Effective Mathematics of the Uncountable

Classical computable model theory is most naturally concerned with countable domains. There are, however, several methods – some old, some new – that have extended its basic concepts to uncountable structures. Unlike in the classical case, however, no single dominant approach has emerged, and different methods reveal different aspects of the computable content of uncountable mathematics. This book contains introductions to eight major approaches to computable uncountable mathematics: descriptive set theory; infinite time Turing machines; Blum-Shub-Smale computability; Sigma-definability; computability theory on admissible ordinals; E-recursion theory; local computability; and uncountable reverse mathematics. This book provides an authoritative and multifaceted introduction to this exciting new area of research that is still in its early stages. It is ideal as both an introductory text for graduate and advanced undergraduate students and a source of interesting new approaches for researchers in computability theory and related areas.

NOAM GREENBERG is a Rutherford Discovery Fellow of the Royal Society of New Zealand where his research interests include computability theory, algorithmic randomness, reverse mathematics, higher recursion theory, computable model theory, and set theory.

JOEL DAVID HAMKINS is professor at the City University of New York and has held visiting positions at the University of California, Berkeley, Kobe University, Carnegie Mellon University, the University of Muenster, the University of Amsterdam, and New York University. His research interests lie in mathematical logic, particularly set theory, focusing on the mathematics and philosophy of the infinite.

DENIS HIRSCHFELDT is a Professor of Mathematics at the University of Chicago and has previously held visiting positions at the University of Wisconsin-Madison, and the University of Notre Dame. He was a recipient of the 1999 Sacks Prize of the Association for Symbolic Logic for the best doctoral dissertation in mathematical logic worldwide, and the 2010 Sacks Prize of the Association for Symbolic Logic for expository writing.

RUSSELL MILLER holds an appointment as Professor of Mathematics jointly between Queens College and the CUNY Graduate Center. His research applies computability to other areas of mathematics, including model theory, set theory, commutative algebra, differential algebra, graph theory, and topology.

LECTURE NOTES IN LOGIC

A Publication for
The Association for Symbolic Logic

This series serves researchers, teachers, and students in the field of symbolic logic, broadly interpreted. The aim of the series is to bring publications to the logic community with the least possible delay and to provide rapid dissemination of the latest research. Scientific quality is the overriding criterion by which submissions are evaluated.

More information, including a list of the books in the series, can be found at http://www.aslonline.org/books-lnl.html

LECTURE NOTES IN LOGIC 41

Effective Mathematics of the Uncountable

Edited by

NOAM GREENBERG
Victoria University of Wellington

JOEL DAVID HAMKINS
City University of New York

DENIS HIRSCHFELDT
University of Chicago

RUSSELL MILLER
City University of New York

ASSOCIATION FOR SYMBOLIC LOGIC

CAMBRIDGE
UNIVERSITY PRESS

CAMBRIDGE
UNIVERSITY PRESS

University Printing House, Cambridge CB2 8BS, United Kingdom

Published in the United States of America by Cambridge University Press, New York

Cambridge University Press is a part of the University of Cambridge.

It furthers the University's mission by disseminating knowledge in the pursuit of education, learning and research at the highest international levels of excellence.

www.cambridge.org
Information on this title: www.cambridge.org/9781107014510

First published 2013

Printed in the United Kingdom by CPI Group Ltd, Croydon CR0 4YY

A catalogue record for this publication is available from the British Library

ISBN 978-1-107-01451-0 Hardback

CONTENTS

PREFACE

Although classical computable model theory is most naturally concerned with countable domains, several methods – some old, some new – have extended its basic concepts to uncountable structures. Unlike in the classical case, however, no single dominant approach has emerged, and different methods reveal different aspects of the computable content of uncountable mathematics. Furthermore, uncountable computable model theory is still in an early stage of development, and, in particular, there has been relatively little work on connecting and comparing the various available approaches. Two *Effective Mathematics of the Uncountable* workshops were held at the CUNY Graduate Center in New York on August 18–22, 2008 and August 17–21, 2009, organized by Noam Greenberg, Joel Hamkins, Denis Hirschfeldt, and Russell Miller, with support from a Templeton Foundation "Exploring the Infinite" program grant. The aim of these workshops was to introduce a variety of approaches to uncountable computable model theory to researchers and students in computability theory and related fields, and to encourage collaboration between those who have developed and studied different facets of the effective content of uncountable mathematics.

Speaking at the EMU workshops were researchers with a wide range of backgrounds and motivations: Nate Ackerman, Wesley Calvert, Samuel Coskey, Noam Greenberg, Joel Hamkins, Denis Hirschfeldt, Julia Knight, Peter Koepke, David Linetsky, Robert Lubarsky, Russell Miller, Antonio Montalbán, Ansten Mørch Klev, Kerry Ojakian, Gerald Sacks, Richard Shore, Alexei Stukachev, and Philip Welch. We would like to thank all our speakers, the Templeton Foundation, those who helped with the local organization of the workshops, and all the participants who contributed to their lively and productive atmosphere.

Following the EMU workshops, we asked several researchers to write introductions to major approaches to uncountable model theory: descriptive set theory, infinite time Turing machines, Blum–Shub–Smale computability, Σ-definability, computability theory on admissible ordinals, E-recursion theory, local computability, and uncountable reverse mathematics. These contributions are gathered in the present volume. Rather than dictate a format, we left authors free to decide how best to present their areas of interest, but encouraged them to aim for self-contained presentations appropriate for researchers and students with a background in classical computability theory and computable mathematics, though not necessarily in computability theory in the uncountable setting. We have also added an introduction, which helps set the stage for the remaining chapters by briefly discussing their contents and raising certain issues relevant to many or all of the approaches described in these chapters. As a result, this book is not a conference proceedings, but a multifaceted introduction to an exciting new area of research, which we hope will help attract to it both established researchers in computability theory and related areas, and students looking for a wide open area with strong potential for fundamental work at both the technical and conceptual levels.

INTRODUCTION

Chang and Keisler [8] famously defined model theory as the sum of logic and universal algebra. In the same spirit, one might describe computable model theory to be the investigation of the constraints on information content imposed by algebraic structure. The analogue of the interplay between syntactical objects and the algebraic structure they define is the connection between definability and complexity. One asks: How complicated are the constructions of model theory and algebra? What kind of information can be coded in structures like groups, fields, graphs, and orders? What mathematical distinctions are unearthed when "boldface" notions such as isomorphism are replaced by their "lightface" analogues such as, say, computable isomorphism?

A special case of the following definition was first rigorously made by Fröhlich and Shepherdson [11], following work of Hermann [17] and van der Waerden [40], which itself built on the constructive tradition of 19th century algebra. It was further developed by Rabin [32, 33] and Mal'cev [27].

Definition. Let $\mathcal{L}$ be a computable signature (language), and let $\mathcal{M}$ be an $\mathcal{L}$-structure whose universe is the set of natural numbers. The *degree of* $\mathcal{M}$ is the Turing degree of the atomic (equivalently, quantifier-free) diagram of $\mathcal{M}$.

A structure is *computable* if its degree is $\mathbf{0}$, the Turing degree of computable sets. Equivalently, a structure $\mathcal{M}$ is computable if, uniformly in the symbols of $\mathcal{L}$, the interpretations in $\mathcal{M}$ of the constant symbols, function symbols, and relation symbols of $\mathcal{L}$ are computable. In the Eastern school of computable model theory, the focus has been on *constructivizations*: in Western terminology, a constructivization of a structure $\mathcal{M}$ is an isomorphism between $\mathcal{M}$ and a computable copy of $\mathcal{M}$. A structure $\mathcal{M}$ is said to be *computably presentable*, or *constructivizable*, if it has some constructivization, that is, if it has a computable copy.

Within computable model theory we identify three research programmes.

1. Pure computable model theory considers the effectiveness of model-theoretic constructions. For example, an examination of the standard proof of the compactness theorem reveals that every complete computable (a.k.a. decidable) theory has a computable model, indeed one whose elementary diagram is computable; such structures are called *decidable* (or *strongly constructivizable*). The countable omitting types theorem can be similarly extended [28]. On the other hand, Millar [29] and Kudaibergenov [25] showed that Vaught's "no two models" theorem fails if we consider only decidable models.

Another example of this line of research is the investigation of the effective properties of "special" models. A typical theorem is the characterization of the decidable complete atomic theories that have decidable prime models (Goncharov and Nurtazin [13] and Harrington [16]), depending on the effective properties of the collection of isolated types of the theory. This result has been extended to an extensive analysis of the degrees of prime, saturated, and homogeneous models of decidable theories by several authors (see e.g. [26]). Similarly, an investigation into the computability of countable models of $\aleph_1$-categorical theories is ongoing (see e.g. [1]).

2. Computable structure theory, a more computability-centric approach, looks at the trace left on computability theory, and in particular on the Turing degrees, by their interaction with model theory. Typical here is Knight's result [22] that if $\mathcal{M}$ is not automorphically trivial and a Turing degree $\mathbf{d}$ computes a copy of $\mathcal{M}$, then $\mathbf{d}$ contains a copy of $\mathcal{M}$. In general, one may ask which sets of Turing degrees are *degree spectra*: the collection of degrees of copies of some structure. For example, Slaman [39] and Wehner [41] showed that the collection of nonzero degrees is a degree spectrum.

This approach, pioneered by Ash and Knight, also asks about the relationship between definability of relations on structures and their complexity. The following is a main result [4, 9]: Let R be a relation on a structure $\mathcal{M}$. Then the property that for every isomorphism $f\colon \mathcal{M} \to \mathcal{N}$, the image of R is c.e. in $\mathcal{N}$ is equivalent to the property that R is definable in $\mathcal{M}$ by an effectively presented infinitary Σ^0_1 formula in the logic $\mathcal{L}_{\omega_1,\omega}$. One investigates not only the degrees of structures, but also how complicated are the isomorphisms between structures. This line of research leads to new notions, motivated by computability, which have no analogue in "boldface" model theory. Central among them are the notions of *computable categoricity*, *relative computable categoricity*, and *computable dimension*. (See [2, 3, 15] for definitions and further discussion of these notions.) These are properties of structures rather than theories. A characterization of relative computable categoricity (Ash, Knight, Manasse, and Slaman [4], Chisholm [9]) in terms of definability of the orbits of $\mathcal{M}$ under the action of the automorphism group of $\mathcal{M}$ is an effective version of Scott's analysis of the isomorphism types of countable structures using infinitary logic.

3. Computable algebra investigates the effective properties of particular classes of structures. In some sense this is applied computable model theory. Researchers have attempted, for example, to characterize, among the class of Abelian p-groups, which Abelian p-groups have computable copies. One also asks about the relationship between the complexity of a structure and the complexity of associated objects; for example, Fröhlich and Shepherdson implemented van der Waerden's construction of a computable field with no splitting algorithm. Similarly, one asks about the complexity of the linear independence relation in computable vector spaces; a definitive answer was given by Shore [37]. One asks how complicated are algebraic constructions: for example, the algebraic closure of a computable field has a computable copy, but the image of the original field in its algebraic closure need not always be computable [33]. Instances of notions from computable model theory may have succinct characterizations: Goncharov and Dzgoev [12] and Remmel [34] showed that a computable linear ordering is computably categorical if and only if it contains only finitely many successor pairs. One also asks what are the degree spectra of structures in particular classes, such as linear orderings. For example, Jockusch and Soare [20] showed that there is a low linear ordering with no computable copy; this result was extended by R. Miller [30], though the question of whether the Slaman–Wehner example mentioned above can be realized by a linear order is still an important open problem. On the other hand, every low_4 Boolean algebra has a computable copy [23].

Computable model theory is also related to reverse mathematics, the project of classifying theorems of mathematics in terms of proof-theoretic strength, often by showing equivalence (over a weak base theory) of these theorems with certain subsystems of second order arithmetic (see [38]). For example, the result that a computable field has a computable algebraic closure translates to a proof in the system RCA_0 of recursive comprehension of the existence of algebraic closure of any given field. Thus, computable algebra is often the key for classifying theorems of algebra within reverse mathematics. Similarly, the investigations of pure model theory yield a reverse mathematical classification of theorems of model theory; see for example [18].

While model theory has interesting things to say about countable models (such as Vaught's theorem, or the Ryll-Nardzewski theorem), the real strength of model theory, and in particular stability theory, is apparent in the realm of uncountable models, with Morley's theorem on uncountable categoricity being both a paragon and the catalyst of modern model theory. It is only natural to wish to find the effective content of this part of mathematics. Yet computable model theory has been restricted to investigating countable models, and its interactions with stability theory has been only at the fringe of the latter, for example using the Baldwin-Lachlan analysis [5] of models of uncountably categorical theories to understand effective properties of *countable* models of such theories. Nonetheless, intuitively one sees "effective" and "non-effective" aspects of uncountable model theory and uncountable mathematics in general, and one would like to formalize them and reason about them.

The source of the restriction to countable structures is the fact the the objects that are manipulated by models of computation are hereditarily finite. Turing machines take as input finite strings over a finite alphabet, register machines store natural numbers, and so on. In other words, the world of computability theory is inherently countable. In order to develop an effective theory of uncountable structures, one needs to generalize the theory of computable functions and sets to include uncountable domains. There is no canonical generalization of this sort, and so the kind of effective theory of uncountable mathematics one gets depends heavily on the choice of the model of computation.

The purpose of this book is to describe eight such choices and the resulting applications to the study of effective properties of uncountable objects. It is intended as both an invitation to uncountable computable mathematics and a resource for researchers in the area who, while working along one or more of these lines, are interested in the possibilities raised by other approaches. We first discuss approaches to uncountable computable model theory in particular, via the choice of some model of computation, and then discuss effective uncountable mathematics in greater generality.

In "Borel structures: a brief survey", Montalbán and Nies take "effective" to mean "Borel". They look at structures whose universe is a Borel subset of a Polish space, and where the relations and functions on the structure are uniformly Borel; and also concentrate on Borel homomorphisms between such structures. More generally, they also accept structures that are quotients of Borel structures by a Borel equivalence relation. In this context, Hjorth and Nies [19] verified the failure of an effective compactness theorem, and Nies and Shore [unpublished] have computed the Borel dimension (number of Borel inequivalent Borel copies) of the field of complex numbers to be $2^{\aleph_0}$.

Coskey and Hamkins ("Infinite time turing machines and an application to the hierarchy of equivalence relations on the reals") show what happens if one lets Turing machines run beyond forever; that is, if computations of Turing machines run for an ordinal amount of time. These machines can then be used to compute subsets of Cantor space 2^{ω}, by writing entire reals on the input tape. The sets of real numbers that can be computed by such machines are all Δ^1_2; all Π^1_1 sets can be so computed. Thus, infinite time Turing machine computation is in a sense an extension of the Borel model. In this context, Hamkins, Miller, Seabold, and Warner [14] showed that the effective version of the completeness theorem is independent of ZFC: it holds if $V = L$, but can be forced to fail, for example in any model in which there are no Σ^1_2 sets of size $\aleph_1$. As in the Borel world, here too there may be a difference between "injective" presentations and presentations that allow for taking a quotient by a computable equivalence relation. Unlike the Borel case, in the context of infinite time Turing machine computability, it is independent of ZFC whether the injective and non-injective notions coincide, that is, whether every structure with a computable presentation has an injective presentation.

Blum, Shub, and Smale [6] introduced a notion of computability over real numbers. In this model, a machine treats a real number as a complete object and does not require an approximation for the number; on the other hand, the machine runs for finitely many steps. Although originally developed with an eye toward modeling numerical analysis, it is natural to consider this notion as a model for computability for sets and functions of real numbers. In "Some results on $\mathbb{R}$-computable structures", Calvert and Porter pursue the development of effective model theory using the Blum–Shub–Smale computation scheme. In particular, they discuss $\mathbb{R}$-vector spaces, 2-manifolds, and homotopy groups.

The concept of Σ-definability was developed by Ershov [10]. We say that a structure $\mathcal{M}$ is *Σ-reducible* to a structure $\mathcal{N}$ if $\mathcal{M}$ is interpretable in the smallest hereditarily finite set containing the elements of $\mathcal{N}$ as ur-elements using existential formulas. In other words, $\mathcal{M} \leqslant_\Sigma \mathcal{N}$ if $\mathcal{N}$ can interpret $\mathcal{M}$ effectively when imbued with the power of arithmetic on the natural numbers. In "Effective model theory: an approach via the Σ-definability", Stukachev surveys the Σ-definability approach. A source of unexpected results in this context is the reducibility of fields to linear orderings. For example, Ershov showed that the field of complex numbers is Σ-reducible to any dense linear ordering of size continuum (but is not Σ-reducible to any set, i.e., to any structure with empty signature), whereas the field of real numbers is not Σ-reducible to any linear ordering.

In "Computable structure theory using admissible recursion theory on ω_1", Greenberg and Knight use admissible recursion theory as a model of computation on infinite cardinals, in particular on ω_1. This model has several equivalent definitions, but the original and shortest definition states that computability is given by definability by existential formulas over the structure $(L_{\omega_1}, \in)$, where L is Gödel's constructible universe. This choice allows for a development of computable model theory for structures of size $\aleph_1$ much along the lines of the development of computable model theory for countable structures. The authors discuss, for example, fields, vector spaces, and linear orderings; and pure computable model theory, with a look into the effective completeness theorem, Scott families, and computable categoricity.

The theory of E-recursion is an extension of admissible recursion theory to inadmissible sets. The *divergence-admissibility split* states that the inadmissible sets L_α that are closed under E-recursive functions are exactly those which admit *divergence witnesses*. Thus computability on these inadmissible *E-closed* domains has new properties that are not mirrored in admissible recursion theory. In "E-recursive intuitions", Sacks discusses how the logic $\mathcal{L}_{\alpha,\omega}$ behaves with respect to the completeness and compactness theorems, when L_α is inadmissible and E-closed.

Miller, in "Local computability and uncountable structures", takes a different approach. Rather than using some theory of computation for uncountable sets, local computability measures the effectiveness of uncountable structures by examining their finitely generated substructures and how embeddings between these lift to containments of substructures of the original uncountable structure. This approach yields distinctions between, for example, the field of real numbers and the field of complex numbers; in a sense, the latter is "more" locally computable than the former. Local computability of the real field relies on Artin's theorem, and so requires not insignificant algebra. Miller also discusses linear orderings and trees.

Finally, in "Reverse mathematics, countable and uncountable: a computational approach", Shore discusses uncountable versions of reverse mathematics. While reverse mathematics is rooted in proof theory, computability theorists often prefer to ignore nonstandard models of (first order) arithmetic, and so concentrate on the ω-models (that is, models with standard first-order part) of theorems of mathematics. This approach is generalized by Shore to uncountable domains, using admissible recursion theory as the computational tool. In this context, Shore analyzes a number of statements of uncountable algebra in terms of their computational strength. For example, he shows that the existence of a basis is equivalent to closure under the Turing jump, and that the existence of prime ideals in a ring is equivalent to an uncountable version of weak König's lemma, the finite character tree property.

We would be remiss if we omitted computable analysis from this discussion. It is the longest-standing and most-developed approach to computability on the real numbers, stemming from Turing's own definition of a computable real number and covering a wide range of topics since then. Over that time, many introductions to the subject have been written and are available to the interested reader. Therefore, we did not feel the need to add another one in this volume, but we recommend [7] as a useful and recent tutorial on computable analysis, very much in the style of the textbook [42], and we encourage the reader to keep computable analysis in mind when considering the fifteen questions at the end of this introduction. Among earlier books on the subject, we would also mention [24] and [31]. The approach taken is to view a real number x as given by a Cauchy sequence of rational approximations $\langle q_n \rangle$, converging effectively to x, i.e. with $|x - q_n| < 2^{-n}$ for all n. A real number is computable if there is a computable Cauchy sequence converging effectively to it. One can then define a computable function f on all real numbers to be given by a Turing functional Φ, which uses as oracle a Cauchy sequence (computable or not) converging effectively to the input x, and, on input n, outputs the nth element of a Cauchy sequence converging effectively to $f(x)$. (Such a function is sometimes called *type-two computable*; there are certain analogies to the infinite-time Turing machines in the chapter of Coskey and Hamkins.)

The methods outlined in the eight papers in this book for developing an effective theory of uncountable mathematics are quite distinct. They yield different collections of computable structures and mappings between them. Nevertheless, we would like to discuss some similarities and particularly noteworthy distinctions between them, and some themes to which most of them relate. Some of these issues are special to uncountable mathematics, and do not have analogues in the countable realm.

The most jarring issue, to some, is independence, and the reliance on set-theoretic hypotheses beyond Zermelo–Fränkel set theory. For example, both Shore and Greenberg–Knight, when dealing with objects of size κ, assume that all bounded subsets of κ are constructible, so $H_\kappa = L_\kappa$; this assumption ensures that every subset of κ is amenable for L_κ. Similarly, we mentioned above that basic results in the infinite time Turing machine model are independent of ZFC, essentially because of the fact that this theory goes beyond the Borel world to Δ^1_2 sets. Traditionally, computation is considered a "down to earth" part of mathematics, absolute between models of set theory, and invariant to the choice of axioms of set theory. (Although independence results do occur in classical computability theory, they do not occur in the basic theory, but rather arise in contexts where there is a mix of set theory with computability theory.) Thus, some may expect any theory of computation, including one on uncountable objects, to be basic enough to maintain this invariance.

Another difference between the various approaches is whether they allow considerations of structures of different cardinalities. A number of approaches – Borel computability, infinite time Turing machines, and the Blum–Shub–Smale model – apply only to structures of size the continuum (although the Blum–Shub–Smale model generalizes to work over any ring). On the other hand, Σ-definability and local computability work, at once, for all cardinals. In the middle, admissible recursion theory can work with any cardinal, but requires us to fix a cardinal. That is, admissible κ-recursion theory is defined for each κ, but for distinct cardinals κ and λ, κ-computability and λ-computability are incompatible. The issue of working with distinct cardinals at once may come up, for example, when considering effective versions of the Löwenheim–Skolem theorems, an avenue that is yet unexplored.

In practice, it turns out that another fundamental distinction between models of computation of uncountable objects is the extent to which they have access to a well-ordering of the universe. Traditionally, the ordering and successor relations on the natural numbers are computable. This fact means that searching for witnesses for a particular computable property is a computable procedure. Indeed, the centrality of this aspect of computability to the general theory is evident from Gödel's definition of the class of partial computable functions by the least number operator. This property has profound implications in classical computable model theory. Consider, for example, a (countable) computable field F. Given a polynomial $f \in F[x]$, if we know that f has a root in F, then such a root can be effectively found by searching over the elements of the field and testing them one by one as inputs for f, until a root is found. An ordering of F in order-type ω, which ensures that such a search will end in finitely many steps, cannot be separated from F itself; computable model theory does not know how to "forget" about this ordering of F, and access F only via its algebraic structure. To some mathematicians, this power is unreasonable: algorithms involving an "explicit" field (in the language of van der Waerden) should rely only on algebra, not on external properties such as an ordering of the field. In that view, finding roots of polynomials should be done algebraically, say using Newton's method, but not via "mindless" search.

Borel computation, Blum–Shub–Smale, and infinite time Turing machines have no access to a well-ordering of the continuum, and so according to the view above are "purer" models of computation. Similarly, Σ-definability has access only to the traditional structure on ω; the elements of a structure $\mathcal{M}$ are considered as urelements and are not effectively ordered. Local computability too does not access a well-ordering of uncountable objects, since it directly manipulates only finitely generated substructures of the given structure.

The consequences are dramatic. They are exemplified by the "lost melody" theorem of infinite time Turing machines, which states the existence of a real $c \in 2^\omega$ that can be recognized by an infinite time Turing machine, but not produced (or enumerated) by any such machine that has no access to c as an oracle. In these models of computation, the traditional equivalence between computable enumerability and semi-decidability is broken. In contrast, admissible recursion theory and E-recursion theory work with a computable well-ordering of their universe, although in E-recursion the utility of such an ordering is limited, compared to the unbridled access given to admissible recursion theory. As a result, admissible recursion reflects many of the properties of countable computability, and many constructions of the latter lift to uncountable constructions in the former; but the theory loses the pure reliance on algebra.

The dichotomy between models with a well-ordering of the universe and models without one is a special case of a larger theme: the choice between an intuitive presentation and computational power of a model of computation. Few would dispute that a desirable property of a model of computation is a presentation that makes it intuitive. We would like to understand easily why the model purports to capture the notion of computability. This desideratum explains why it was Turing's machine model, rather than, say, Gödel's equivalent definition of partial recursive functions, that convinced mathematicians that this class of functions correctly captures the notion of computability of functions of natural numbers.

Often, the price of a clear intuitive definition is a weaker theory of effective mathematics. An alternative approach would allow for intuition to develop via usage. Thus, for example, admissible recursion theory has a machine definition via ordinal register machines, or by Turing machines with an ordinal-length tape; but calling these definitions "intuitive" relies, at least, on initially being comfortable with ordinals as basic building blocks. In any case, this is a matter of taste and personal judgment, and so we leave it to the readers.

The above discussion points to some issues that the reader might wish to keep in mind when reading the ensuing papers. Of course, each approach to uncountable computable model theory also has its own specific set of open problems and research directions, discussion of which is best left to the individual papers.

However, there are several other general questions relevant to many or all of these approaches. To finish this introduction, we will list some here, including ones suggested by our previous discussion. In some cases, the answers for particular approaches may be clear or already known, but in many cases, thorough accounts are still challenges for the future of this fascinating area of research.

Here then are some of the questions one might ask when considering an approach to uncountable computable model theory such as the ones discussed in this book.

1. What are the effective versions, under the given approach, of the most basic model-theoretic notions and results, such as the notion of isomorphism and the Completeness Theorem? Moving beyond these, we may consider areas of investigation that have met with great success in the countable case, for example the effective content of the study of special models, such as atomic and homogeneous models, or structure/nonstructure theorems that point out differences between various classes of structures, as in the work of Richter [35, 36].
2. Beyond the above areas, what does the approach have to say about modern concerns in model theory, such as stability theory, which are in many ways beyond the ken of classical computable model theory?
3. Which particular areas of "classical" mathematics is the approach well suited to investigating?
4. There are certain uncountable structures, such as $\mathbb{R}$, which most people would agree are "intuitively effective". Does the given approach make such structures formally effective, or if not, is there a good reason for the mismatch between our intuitions and the formal definitions? The same question may be asked about intuitively effective constructions and theorems.
5. A great part of the success of classical computable mathematics comes from the close connections between computability theory and definability. How well does the given approach interact with definability?
6. How well does the notion of relative computability generalize under this approach, and what impact does the answer to this question have to generalizing notions that such as degree spectra of structures and relations on structures, which in the countable setting rely on the structure of the Turing degrees?
7. In the countable setting, one of the most important tools in analyzing the effective content of mathematics is the theory of classes of degrees, such as low degrees, PA degrees, hyperimmune-free degrees, and so on, which, while usually defined in computability-theoretic terms, have deep connections with combinatorial principles that often arise in the deep analysis of mathematical concepts and constructions. What are the useful analogues to such notions under the given approach, and how may they be applied to the study of uncountable effective mathematics?

8. One reason to be interested in computable mathematics, and the closely related program of reverse mathematics, is that computable procedures (or, in the context of reverse mathematics, the roughly corresponding weak base system RCA_0) may be seen as a mathematically precise version of Hilbert's finitistic procedures (or at least a concept that can play a similar foundational role to finitistic procedures). To what extent does the given approach provide a notion of effective procedure that can reasonably fit the same foundational role?
9. How dependent is the approach on set-theoretic assumptions, and how is it affected by varying these assumptions? Do the differences thus obtained say something of interest about the effective aspects of uncountable mathematics?
10. Is the approach limited to particular cardinalities? If so, can it be generalized to other cardinalities?
11. Is there a way under this approach to consider the effectiveness of structures of different cardinalities at once?
12. To what extent are well-orderings of the universe important to this approach? If such well-orderings of uncountable universes are available, what effects do they have on the results obtained under the approach?
13. Are there ways of extending the approach to address questions of efficiency raised by complexity theory? In this connection, a particularly active area of current research in the countable setting is that of automatic structures (see e.g. [21]).
14. Is there anything the approach can tell us about the countable setting?
15. What are the connections between this approach and other ones? Here we mean both "hard" mathematical connections, allowing us perhaps to transfer results obtained via one approach to another, and comparisons of results. For instance, does effectiveness (of a particular structure or construction, say) under one approach tend to imply effectiveness under the other, at least heuristically? If certain constructions turn out to be effective under one approach but not the other, what does this situation tell us about the nature of these constructions? Turning things around, can such differences in the resulting theories of effective mathematics be used to improve our understanding of the differences between various approaches to "pure" uncountable computability theory?

REFERENCES.

[1] U. Andrews, A new spectrum of recursive models using an amalgamation construction. *J. Symbolic Logic* **76** (2011), 883–896.

[2] C.J. Ash, Isomorphic recursive structures. In *Handbook of Recursive Mathematics*, vol. 1 (Yu.L. Ershov, S.S. Goncharov, A. Nerode, and J.B. Remmel, eds., V. W. Marek, assoc. ed.), Elsevier, 1998, 167–181.

[3] C.J. Ash and J.F. Knight, *Computable Structures and the Hyperarithmetical Hierarchy*, Amsterdam, 2000.

[4] C. Ash, J. Knight, M. Manasse, and T. Slaman, Generic copies of countable structures, *Ann. Pure Appl. Logic* **42** (1989), 195–205.

[5] J.T. Baldwin and A.H. Lachlan, On strongly minimal sets, *J. Symbolic Logic* **36** (1971), 79–96.

[6] L. Blum, M. Shub, and S. Smale, On a theory of computation and complexity over the real numbers: NP-completeness, recursive functions and universal machines, *Bull. Amer. Math. Soc.* (N.S.) **21** (1989), 1–46.

[7] V. Brattka, P. Hertling, and K. Weihrauch, A tutorial on computable analysis. In *New Computational Paradigms: Changing Conceptions of What is Computable*, (S.B. Cooper, B. Löwe, and A. Sorbi, eds.), Springer, 2008, 425–491.

[8] C.C. Chang and H.J. Keisler, *Model Theory*, third edition; Studies in Logic and the Foundations of Mathematics **73**, North-Holland, 1990.

[9] J. Chisholm, Effective model theory vs. recursive model theory, *J. Symbolic Logic* **55** (1990), 1168–1191.

[10] Yu.L. Ershov, Σ-definability in admissible sets, *Sov. Math. Dokl.* **32** (1985), 767–770.

[11] A. Fröhlich and J.C. Shepherdson, Effective procedures in field theory, *Philos. Trans. Roy. Soc. London*. Ser. A. **248** (1956), 407–432.

[12] S.S. Goncharov and V.D. Dzgoev, Autostability of models, *Algebra Logic* **19** (1980), 45–58 (Russian), 28–37 (English translation).

[13] S.S. Goncharov and A.T. Nurtazin, Constructive models of complete decidable theories, *Algebra Logic* **12** (1973), 125–142, 243 (Russian), 67–77 (English translation).

[14] J.D. Hamkins, R. Miller, D. Seabold, and S. Warner, Infinite time computable model theory. In *New Computational Paradigms: Changing Conceptions of What is Computable* (S.B. Cooper, B. Löwe, and A. Sorbi, eds.), Springer, 2008, 521–557.

[15] V.S. Harizanov, Pure computable model theory. In *Handbook of Recursive Mathematics*, vol. 1 (Yu.L. Ershov, S.S. Goncharov, A. Nerode, and J.B. Remmel, eds., V.W. Marek, assoc. ed.), Elsevier, 1998, 3–114.

[16] L. Harrington, Recursively presentable prime models, *J. Symbolic Logic* **39** (1974), 305–309.

[17] G. Hermann, Die Frage der endlich vielen Schritte in der Theorie der Polynomideale, *Math. Ann.* **95** (1926), 736–788.

[18] D.R. Hirschfeldt, R.A. Shore, and T.A. Slaman, The atomic model theorem and type omitting, *Trans. Amer. Math. Soc.* **361** (2009), 5805–5837.

[19] G. Hjorth and A. Nies, Borel structures and Borel theories. *J. Symbolic Logic* **76** (2011), 461–476.

[20] C.G. Jockusch, Jr. and R.I. Soare, Degrees of orderings not isomorphic to recursive linear orderings, *Ann. Pure Appl. Logic* **52** (1991), 39–64.
[21] B. Khoussainov and M. Minnes, Three lectures on automatic structures. In *Logic Colloquium '07* (F. Delon, U. Kohlenbach, P. Maddy, and F. Stephan, eds.), Lecture Notes in Logic **35**, Cambridge University Press, and Association for Symbolic Logic, 2010, 132–176.
[22] J.F. Knight, Degrees coded in jumps of orderings, *J. Symbolic Logic* **51** (1986), 1034–1042.
[23] J.F. Knight and M. Stob, Computable Boolean algebras, *J. Symbolic Logic* **65** (2000), 1605–1623.
[24] K. Ko, *Computational Complexity of Real Functions*, Birkhauser Boston, 1991.
[25] K.Zh. Kudaibergenov, A theory with two strongly constructible models, *Algebra Logic* **18** (1979), 176–185, 253 (Russian), 111–117 (English translation).
[26] K. Lange and R.I. Soare, Computability of homogeneous models, *Notre Dame J. Formal Logic* **48** (2007), 143–170.
[27] A.I. Mal'cev, Constructive algebras I, *Uspekhi Mat. Nauk* **16** (1961), 3–60.
[28] T. Millar, Omitting types, type spectrums, and decidability, *J. Symbolic Logic* **48** (1983), 171–181.
[29] T.S. Millar, A complete, decidable theory with two decidable models, *J. Symbolic Logic* **44** (1979), 307–312.
[30] R. Miller, The Δ^0_2-spectrum of a linear order, *J. Symbolic Logic* **66** (2001), 470–486.
[31] M. Pour-El and I. Richards, *Computability in Analysis and Physics*, Springer-Verlag, 1989.
[32] M.O. Rabin, Computable algebraic systems. In *Summer Institute for Symbolic Logic*, Cornell University, Institute for Defence Analyses, 1957, 134–138.
[33] M.O. Rabin, Computable algebra, general theory, and theory of computable fields, *Trans. Amer. Math. Soc.* **95** (1960), 341–360.
[34] J.B. Remmel, Recursively categorical linear orderings, *Proc. Amer. Math. Soc.* **83** (1981), 387–391.
[35] L.J. Richter, *Degrees of Structures*, PhD Dissertation, University of Illinois at Urbana–Champaign, 1979.
[36] L.J. Richter, Degrees of structures, *J. Symbolic Logic* **46** (1981), 723–731.
[37] R.A. Shore, Controlling the dependence degree of a recursively enumerable vector space, *J. Symbolic Logic* **43** (1978), 13–22.
[38] S.G. Simpson, *Subsystems of Second Order Arithmetic*, second edition, Perspectives in Logic, Cambridge University Press, and Association for Symbolic Logic, 2009.

[39] T.A. Slaman, Relative to any nonrecursive set, *Proc. Amer. Math. Soc.* **126** (1998), 2117–2122.

[40] B.L. van der Waerden, Eine Bemerkung über die Unzerlegbarkeit von Polynomen, *Math. Ann.* **102** (1930), 738–739.

[41] S. Wehner, Enumeration, countable structures and Turing degrees, *Proc. Amer. Math. Soc.* **126** (1998), 2131–2139.

[42] K. Weihrauch, *Computable Analysis: An Introduction*, Springer, Berlin, 2000.

Noam Greenberg, Victoria University of Wellington

Joel David Hamkins, City University of New York

Denis R. Hirschfeldt, University of Chicago

Russell G. Miller, City University of New York

SOME RESULTS ON $\mathbb{R}$-COMPUTABLE STRUCTURES

WESLEY CALVERT AND JOHN E. PORTER

§1. Introduction. The theory of effectiveness properties on countable structures whose atomic diagrams are Turing computable is well-studied (see, for instance, [1, 15]). Typical results describe which structures in various classes are computable (or have isomorphic copies that are) [19], or the potential degree of unsolvability of various definable subsets of the structure [16]. The goal of the present paper is to survey some initial results investigating similar concerns on structures which are effective in a different sense.

A rather severe limitation of the Turing model of computability is its traditional restriction to the countable. Of course, many successful generalizations have been made (see, for instance, [28, 12, 13, 23, 24, 26] and the other chapters in the present volume). The generalization that will be treated here is based on the observation that while there is obviously no Turing machine for addition and multiplication of real numbers, there is strong intuition that these operations are "computable." The BSS model of computation, first introduced in [5], approximately takes this to be the definition of computation on a given ring (a more formal definition is forthcoming). This allows several problems of computation in numerical analysis and continuous geometry to be treated rigorously. The monograph [4] gives the examples of the "decision problem" of the points for which Newton's method will converge to a root, and determining whether a given point is in the Mandelbrot set.

1.1. Basic Definitions. The definition of a BSS machine comes from [4]. Such a machine should be thought of as the analogue of a Turing machine (indeed, the two notions coincide where $R = \mathbb{Z}$). Let R be a ring with 1. Let R^∞ be the set of finite sequences of elements from R, and R_∞ the bi-infinite direct sum

$$\bigoplus_{i \in \mathbb{Z}} R.$$

Definition 1.1. A machine M over R is a finite connected directed graph, containing five types of nodes: input, computation, branch, shift, and output, with the following properties:

1. The unique input node has no incoming edges and only one outgoing edge.
2. Each computation and shift node has exactly one output edge and possibly several input branches.

[1]The first author is grateful for the support of Grant #13397 from the Templeton Foundation.

3. Each output node has no output edges and possibly several input edges.
4. Each branch node η has exactly two output edges (labeled 0_η and 1_η) and possibly several input edges.
5. Associated with the input node is a linear map $g_I : R^\infty \to R_\infty$.
6. Associated with each computation node η is a rational function $g_\eta : R_\infty \to R_\infty$.
7. Associated with each branch node η is a polynomial function $h_\eta : R_\infty \to R$.
8. Associated with each shift node is a map $\sigma_\eta \in \{\sigma_l, \sigma_r\}$, where $\sigma_l(x)_i = x_{i+1}$ and $\sigma_r(x)_i = x_{i-1}$.
9. Associated with each output node η is a linear map $O_\eta : R_\infty \to R^\infty$.

A machine may be understood to compute a function in the following way:

Definition 1.2. Let M be a machine over R.

1. A *path* through M is a sequence of nodes $(\eta_i)_{i=0}^n$ where η_0 is the input node, η_n is an output node, and for each i, we have an edge from η_i to η_{i+1}.
2. A *computation* on M is a sequence of pairs $((\eta_i, x_i))_{i=0}^n$ with a number x_{n+1}, where $(\eta_i)_{i=0}^n$ is a path through M, where $x_0 \in R^\infty$, and where, for each i, the following hold:
   (a) If η_i is an input node, $x_{i+1} = g_I(x_i)$.
   (b) If η_i is a computation node, $x_{i+1} = g_{\eta_i}(x_i)$.
   (c) If η_i is a branch node, $x_{i+1} = x_i$ and η_{i+1} is determined by h_{η_i} so that if $h_{\eta_i}(x_i) \geq 0$, then η_{i+1} is connected to η_i by 1_{η_i} and if $h_{\eta_i}(x_i) < 0$, then η_{i+1} is connected to η_i by 0_{η_i}. (Note that in all other cases, η_{i+1} is uniquely determined by the definition of path.)
   (d) If η_i is a shift node, $x_{i+1} = \sigma_{\eta_i}(x_i)$
   (e) If η_i is an output node, $x_{i+1} = O_{\eta_i}(x_i)$.

The proof of the following lemma is an obvious from the definitions.

Lemma 1.3. *Given a machine M and an element $z \in R^\infty$, there is at most one computation on M with $x_0 = z$.*

Definition 1.4. The function $\varphi_M : R^\infty \to R^\infty$ is defined in the following way: For each $z \in R^\infty$, let $\varphi_M(z)$ be x_{n+1}, where $\left(((\eta_i, x_i))_{i=0}^n, x_{n+1}\right)$ is the unique computation, if any, where $x_0 = z$. If there is no such computation, then φ_M is undefined on z.

Since a machine is a finite object, involving finitely many real numbers as parameters, it may be coded by a member of R^∞ — indeed, by a an element $\sigma \in R^\infty$ whose first element is the length of σ.

Definition 1.5. If σ is a code for M, we define $\varphi_\sigma = \varphi_M$.

We can now say that a set is computable if and only if its characteristic function is φ_M for some M.

EXAMPLE 1.6. Let $R = \mathbb{Z}$. Now the R-computable functions are exactly the classical Turing-computable functions.

EXAMPLE 1.7. Let $R = \mathbb{R}$. Then the Mandelbrot set is not R-computable (see Chapter 2 of [4]).

DEFINITION 1.8. A *machine over R with oracle X* is exactly like a machine over R, except that it has an additional type of nodes, the oracle nodes. Each oracle node is exactly like a computation node, except that $g_\eta = \chi_X$. Computations in oracle machines are defined in the obvious way.

We say that a set S is decidable (respectively, X-decidable) over R if and only if S is both the halting set of an R-machine (respectively, with oracle X) and the complement of the halting set of an R-machine (respectively, with oracle X). We also say that S is semi-decidable if and only if S is the domain of an R-computable function (if R is a real closed field, it is equivalent to say that S is the range of an R-computable function [4]). Ziegler [30] gives a specialized but recent survey of results on $\mathbb{R}$-computation. The following result, first presented by Michaux, but proved in detail in [10], is useful in characterizing the decidable and semi-decidable sets:

PROPOSITION 1.9. *Let $S \subseteq \mathbb{R}^\infty$. Then S is semi-decidable if and only if S is the union of a countable family of semialgebraic sets defined over a single finitely generated extension of $\mathbb{Q}$.*

The "only if" part of this statement is the upshot of an earlier theorem described in [4], called the Path Decomposition Theorem. We can now proceed to define computable structures.

DEFINITION 1.10. Let $\mathcal{L} = (\{P_i\}_{i\in I_P}, \{f_i\}_{i\in I_f}, \{c_i\}_{i\in I_C})$ be a language with relation symbols $\{P_i\}_{i\in I_P}$, function symbols $\{f_i\}_{i\in I_f}$, and constant symbols $\{c_i\}_{i\in I_C}$. Let $\mathcal{A}$ be an $\mathcal{L}$-structure with universe $A \subseteq R^\infty$.

1. We say that $\mathcal{L}$ is R-computable if the sets of relations, functions, and constants are each decidable over R, and if, in addition, there are R-machines which will tell, given P_i (respectively, f_i), the arity of P_i (respectively, f_i).
2. We identify $\mathcal{A}$ with its atomic diagram; in particular,
3. We say that $\mathcal{A}$ is computable if and only if the atomic diagram of $\mathcal{A}$ is R-decidable.

The obstructions to a direct parallel between the theory of $\mathbb{R}$-computable structures and that of Turing computable structures which we have encountered so far are two in number (one for the parsimonious):

1. The real numbers do not admit an ω-like well-ordering to facilitate searching or priority constructions, and in particular
2. There exist $\mathbb{R}$-computable injective functions whose inverses are not $\mathbb{R}$-computable.

1.2. Plan of the Paper. In the present paper, we will survey recent work on the theory of ℝ-computable structures. In Section 2, we give some basic calculations, showing some parallels with the classical theory, including computable ordinals (Section 2.1), satisfaction of computable infinitary formulas (Section 2.2), and the use of forcing to carry out a simple priority construction (Section 2.3). In Section 3, we explore effective categoricity, using vector spaces as an example. In Section 4, we describe some recent results in effective geometry and topology from the perspective of ℝ-computation. In Section 5 we address the relationship of ℝ-computation with other models of effective mathematics for uncountable structures. In Section 6 we summarize the state of ℝ-computable model theory and describe some directions for future research.

§2. Basic Results.

2.1. ℝ-computable Ordinals. The Turing computable ordinals constitute a proper initial segment of the countable ordinals [29, 20]. This initial segment includes, for instance, the ordinal $\omega^{\omega^{\omega^{\cdot^{\cdot^{\cdot}}}}}$. In the present section, we will establish the following theorem:

THEOREM 2.1. *A well-ordering $(L, <)$ has an isomorphic copy which is ℝ-computable if and only if L is countable.*

PROPOSITION 2.2. *Every countable well-ordering $(M, \prec)$ has an isomorphic copy $(L, <)$ which is ℝ-computable.*

PROOF. Since $(M, \prec)$ is countable, it has an isomorphic copy with universe ω. Now $D(M) = \{(a, b) \in M^2 | a \prec b\}$ is a subset of ω^2. Now we define a real number ℓ in the following way:

$$\ell = \sum_{i \in \omega} 10^{-i} \chi_{D(M)}(i).$$

There is a ℝ machine which, given a pair $(a, b) \in \omega^2$ will return the $10^{-\langle a,b \rangle}$ place of ℓ if that place is 1 and will diverge if that place is 0. This shows that $D(M)$ is the halting set of a ℝ-computable function, as required. A symmetric argument shows that the complement of $D(M)$ is the halting set of a ℝ-computable function. ⊣

PROPOSITION 2.3. *Suppose $(L, <)$ is a ℝ-computable well-ordering. Then $|L| \leq \aleph_0$.*

PROOF. Since $(L, <)$ is ℝ-computable, the set $L_< := \{(a, b) \subseteq L^2 : a < b\}$ is the halting set of a ℝ-machine. By Path Decomposition, it must be a disjoint union of semialgebraic sets, and consequently Borel. By the Kunen–Martin Theorem (Theorem 31.5 of [18]), analytic (and hence Borel) well-orderings are countable. ⊣

A rather different proof of Proposition 2.3, using Fubini's Theorem, is possible and enlightening.

Proof. Assume $L_<$ is uncountable. Then since $L_<$ is uncountable and Borel, $|L_<| = 2^{\aleph_0}$. This implies L is Borel with $|L| = 2^{\aleph_0}$. Without loss of generality, we suppose that L is order isomorphic to the cardinal $2^{\aleph_0}$; otherwise, an initial segment of L which was isomorphic to this cardinal would also be $\mathbb{R}$-computable. In particular, L contains a Cantor set C. Fix a Borel measure μ on C such that $\mu(C) = 1$ and extend μ to L by setting $\mu(L \setminus C) = 0$.

We define two auxiliary sets:

$$L_x = \{b \in L : (x, b) \in L_<\}$$
$$L^y = \{a \in L : (a, y) \in L_<\}$$

Each of these is a Borel set. For any y, we have $|L^y| < 2^{\aleph_0}$, since $2^{\aleph_0}$ is a cardinal and L^y is isomorphic to an ordinal less than $2^{\aleph_0}$. Since L^y is Borel, we have $|L^y| = \aleph_0$. This implies the set L_y is co-countable for any y.

Since $L_<$ is Borel, we can apply Fubini's theorem to calculate $\int_{L_<} 1 d\lambda$, where λ is the product measure $\mu \times \mu$. On the one hand,

$$\int_{L_<} 1 d\lambda = \int_L \int_{L_x} 1(d\mu)(d\mu) = \int_L \mu(L_x) d\mu = \int_L 1 d\mu = \mu(L) = 1$$

since L_x is co-countable for each x, and thus of full measure. On the other hand, since L^y is countable for each y, we have $\int_{L_y} 1 dy = 0$. Hence

$$\int_{L_<} 1 d\lambda = \int_L \int_{L^y} 1(d\mu)(d\mu) = \int_L \mu(L^y) d\mu = \int_L 0 d\mu = 0,$$

which is a contradiction. ⊣

2.2. The Complexity of Satisfaction. We define the class of $\mathbb{R}$-computable infinitary formulas. The definition is by analogy with the (Turing) computable infinitary formulas already in broad usage, described in [1]. The choice of computable infinitary formulas is nontrivial, since there are uncountably many $\mathbb{R}$-machines. One natural approach, not pursued here, would be to work in the $\mathbb{R}$-computable fragment of $\mathcal{L}_{(2^{\aleph_0})^+,\omega}$. This would certainly be an interesting logic to understand, but the present authors found it more desirable at first to understand the more familiar $\mathbb{R}$-computable fragment of $\mathcal{L}_{\omega_1\omega}$. At issue is which conjunctions and disjunctions are allowed in a "computable" formula. The logic $\mathcal{L}_{\omega_1\omega}$ allows countable conjunctions and disjunctions, while $\mathcal{L}_{(2^{\aleph_0})^+,\omega}$ allows any of size at most $2^{\aleph_0}$. However, the difficulty of describing what is meant by, for instance, an interval of formulas is a motivation (beyond the avoidance of set-theoretic independence) to consider first the countably long formulas.

Definition 2.4. Let $\mathcal{L}$ be an R-computable language.

1. The Σ_0 formulas of $\mathcal{L}$ are exactly the finitary quantifier-free formulas. The Π_0 formulas are the same.

2. For any ordinal $\alpha = \beta + 1$, the Σ^0_α formulas are those of the form

$$\bigvee\!\!\!\!\bigvee_{i \in S} \exists \bar{y}[\varphi_i(\bar{x}\bar{y})]$$

where S is countable and is the halting set of an R-machine, and there is a finitely generated field $F \subset \mathbb{Q}$ such that all parameters in ϕ_i are in F.
3. For any ordinal $\alpha = \beta + 1$, the Π^0_α formulas are those of the form

$$\bigwedge\!\!\!\!\bigwedge_{i \in S} \forall \bar{y}[\varphi_i(\bar{x}\bar{y})]$$

where S is countable and is the halting set of an R-machine, and there is a finitely generated field $F \subset \mathbb{Q}$ such that all parameters in ϕ_i are in F.
4. Suppose $\alpha = \lim_n \beta_n$ where β_n is a bounded R-computable sequence of ordinals, and there is a finitely generated field $F \subset \mathbb{Q}$ such that all parameters in ϕ_i are in F.
   (a) The Σ_α formulas are those of the form

   $$\bigvee\!\!\!\!\bigvee_{n \in S} \varphi_n,$$

   where for each n the formula φ_n is a Σ_{β_n} formula and S is countable and is the halting set of an R-machine.
   (b) The Π_α formulas are those of the form

   $$\bigwedge\!\!\!\!\bigwedge_{n \in S} \varphi_n,$$

   where for each n the formula φ_n is a Π_{β_n} formula and S is countable and is the halting set of an R-machine.

The ℝ-computable infinitary formulas will be exactly the formulas which belong to either Σ_α or Π_α for some countable (i.e. ℝ-computable) α. Ash showed that Turing computable Σ_α formulas defined sets which were Σ^0_α [1].

We will say that a set is *semantically* ℝ-Σ_α if and only if it is the set of solutions to a ℝ-computable Σ_α formula, and similarly for Π_α. We will say that a set is *topologically* Σ^0_α if it is of that level in the standard Borel hierarchy using the order topology on ℝ.

THEOREM 2.5. *We characterize the topological structure of sets in the semantic hierarchy:*

1. *The semantically* ℝ-Σ_0 *sets are topologically* Δ^0_2.
2. *If* $0 < \alpha < \omega$, *then the semantically* ℝ-Σ_α *sets are included among the topologically* $\Sigma^0_{\alpha+1}$ *sets.*
3. *If* $\alpha \geq \omega$, *then the semantically* ℝ-Σ_α *sets are included among the topologically* Σ^0_α *sets.*

Proof. Since $\mathcal{A}$ is a $\mathbb{R}$-computable structure, the semantically $\mathbb{R}$-Σ_0 sets are all countable unions of semialgebraic sets, and the complements of semantically $\mathbb{R}$-Σ_0 are all countable unions of semialgebraic sets. Since all semialgebraic sets are topologically Δ_2^0 (that is, both topologically Σ_2^0 and Π_2^0), the countable unions of them are all topologically Σ_2^0. Now if the statement holds for $n \leq k$, it clearly holds for $n = k + 1$ by the definitions of the various classes involved.

Toward the final statement, notice that the semantically $\mathbb{R}$-Σ_ω sets are countable unions of sets at lower levels, and are all topologically Σ_ω. Above that level, the induction follows exactly as before. ⊣

At the finite levels, Cucker proved [10] that the union of all the semantically $\mathbb{R}$-Σ_n for $n < \omega$ is the class of Borel sets of finite order. Cucker [10] defined another arithmetical hierarchy: we call a set *computationally* $\Sigma_{\alpha+1}$ if it can be enumerated by a real machine with a computationally Σ_α oracle. In particular, the semi-decidable sets are the computationally Σ_1 sets. Cucker proved that for all $k < \omega$, the computationally Σ_k sets are exactly the semantically $\mathbb{R}$-Σ_k sets. It seems likely that this result could be generalized for transfinite α, but we do not have a proof of this.

2.3. Forcing as a Construction Technique. Aside from the lack of inverse functions, the most difficult part of classical computability theory to get by without is the priority construction. Unless this niche can be filled, we are not optimistic concerning the parallel between Turing-computable structures and $\mathbb{R}$-computable structures. Consequently, although there are ad-hoc methods to construct $\mathbb{R}$-incomparable sets [22], we give an example in this section that is potentially more easily generalized.

Proposition 2.6. *There exist sets A_0 and A_1 such that neither is computable by a $\mathbb{R}$-machine using the other as an oracle.*

Proof. The proof will closely follow the second proof given for the classical case by Lerman [21]. Let $(F, \leq)$ be the set of pairs of partial functions from $\mathbb{R}$ to 2 such that the cardinality of each component is less than 2^ω, partially ordered by extension in the sense that $(p_0, p_1) \leq (q_0, q_1)$ if and only if p_i extends q_i for each i. It suffices to satisfy the following requirements for every $e \in \mathbb{R}^\infty$:

$$P_{e,i} : M_e^{A_i} \neq A_{1-i}.$$

We say that $(p_0, p_1) \Vdash P_{e,i}$ if there is some x such that either $M_e^{p_i}(x) \downarrow \neq p_{1-i}(x)$ and the latter is defined, or for every p extending p_i, we have $M_e^p(x) \uparrow$.

Lemma 2.7 (Density Lemma). *For any pair (e, i), the set $\{p \in F : p \Vdash P_{e,i}\}$ is dense.*

Proof. Let $q = (q_0, q_1) \in F$. We will show that there is some $p \leq q$ such that $p \Vdash P_{e,i}$. Let x be outside the domain of q_{1-i}. If there is no pair $r = (r_0, r_1)$ such that r_i extends q_i and $M_e^{r_i}(x) \downarrow$, then $q \Vdash P_{e,i}$, so assume that such an r exists. Now we extend q_{1-i} by setting $p_{1-i} = q_{1-i} \cup \{(x, 1 - M_e^{q_i}(x))\}$. ⊣

The following Lemma is the only part of the construction which becomes genuinely more difficult in the uncountable case.

LEMMA 2.8 (Existence of a Generic). *Let $\mathcal{C}$ be the collection of all sets of the form $\{p \in F : p \Vdash P_{e,i}\}$. There exists a $\mathcal{C}$-generic set; that is, a pair of functions $G = (G_0, G_1)$ where $G_i : \mathbb{R} \to 2$ such that for each pair (e, i), the function G extends some element of F which forces $P_{e,i}$.*

PROOF. Let $G^0 := (\emptyset, \emptyset)$, and well-order the requirements (note that they will have order type 2^ω. We define $G^{\alpha+1}$ to be the extension of G^α which forces the αth requirement. For limit ordinals γ, we define $G^\gamma := \bigcup_{\beta<\gamma} G^\beta$. The union of all the G^α is a $\mathcal{C}$-generic. ⊣

Of course, we may take G to be total, by setting all undefined values to 0. Now we take A_0 to be the set whose characteristic function is G_0 and A_1 the set with characteristic function G_1. ⊣

§3. Effective Categoricity.

It is often possible to produce two classically computable structures which are isomorphic, but for which the isomorphism is not witnessed by a computable function. Any theory of effective mathematics must take account of this phenomenon.

DEFINITION 3.1. A computable structure $\mathcal{M}$ is said to be *computably categorical* if and only if for any computable structure $\mathcal{N} \simeq \mathcal{M}$ there is a computable function $f : \mathcal{N} \xrightarrow{\simeq} \mathcal{M}$. The number of equivalence classes under computable isomorphism contained in an isomorphism type is called its *computable dimension*.

In the present section, we describe progress toward a parallel to the following classical result:

THEOREM 3.2 (see [27], although it was almost certainly known earlier). *If V is a countable vector space over $\mathbb{Q}$,*

1. *There is a Turing-computable copy of V,*
2. *The categoricity properties are as follows:*
   (a) *If $dim(V)$ is finite, then V is computably categorical, and*
   (b) *If $dim(V) = \omega$, then the computable dimension of V is ω.*

The existence part of the theorem is still true without serious modification.

PROPOSITION 3.3. *Let $n \in \aleph_0 \cup \{\aleph_0, 2^{\aleph_0}\}$. Then there is a $\mathbb{R}$-computable vector space V^n of dimension n. Further, V^n has a $\mathbb{R}$-computable basis.*

PROOF. Consider the language of real vector spaces (addition, plus one scaling operation for each element of $\mathbb{R}$). Let $\{b_i : i \in I\}$ be a $\mathbb{R}$-computable set of constants, where $|I| = n$. The set of closed terms with constants from $\{b_i : i \in I\}$, modulo provable equivalence (in the theory of vector spaces) is a model of the theory of vector spaces, and has dimension n. ⊣

Of course, the categoricity result highlights an additional concern with $\mathbb{R}$-computation: It may happen that there is a $\mathbb{R}$-computable isomorphism with no $\mathbb{R}$-computable inverse. Thus, while the following result establishes, according to Definition 3.1, something very close to part 2a of Theorem 3.2, it falls short of full analogy.

Proposition 3.4. *Let $n < \aleph_0$. Then for any $\mathbb{R}$-computable real vector space W of dimension n, there is a computable isomorphism $f : V^n \to W$.*

Proof. Let $\{a_1, \dots, a_n\}$ be a basis of W. Each member of V^n is a $\mathbb{R}$-linear combination $\sum_{i=1}^{n} \lambda_i b_i$. We map $\sum_{i=1}^{n} \lambda_i b_i$ to $\sum_{i=1}^{n} \lambda_i a_i$. ⊣

The classical way to prove part 2b of Theorem 3.2 is to produce a computable vector space with a computable basis, and an isomorphic (i.e. same dimension) vector space with no computable basis. Without recourse to priority constructions, this strategy seems, for the present, very difficult in the $\mathbb{R}$-computable context.

§4. Geometry and Topology. In the talk by the first author at EMU 2008, an early slide asked for a context in which one could formulate effectiveness questions for results like Thom's Theorem on cobordism or the classification of compact 2-manifolds. Some work in the intervening months, which began at that meeting, has yielded interesting results in $\mathbb{R}$-computable topology.

An n-manifold is a topological space which is locally homeomorphic to $\mathbb{R}^n$, satisfying some fairly obvious regularity conditions on the intersections of the neighborhoods on which homeomorphism holds. The following definition is given in [9].

Definition 4.1. A *real-computable d-manifold M* consists of real-computable i, j, j', k, the *inclusion functions*, satisfying the following conditions for all $m, n \in \omega$.

- If $i(m, n) \downarrow= 1$, then $\phi_{j(m,n)}$ is a total real-computable homeomorphism from $\mathbb{R}^d$ into $\mathbb{R}^d$, and $\phi_{j'(m,n)} = \phi^{-1}_{j(m,n)}$, and $k(m, n) \downarrow= k(n, m) \downarrow= m$.
- If $i(m, n) \downarrow= 0$, then $k(m, n) \downarrow= k(n, m) \downarrow\in \omega$ with $i(k(m, n), m) = i(k(m, n), n) = 1$ and for all $p \in \omega$, if $i(p, m) = i(p, n) = 1$, then $i(p, k(m, n)) = 1$, and for all $q \in \omega$, if $i(m, q) = i(n, q) = 1$, then $i(k(m, n), q) = 1$ with
$$\mathrm{range}(\phi_{j(m,q)}) \cap \mathrm{range}(\phi_{j(n,q)}) = \mathrm{range}(\phi_{j(k(m,n),q)}).$$
- If $i(m, n) \notin \{0, 1\}$, then $i(m, n) \downarrow= i(n, m) \downarrow= -1$, and
$$(\forall p \in \omega)[i(p, m) \neq 1 \text{ or } i(p, n) \neq 1],$$
and for all $q \in \omega$, if $i(m, q)$ and $i(n, q)$ both lie in $\{0, 1\}$, then
$$\mathrm{range}(\phi_{j(k(m,q),q)}) \cap \mathrm{range}(\phi_{j(k(n,q),q)}) = \emptyset.$$

◦ For all $q \in \omega$, if $i(m, n) = i(n, q) = 1$, then $i(m, q) = 1$ and

$$\phi_{j(n,q)} \circ \phi_{j(m,n)} = \phi_{j(m,q)}.$$

In essence, each natural number m represents a chart U_m. The functions $i(m, n)$ tell whether U_m is a subset of U_n and whether U_n is a subset of U_m. The function $j(m, n)$ is the index for a computable map giving the inclusion of U_m in U_n.

4.1. Classifying Compact 2-Manifolds. Classification of n-manifolds up to homeomorphism in general is quite difficult. However, a well-known theorem of disputed priority offers the following classification of compact connected 2-manifolds.

THEOREM 4.2. *Let X be a compact connected 2-manifold. Then X is homeomorphic to a connected sum of 2-spheres, copies of $\mathbb{RP}^2$, copies of $S^1 \times S^1$, and copies of the Klein Bottle.*

Unpublished work by the first author and Montalban, inspired in part by discussions with R. Miller, gives an effective version of this result.

THEOREM 4.3 (Calvert–Montalban). *Let M and N be $\mathbb{R}$-computable compact 2-manifolds which are classically homeomorphic. Then there is a $\mathbb{R}$-computable homeomorphism $f : M \to N$.*

PROOF OUTLINE. We can triangulate each of M and N to form a finite simplicial complex. The function f consists of a mapping on the complexes, with a smoothing effect. ⊣

COROLLARY 4.4. *Let X be a compact connected $\mathbb{R}$-computable 2-manifold. Then X is homeomorphic by a $\mathbb{R}$-computable function to a connected sum of 2-spheres, copies of $\mathbb{RP}^2$, copies of $S^1 \times S^1$, and copies of the Klein Bottle.*

4.2. Computing Homotopy Groups. One standard set of topological invariants for a manifold M is the sequence of groups $(\pi_n(M))_{n\in\omega}$, where $\pi_n(M)$ is the group of continuous mappings from S^n to M, up to homotopy equivalence. Under the classical model of computation, manifolds are often represented by simplicial complexes in order to discuss the possibility of computing various topological invariants. Brown showed [6] that there is a procedure which will, given a finite simplicial complex M, compute a set of generators and relations for each of the groups $\pi_n(M)$. It is natural to ask, now that we have a notion of computation that gives us algorithmic access to the manifolds themselves, whether this can be computed directly from the manifolds. We restrict attention here to the case of π_1, studied in detail in [9], although it is likely that similar results could be established for π_n.

LEMMA 4.5 (Calvert–Miller [9]). *Every loop f in a computable manifold M is homotopic to a computable loop in M whose only real parameters are the base point and the inclusion functions necessary to define M.*

Nevertheless, the answer to the question of computing a fundamental group from a manifold is largely negative:

THEOREM 4.6 (Calvert–Miller [9]). *Let M be a $\mathbb{R}$-computable manifold which is connected but not simply connected. Then there is no algorithm to decide whether a given loop is nullhomotopic.*

THEOREM 4.7 (Calvert–Miller [9]). *There is no $\mathbb{R}$-computable function which will decide, given a $\mathbb{R}$-computable manifold, whether that manifold is simply connected.*

Nevertheless, there is a canonical family of loops, sufficient to represent the whole (but not recoverable by a uniform procedure) from which we could make the necessary computations for a fundamental group.

LEMMA 4.8 (Calvert–Miller [9]). *Let M be a $\mathbb{R}$-computable manifold. Then there is a $\mathbb{R}$-computable function S_M, defined on the naturals, such that the set $S_M(n)$ consists of a set of indices for loops and contains exactly one representative from each homotopy equivalence type.*

While we cannot effectively pass from an index for M to an index for S_M, this step includes all of the difficulty in computing $\pi_1(M)$:

THEOREM 4.9 (Calvert–Miller [9]). *Let M be a $\mathbb{R}$-computable manifold. Then there is a uniform procedure to pass from an index for S_M to an index for a real-computable presentation of the group $\pi_1(M)$.*

§5. Relations with Other Models.

5.1. Local Computability. Let T be a $\forall$-axiomatizable theory in a language with n symbols.

DEFINITION 5.1. A *simple cover* of $\mathcal{S}$ is a (finite or countable) collection $\mathfrak{U}$ of finitely generated models $\mathcal{A}_0, \mathcal{A}_1, \ldots$ of T, such that:

- every finitely generated substructure of $\mathcal{S}$ is isomorphic to some $\mathcal{A}_i \in \mathfrak{U}$; and
- every $\mathcal{A}_i \in \mathfrak{U}$ embeds isomorphically into $\mathcal{S}$.

A simple cover $\mathfrak{U}$ is *computable* if every $\mathcal{A}_i \in \mathfrak{U}$ is a computable structure whose domain is an initial segment of ω. $\mathfrak{U}$ is *uniformly computable* if the sequence $\langle(\mathcal{A}_i, \overline{a}_i)\rangle_{i\in\omega}$ can be given uniformly: there must exist a computable function which, on input i, outputs a tuple of elements $\langle e_1, \ldots, e_n, \langle a_0, \ldots, a_k\rangle\rangle \in \omega^n \times \mathcal{A}_i^{<\omega}$ such that $\{a_0, \ldots, a_{k_i}\}$ generates $\mathcal{A}_i$ and ϕ_{e_j} computes the j-th function, relation, or constant in $\mathcal{A}_i$.

Definition 5.2. An embedding $f : \mathcal{A}_i \hookrightarrow \mathcal{A}_j$ *lifts* to the inclusion $\mathcal{B} \subset \mathcal{C}$, via isomorphisms $\beta : \mathcal{A}_i \twoheadrightarrow \mathcal{B}$ and $\gamma : \mathcal{A}_j \twoheadrightarrow \mathcal{C}$, if the diagram below commutes:

$$\begin{array}{ccc} \mathcal{B} & \xrightarrow[\subseteq]{} & \mathcal{C} \\ \beta \uparrow \cong & & \gamma \uparrow \cong \\ \mathcal{A}_i & \xrightarrow{f} & \mathcal{A}_j \end{array} \qquad \text{with } \gamma \circ f = \beta$$

A *cover* of $\mathcal{S}$ consists of a simple cover $\mathfrak{U} = \{\mathcal{A}_0, \mathcal{A}_1, ...\}$ of $\mathcal{S}$, along with sets $I_{ij}^{\mathfrak{U}}$ (for all $\mathcal{A}_i, \mathcal{A}_j \in \mathfrak{U}$) of injective homomorphisms $f : \mathcal{A}_i \hookrightarrow \mathcal{A}_j$, such that:

1. for all finitely generated substructures $\mathcal{B} \subseteq \mathcal{C}$ of $\mathcal{S}$, there exists $i, j \in \omega$ and an $f \in I_{ij}^{\mathfrak{U}}$ which lifts to $\mathcal{B} \subseteq \mathcal{C}$ via some isomorphisms $\beta : \mathcal{A}_i \twoheadrightarrow \mathcal{B}$ and $\gamma : \mathcal{A}_j \twoheadrightarrow \mathcal{C}$; and
2. for every i and j, every $f \in I_{ij}^{\mathfrak{U}}$ lifts to an inclusion $\mathcal{B} \subseteq \mathcal{C}$ in $\mathcal{S}$ via some isomorphism β and γ.

This cover is *uniformly computable* if $\mathfrak{U}$ is a uniformly computable simple cover of $\mathcal{S}$ and there exists a c.e. set W such that for all $i, j \in \omega$

$$I_{ij}^{\mathfrak{U}} = \{\phi_e \upharpoonright \mathcal{A}_i : \langle i, j, e\rangle \in W\}.$$

A structure $\mathcal{B}$ is *locally computable* if it has a uniformly computable cover.

Proposition 5.3 ([23]). *A structure $\mathcal{S}$ is* locally computable *if and only if it has a uniformly computable simple cover.*

Proposition 5.4 ([23]). *The ordered field of real numbers is not locally computable.*

However, the ordered field of real numbers is trivially $\mathbb{R}$-computable. It appears at first that the ordering might be essential in escaping local computability.

Definition 5.5. A $\mathbb{R}$-machine is said to be *equational* if and only if each branch node is decided by a polynomial *equation*. We call a structure *equationally $\mathbb{R}$-computable* if its diagram is computable by an equational $\mathbb{R}$-machine.

Lemma 5.6 (Path Decomposition for Equational Machines). *Let M be an equational $\mathbb{R}$-machine. Then the halting set of M is a countable disjoint union of algebraic sets.*

Proof. The proof is exactly the same as for normal $\mathbb{R}$-machines. ⊣

Corollary 5.7. *The ordered field of real numbers is not equationally $\mathbb{R}$-computable.*

Theorem 5.8. *There is an equationally $\mathbb{R}$-computable structure which is not locally computable.*

Proof. Let S be a noncomputable set of natural numbers, and denote by C_n a cyclic graph on n vertices (i.e. an n-gon). Now let $\mathcal{G}$ be the structure given by

$$\left(\dot{\bigcup_{n\in S}} C_{2n}\right) \dot{\cup} \left(\dot{\bigcup_{n\notin S}} C_{2n+1}\right).$$

To show that $\mathcal{G}$ is equationally $\mathbb{R}$-computable, we observe that the disjoint union of two $\mathbb{R}$-computable structures is $\mathbb{R}$-computable (since the same is true of the cardinal sum). However, each of the graphs C_k has a $\mathbb{R}$-computable copy by Lagrange interpolation.

Suppose f is a uniform computable enumeration of the finitely generated substructures of $\mathcal{G}$. Then we could compute whether $n \in S$ by searching the structures indexed by $f(t)$ for successive t until we see a substructure of type C_{2n} or of type C_{2n+1}. Since S is noncomputable, no such f can exist, so that $\mathcal{G}$ is not locally computable. ⊣

Theorem 5.9. *There is a locally computable structure which is not $\mathbb{R}$-computable.*

Proof. Let X be the set of all countable graphs with universe ω, and let E be the isomorphism relation on X. Now E is complete analytic [14], so $F = E^c$ is complete co-analytic. Now for any $x \in X$, we have $\neg xFx$, and for any $x, y \in X$ we have xFy if and only if yFx. Thus, F defines the adjacency relation of a graph on X. Let $\mathcal{X}$ denote the graph (X, F).

Now $\mathcal{X}$ is not real-computable, since its diagram is complete co-analytic (contradicting path decomposition). We will show that $\mathcal{X}$ is locally computable. Now the finitely generated substructures of $\mathcal{X}$ are all finite graphs, and it only remains to determine which finite graphs are included. Let T be the following graph:

Let $\mathcal{G}$ be a finite T-free graph. We will show that $\mathcal{G}$ embeds in $\mathcal{X}$. Let $\mathcal{G} = (\{0, \dots, n\}, G)$. We will define an equivalence structure R with universe $N = \{0, \dots, n\}$. For $x, y \in N$, we say that xRy if and only if $\neg xGy$. This relation R will be reflexive and symmetric. Since G is T-free, R will also be transitive. Now since the isomorphism relation is Borel complete [14], there is a function $f : N \to X$ such that xRy if and only if $f(x)Ef(y)$. This function can also be required to be injective [17].

Now let Φ be a computable Friedberg enumeration of finite graphs up to isomorphism (i.e. a total computable function whose range consists of an index for exactly one representative from each isomorphism class of finite graphs). Such an enumeration was given in [8]. We will define a Friedberg enumeration Ψ of finite T-free graphs up to isomorphism as follows: $\Psi(x)$ will be $\Phi(x')$ for the least x'

such that $\Phi(x')$ is T-free and $\Phi(x') \notin ran(\Psi|_x)$. Since all of the graphs are finite, we can effectively check whether each is T-free, so that Ψ is computable. Now Ψ provides a uniform simple computable cover for $\mathcal{X}$. ⊣

COROLLARY 5.10. *There is another structure $\tilde{\mathcal{X}}$ with the same uniform simple computable cover as $\mathcal{X}$, such that $\tilde{\mathcal{X}}$ is $\mathbb{R}$-computable.*

PROOF. Let $\tilde{\mathcal{X}}$ be the disjoint union

$$\dot{\bigcup_{x \in \omega}} \Psi(x).$$

Now $\tilde{\mathcal{X}}$ is countable, and so is trivially $\mathbb{R}$-computable. ⊣

One can say more about the structure described in Theorem 5.9. The structure satisfies a stronger condition called *perfectly local computability*. We recall the definition of perfectly locally computable and leave the details to the reader.

DEFINITION 5.11. Let $\mathfrak{U}$ be a uniformly computable cover for a structure $\mathcal{S}$. A Set M is a *correspondence system* for $\mathfrak{U}$ and $\mathcal{S}$ if it satisfies all of the following:

1. Each element of M is an embedding of some $\mathcal{A}_i \in \mathfrak{U}$ into $\mathcal{S}$; and
2. Every $\mathcal{A}_i \in \mathfrak{U}$ is the domain of some $\beta \in M$; and
3. Every generated $\mathcal{B} \subset \mathcal{S}$ is the image of some $\beta \in M$; and
4. For every i and j and every $\beta \in M$ with domain $\mathcal{A}_i$, every $f \in I_{ij}^{\mathfrak{U}}$ lifts to an inclusion $\beta(\mathcal{A}_i) \subset \gamma(\mathcal{A}_i)$ via β and some $\gamma \in M$; and
5. For every i, every $\beta \in M$ with domain $\mathcal{A}_i$, and every finitely generated $\mathcal{C} \subset \mathcal{S}$ containing $\beta(\mathcal{A}_i)$, there exist a j and an $f \in I_{ij}^{\mathfrak{U}}$ which lifts to $\beta(\mathcal{A}_i) \subset \mathcal{C}$ via β and some $\gamma : \mathcal{A}_j \twoheadrightarrow \mathcal{C} \in M$.

The correspondence system is *perfect* if it also satisfies

6. For every finitely generated $\mathcal{B} \subset \mathcal{S}$, if $\beta : \mathcal{A}_i \twoheadrightarrow \mathcal{B}$ and $\gamma : \mathcal{A}_j \twoheadrightarrow \mathcal{B}$ both lie in M and have image $\mathcal{B}$, then $\gamma^{-1} \circ \beta \in I_{ij}^{\mathfrak{U}}$.

If a perfect correspondence system exists, then its elements are called *perfect matches* between their domains and their images. We then say that $\mathcal{S}$ is *perfectly locally computable* with *perfect cover* $\mathfrak{U}$.

There is one point about the structure in Theorem 5.9 that is unsatisfactory. While $\mathcal{X}$ itself is certainly not $\mathbb{R}$-computable, our proof does not exclude the possibility that there might be some isomorphic copy of $\mathcal{X}$ which is $\mathbb{R}$-computable. A referee has conjectured that there is such a copy, but we leave both this and the following broader question open:

QUESTION 5.12. Is there a locally computable structure of size continuum which has no isomorphic copy which is $\mathbb{R}$-computable?

5.2. Σ-Definability. The following definition is standard, and appears in equivalent forms in [3] and [11].

DEFINITION 5.13. Given a structure $\mathcal{M}$ with universe M, we define a new structure $HF(\mathcal{M})$ as follows.

1. The universe of $HF(\mathcal{M})$ is the union of the chain $HF_n(M)$ defined as follows:
   (a) $HF_0(M) = M$
   (b) $HF_{n+1}(M) = \mathcal{P}^{<\omega}(M \cup HF_n(M))$, where $\mathcal{P}^{<\omega}(S)$ is the set of all finite subsets of S
2. The language for $HF(\mathcal{M})$ consists of a unary predicate U for $HF_0(M)$, as well as a predicate $\in$ interpreted as membership, plus a symbol σ^* for each symbol σ of the language of $\mathcal{M}$, given the interpretation of σ on $M = HF_0(M)$.

Ershov gave a definition [11] of a notion generalizing computability to structures other than $\mathfrak{N}$. We will first give Barwise's definition [3] of the class of Σ-formulas.

DEFINITION 5.14. The class of Σ-formulas are defined by induction.

1. Each Δ_0 formula is a Σ-formula.
2. If Φ and Ψ are Σ-formulas, then so are $(\Phi \wedge \Psi)$ and $(\Phi \vee \Psi)$.
3. For each variable x and each term t, if Φ is a Σ-formula, then the following are also Σ-formulas:
   (a) $\exists(x \in t)\ \Phi$
   (b) $\forall(x \in t)\ \Phi$, and
   (c) $\exists x\Phi$.

A predicate S is called a *Δ-predicate* if both S and its complement are defined by Σ-formulas.

DEFINITION 5.15. Let $\mathcal{M}$ and $\mathcal{N} = (N, P_0, P_1, \dots)$ be structures. We say that $\mathcal{N}$ is Σ-definable in $HF(\mathcal{M})$ if and only if there are Σ-formulas $\Psi_0, \Psi_1, \Psi_1^*, \Phi_0, \Phi_0^*, \Phi_1, \Phi_1^*, \dots$ such that

1. $\Psi_0^{HF(\mathcal{M})} \subseteq HF(\mathcal{M})$ is nonempty,
2. Ψ_1 defines a congruence relation on $\left(\Psi_0^{HF(\mathcal{M})}, \Phi_0^{HF(\mathcal{M})}, \Phi_1^{HF(\mathcal{M})}, \dots\right)$,
3. $(\Psi_1^*)^{HF(\mathcal{M})}$ is the relative complement in $(\Psi_0^{HF(\mathcal{M})})^2$ of $\Psi_1^{HF(\mathcal{M})}$,
4. For each i, the set $(\Phi_i^*)^{HF(\mathcal{M})}$ is the relative complement in $\Psi_0^{HF(\mathcal{M})}$ of $\Phi_i^{HF(\mathcal{M})}$, and
5. $\mathcal{N} \simeq \left(\Psi_0^{HF(\mathcal{M})}, \Phi_0^{HF(\mathcal{M})}, \Phi_1^{HF(\mathcal{M})}, \dots\right) / _{\Psi_1^{HF(\mathcal{M})}}$.

THEOREM 5.16 (Calvert [7]). *The structures which have isomorphic copies Σ-definable over $HF(\mathbb{R})$ are exactly the ones which have isomorphic copies which are $\mathbb{R}$-computable.*

An interesting consequence of this (an immediate corollary of Theorem 5.16 and a result of Morozov and Korovina [24]) gives a sense in which some ℝ-computable structures can be approximated by classically computable structures.

DEFINITION 5.17. Let $\mathcal{A}$ and $\mathcal{B}$ be structures in a common signature. We write that $\mathcal{A} \leq_1 \mathcal{B}$ if $\mathcal{A}$ is a substructure of $\mathcal{B}$, and for all existential formulas $\varphi(\bar{x})$ and for all tuples $\bar{a} \subseteq \mathcal{A}$, we have

$$\mathcal{B} \models \varphi(\bar{a}) \Rightarrow \mathcal{A} \models \varphi(\bar{a}).$$

COROLLARY 5.18. *For any ℝ-computable structure $\mathcal{M}$ whose defining machine involves only rationals as parameters, there is a computable structure $\mathcal{M}^*$ such that $\mathcal{M}^* \leq_1 \mathcal{M}$.*

Actually, the necessary result of Morozov and Korovina calls for structures Σ-definable over ℝ without parameters, but since rational parameters may be defined by Σ-formulas, this extension is possible.

5.3. F-Parameterizability. Morozov introduced a concept that he called *F-parameterizability* in order to understand the elementary substructure relation on both automorphism groups and the structure of hereditarily finite sets over a given structure [26]. In a talk at Stanford University, though, he identified this notion as one "which generalizes the notion of computable" [25].

DEFINITION 5.19 ([26]). Let $\mathcal{M}$ be a structure in a finite relational language $\left(P_n^{k_n}\right))_{n\leq k}$. We say that $\mathcal{M}$ is F-parameterizable if and only if there is an injection $\xi : \mathcal{M} \to \omega^\omega$ with the following properties:

1. The image of ξ is analytic in the Baire space, and
2. For each n, the set $\left\{(\xi(a_i))_{i\leq k_n} : \mathcal{M} \models P_n(\bar{a})\right\}$ is analytic.

The function ξ is called an F-parameterization of $\mathcal{M}$. Morozov also introduced the following stronger condition, essentially requiring that $\mathcal{M}$ be able to define its own F-parameterization.

DEFINITION 5.20 ([26]). Let $\mathcal{M}$ be an F-parameterizable structure. We say that $\mathcal{M}$ is weakly selfparameterizable if and only if there are functions $\Xi, p : \mathcal{M} \times \omega \to \omega$, both definable without parameters in $HF(\mathcal{M})$, with the following properties:

1. For all $x \in \mathcal{M}$ and all $m \in \omega$, we have $\Xi(x, m) = \xi(x)[m]$, and
2. For all $f \in \omega^\omega$ there is some $x \in \mathcal{M}$ such that for all $n \in \omega$ we have $p(x, n) = f(n)$.

In making sense of effectiveness on uncountable structures, a major motivation is to describe a sense in which real number arithmetic — an operation that, while not Turing computable, does not seem horribly ineffective — can be considered to be effective.

PROPOSITION 5.21 (Morozov [26]). *The real field is weakly F-selfparameterizable.*

Outline of proof. Define a function $\xi : \mathbb{R} \to \omega^\omega$ mapping x to its decimal expansion. This function is definable without parameters in $HF(\mathbb{R})$, in the sense required by Definition 5.20. ⊣

Theorem 5.22 (Calvert [7]). *Every $\mathbb{R}$-computable structure is F-parameterizable. On the other hand, the structure $(\mathbb{R}, +, \cdot, 0, 1, e^x)$ is weakly F-selfparameterizable but not $\mathbb{R}$-computable.*

§6. Conclusion. We state here some open problems arising from issues discussed in the present paper. The first is perhaps the most vital.

Problem 6.1. Develop a substitute for the priority method which is capable of handling constructions with injury.

After this paper was submitted, we became aware of a paper of Ashaev [2] which may answer the challenge of Problem 6.1, at least in part. However, the extent of the applicability of his method is still unknown.

Question 6.2. Is it true that for any $\mathbb{R}$-computable finite dimensional $\mathbb{R}$-vector spaces M and N with the same dimension, there is a $\mathbb{R}$-computable isomorphism from M to N?

Conjecture 6.3. *A $\mathbb{R}$-computable $\mathbb{R}$-vector space of dimension greater than $\aleph_0$ is not $\mathbb{R}$-computably categorical.*

We would also like to know about the categoricity of vector spaces of dimension $\aleph_0$, but are not ready to hazard a conjecture at this time.

Question 6.4. Does there exist a $\mathbb{R}$-computable Banach space of infinite dimension in the language of vector spaces, augmented by a sort for $\mathbb{R}$ and a function interpreted as the norm?

Question 6.5. Does there exist a $\mathbb{R}$-computable Hilbert space of infinite dimension in the language of vector spaces, augmented by a sort for $\mathbb{R}$ and a binary function interpreted as the inner product?

On each of the previous two questions, the authors had difficulty guaranteeing completeness. In any case, the sense intended is that the norm and inner product, as well as the vector space operations, should be $\mathbb{R}$-computable.

REFERENCES.

[1] C.J. Ash and J.F. Knight, *Computable Structures and the Hyperarithmetical Hierarchy*, Studies in Logic and the Foundations of Mathematics, vol. 144, Elsevier, 2000.

[2] I.V. Ashaev, Priority method in generalized computation. In *Recursion Theory and Complexity: Proceedings of the Kazan '97 Workshop*, de Gruyter, 1999, 1–13.

[3] J. Barwise, *Admissible Sets and Structures: an Approach to Definability Theory*, Springer, 1975.
[4] L. Blum, F. Cucker, M. Shub, and S. Smale, *Complexity and Real Computation*, Springer, 1997.
[5] L. Blum, M. Shub, and S. Smale, On a theory of computation and complexity over the real numbers, *Bulletin of the American Mathematical Society* (New Series) **21** (1989), 1–46.
[6] E.H. Brown, Computability of Postnikov complexes, *Annals of Mathematics* **65** (1957), 1–20.
[7] W. Calvert, On three notions of effective computation over ℝ, preprint, 2009.
[8] W. Calvert, D. Cummins, S. Miller, and J.F. Knight, Comparing classes of finite structures, *Algebra and Logic* **43** (2004), 374–392.
[9] W. Calvert and R. Miller, Real computable manifolds and homotopy groups. In *Unconventional Computation 2009*, Lecture Notes in Computer Science, no. 5715, Springer, 2009, 98–109.
[10] F. Cucker, The arithmetical hierarchy over the reals, *Journal of Logic and Computation* **2** (1992), 375–395.
[11] Ю.Л. Ершов, Определимость и Вичислимость, Сибирская Школа Алгебры и Логики, Научная Книга, 1996.
[12] J.E. Fenstad and P.G. Hinman (editors), *Generalized Recursion Theory*, Studies in Logic and the Foundations of Mathematics, no. 79, North-Holland, 1974.
[13] J.E. Fenstad, R.O. Gandy, and G.E. Sacks (editors), *Generalized Recursion Theory ii*, Studies in Logic and the Foundations of Mathematics, no. 94, North-Holland, 1978.
[14] H. Friedman and L. Stanley, A Borel reducibility theory for classes of countable structures, *Journal of Symbolic Logic* **54** (1989), 894–914.
[15] V. Harizanov, Pure computable model theory. In *Handbook of Recursive Mathematics*, vol. 1, Studies in Logic and the Foundations of Mathematics, no. 138, North-Holland, 1998, 3–114.
[16] ———, Turing degrees of certain isomorphic images of computable relations, *Annals of Pure and Applied Logic* **93** (1998), 103–113.
[17] G. Hjorth and A.S. Kechris, Analytic equivalence relations and Ulm-type classifications, *Journal of Symbolic Logic* **60** (1995), 1273–1300.
[18] A.S. Kechris, *Classical Descriptive Set Theory*, Graduate Texts in Mathematics, no. 156, Springer, 1995.
[19] N.G. Khisamiev, Constructive Abelian groups. In *Handbook of Recursive Mathematics*, vol. 2, Studies in Logic and the Foundations of Mathematics, no. 139, North-Holland, 1998, 1177–1231.
[20] S.C. Kleene, On notation for ordinal numbers, *Journal of Symbolic Logic* **3** (1938), 150–155.

[21] M. Lerman, *Degrees of Unsolvability*, Perspectives in Mathematical Logic, Springer, 1983.

[22] K. Meer and M. Ziegler, An explicit solution to Post's problem over the reals. In *Proceedings of the 15th International Symposium on Fundamentals of Computation Theory*, Lecture Notes in Computer Science, vol. 3623, Springer, 2005, 467–478.

[23] R. Miller, Locally computable structures. *Computation and Logic in the Real World*, Lecture Notes in Computer Science, vol. 4497, Springer, 2007, 575–584.

[24] A. Morozov and M. Korovina, On Σ-definability of countable structures over real, complex numbers, and quaternions, *Algebra and Logic* **47** (2008), 193–209.

[25] A.S. Morozov, On elementary submodels of F-parameterizable models, *Bulletin of Symbolic Logic* **12** (2006), 146, Abstract of invited talk in 2005 Annual Meeting of the Association for Symbolic Logic, Stanford University, Stanford, California, March 19–22, 2005.

[26] А.С. Морозов, Элементарные подмодели параметризуемых моделей, Сибирский Математический Журнал **47** (2006), 595–612.

[27] A. Nurtazin, Strong and weak constructivizations and computable families, *Algebra and Logic* **13** (1974), 177–184.

[28] G.E. Sacks, *Higher Recursion Theory*, Perspectives in Mathematical Logic, Springer, 1990.

[29] C. Spector, Recursive well orderings, *Journal of Symbolic Logic* **20** (1955), 151–163.

[30] M. Ziegler, (Short) survey of real hypercomputation. In *Computation and Logic in the Real World*, Lecture Notes in Computer Science, vol. 4497, Springer, 2007, 809–824.

Department of Mathematics,
Southern Illinois University,
Carbondale, IL 62901, USA
E-mail: wcalvert@siu.edu

Department of Mathematics & Statistics,
Murray State University,
Murray, KY 42071, USA
E-mail: ted.porter@murraystate.edu

INFINITE TIME TURING MACHINES AND AN APPLICATION TO THE HIERARCHY OF EQUIVALENCE RELATIONS ON THE REALS

SAMUEL COSKEY AND JOEL DAVID HAMKINS

Abstract We describe the basic theory of infinite time Turing machines and some recent developments, including the infinite time degree theory, infinite time complexity theory, and infinite time computable model theory. We focus particularly on the application of infinite time Turing machines to the analysis of the hierarchy of equivalence relations on the reals, in analogy with the theory arising from Borel reducibility. We define a notion of infinite time reducibility, which lifts much of the Borel theory into the class $\utilde{\Delta}^1_2$ in a satisfying way.

Infinite time Turing machines fruitfully extend the operation of ordinary Turing machines into transfinite ordinal time and by doing so provide a robust theory of computability on the reals. In a mixture of methods and ideas from set theory, descriptive set theory and computability theory, the approach provides infinitary concepts of computability and decidability on the reals, which climb nontrivially into the descriptive set-theoretic hierarchy (at the level of Δ^1_2) while retaining a strongly computational nature. With infinite time Turing machines, we have infinitary analogues of numerous classical concepts, including the infinite time Turing degrees, infinite time complexity theory, infinite time computable model theory, and now also the infinite time analogue of the theory of Borel equivalence relations under Borel reducibility.

In this article, we shall give a brief review of the machines and their basic theory, and then explain in a bit more detail our recent application of infinite time computability to an analogue of Borel equivalence relation theory, a full account of which is given in [2]. The basic idea of this application is to replace the concept of Borel reducibility commonly used in that theory with forms of infinite time computable reducibility, and study the accompanying hierarchy of equivalence relations. This approach retains much of the Borel analysis and results, while also illuminating a part of the hierarchy of equivalence relations that seems beyond the reach of the Borel theory, including many highly canonical equivalence relations that are infinite time computable but not Borel, such as the isomorphism relations for diverse classes of countable structures.

---

[1]The second author's research has been supported in part by grants from the Research Foundation of CUNY and from the National Science Foundation.

[2]Math Subject Codes: 03D30, 03D60, 03E15. Keywords: infinite time Turing machines, infinitary computability, ordinal computation.

Major parts of this article are adapted from the surveys [8] and [7] and from our article [2] on infinite time computable equivalence relations. Infinite time Turing machines were first studied by Hamkins and Kidder in 1989, with the core introduction provided by Hamkins and Lewis [9]. The theory has now been extended by many others, including Philip Welch, Peter Koepke, Benedikt Löwe, Daniel Seabold, Ralf Schindler, Vinay Deolalikar, Russell Miller, Steve Warner, Giacomo Lenzi, Erich Monteleone, Samuel Coskey and others. Numerous precursors to the theory include Blum–Shub–Smale machines (1980s), Büchi machines (1960s) and accompanying developments, Barry Burd's model of Turing machines with "blurs" at limits (1970s), the extensive development of α-recursion and E-recursion theory, a part of higher recursion theory (since the 1970s), Jack Copeland's accelerated Turing machines (1990s), Ryan Bissell-Siders' ordinal machines (1990s), and more recently, Peter Koepke's ordinal Turing machines and ordinal register machines (2000s). The expanding literature involving infinite time Turing machines includes [9], [28], [29], [30], [23], [15], [10], [26], [16], [5], [6], [25], [3], [14], [7], [32], [31], [20], [8], [13], [12], [11] and others.

§1. A Brief Review of Infinite time Turing machines. Infinite time Turing machines have exactly the same hardware as their classical finite time counterparts, with a head moving back and forth on a semi-infinite paper tape, writing 0s and 1s according to the rigid instructions of a finite program with finitely many states. What is new about the infinite time Turing machines is that their operation is extended into transfinite ordinal time. For convenience, the machines are implemented with a three-tape model, with separate tapes for input, scratch work and output. The machine operates at successor stages of computation in exactly the

			q						
input:	0	0	1	1	1	1	0	0	...
scratch:	1	1	0	1	0	0	1	0	...
output:	0	0	1	0	1	0	1	1	...

classical manner, according to the program instructions. Computation is extended to limit ordinal stages simply by defining the limit configuration of the machines. The idea is to try to preserve as much as possible the information that the computation has been creating up to that stage, preserving it in the limit configuration as a kind of limit of the earlier configurations. Specifically, at any limit ordinal stage ξ, the machine enters what we call the *limit* state, one of the distinguished states along with the *start* and *halt* states; the head is reset to the first cell at the left; and each cell of the tape is updated with the lim sup of the values previously displayed in that cell. Having thus specified the complete configuration of the machine at stage ξ, the computation may now continue to stage $\xi + 1$ and so on.

Computational output is given only when the machine explicitly enters the *halt* state, and computation ceases when this occurs.

Since the tapes naturally accommodate infinite binary strings – and there is plenty of time for the head to inspect every cell – the natural context for input and output to the machines is Cantor space 2^ω, which we denote by $\mathbb{R}$ and refer to as the reals. Thus, the machines provide an infinitary notion of computability on the reals. A program p computes the partial function $\varphi_p : \mathbb{R} \to \mathbb{R}$, defined by $\varphi_p(x) = y$ if program p on input x yields output y, where the output of a computation is the content of the output tape when the machine enters the *halt* state. A subset $A \subseteq \mathbb{R}$ is *infinite time decidable* if the characteristic function of A is infinite time computable. The set A is *infinite time semi-decidable* if the constant partial function $1 \upharpoonright A$ is computable. This is equivalent to A being the domain of an infinite time computable function (but not necessarily equivalent to A being the range of such a function). Elementary results in [9] show that the arithmetic sets are exactly those that are decidable in time uniformly less than ω^2 and the hyperarithmetic sets are those that are decidable in time less than some recursive ordinal. The power of the machines, however, reaches much higher than this into the descriptive set theoretic hierarchy.

For example, every Π^1_1 and Σ^1_1 set is infinite time decidable. To see this, it suffices to show that the complete Π^1_1 set WO, consisting of reals coding a well-ordered relation on ω, is infinite time computable. This is accomplished by the *count-through* argument of [9, Theorem 2.2], which we should like to sketch here. Given a real x, we view it as coding the relation $\lhd$ on ω for which $n \lhd m$ if and only if the $\langle n, m\rangle$ bit of x is 1. The assertion that $\lhd$ is a linear order is arithmetic in x, and therefore easily determined by the machines. After this, the machine will check for well-foundedness essentially by counting through the order, relying on the fact that the computational steps are themselves well-ordered. Specifically, the machine places an initial guess for the current minimal element in the relation $\lhd$, updating it with better guesses as they are encountered. At each revision, the machine flashes a certain master flag, so that at the limit stage the machine can know if the guess was changed infinitely often, indicating ill-foundedness (the machine should reset the master flag at limits of limit stages). Otherwise, the true current minimal element has been found, and so the machine can delete all mention of it from the field of the relation coded by x. Iterating this, the algorithm in effect systematically erases the well-founded initial segment of the relation coded by the input real, until either nothing is left or the ill-founded part is discovered, either of which can be determined. In this way, membership in WO is infinite time decidable. It follows that every Π^1_1 and Σ^1_1 set is infinite time decidable, and so the machines climb properly into Δ^1_2. Meanwhile, the class of infinite time decidable sets is easily observed to be contained in Δ^1_2, and in fact the class Δ^1_2 is closed under the infinite time jump operations and is therefore stratified by a significant part of the infinite time Turing degrees.

Although transfinite, computations are nevertheless inherently countable, since an easy cofinality argument establishes that every computation either halts or repeats by some countable ordinal stage. An ordinal α is said to be *clockable*, if there is a computation $\varphi_p(0)$ halting on exactly the αth step. A real x is *writable* if it is the output of a computation $\varphi_p(0)$, and an ordinal is writable if it is coded by such a real. Because there are only countably many programs, it follows that there are only countably many clockable and writable ordinals. The clockable and writable ordinals extend through all the recursive ordinals and far beyond; their supremum is recursively inaccessible and more. The writable ordinals form an initial segment of the ordinals, since whenever an ordinal is writable, then the algorithm writing it can be easily modified to write a code for any smaller ordinal. But the same is not true for the clockable ordinals; in the midst of the clockable ordinals, there are increasingly complex forbidden regions at which no (parameter-free) infinite time Turing machine can halt.

Let us quickly sketch the argument that such gaps in the clockable ordinals exist, since this is an interesting exercise in ordinal reflection that constitutes a basic method of many later constructions in the theory. Consider the algorithm that simulates all programs on input 0 simultaneously, by some bookkeeping method that reserves and manages sufficient separate space for each, simulating ω many steps of computation for each program in each ω many steps of actual computation. Our algorithm might keep careful track of which programs have halted, and pay attention to find a stage at which none of the programs halt. Since such a stage exists above the supremum of all clockable ordinals, we will definitely find such a stage eventually. Since our algorithm can recognize the first such stage, we can arrange that it halts immediately after this discovery. So we have described a computational procedure that will halt at an ordinal stage that is larger than a stage at which no computations halted, and so there are gaps in the clockable ordinals, as desired. A careful analysis of the algorithm shows that the first gap after any clockable ordinal has order type ω, essentially because it takes ω many additional steps to realize that a gap has been reached. Modified algorithms search for longer gaps and show that there must be increasingly complex gaps at increasingly complex admissible limit stages – for any clockable or writable ordinal α, there are gaps of size at least α. The structure of these gaps exhibits the same complexity as the infinite time halting problem.

Although it was established in [9] that the clockable and writable ordinals have the same order type, perhaps the main question left open in that paper was whether the supremum of these ordinals was the same. This was settled in the affirmative by Philip Welch in [30]. Another way to describe the result is that whenever program p on input x yields a halting computation, then there is another computation that writes out a certificate of this computation, a real coding the entire computation history including a well-ordered relation whose order type is the length of the computation. This important fact, far from obvious, relies on a subtle treatment of eventual writability and constitutes a foundation of many further developments of the theory, including the applications we mention in this article.

The reflective aspect of the count-through argument described above consists of the observation that any decidable property that holds of a real that might be encountered during the course of a computation must hold of a writable real, since we may embark on the computational search to find such a witness and output it when it is found. This idea is greatly extended by the λ-ζ-Σ theorem of Philip Welch. Specifically, [9] defines that a real x is *eventually writable* if there is a computation $\varphi_p(0)$ for which x appears on the output tape from some point on (even if the computation does not halt), and x is *accidentally writable* if it appears on any of the tapes at any stage during a computation $\varphi_p(0)$. By coding ordinals with reals, we obtain the notions of eventually and accidentally writable ordinals. If λ is the supremum of the clockable or writable ordinals, ζ is the supremum of the eventually writable ordinals and Σ is the supremum of the accidentally writable ordinals, then [9] establishes $\lambda < \zeta < \Sigma$. The λ-ζ-Σ theorem of Welch [29] asserts moreover that $L_\lambda \prec_{\Sigma_1} L_\zeta \prec_{\Sigma_2} L_\Sigma$, using the initial segments of Gödel's constructible universe, and furthermore, that these ordinals are characterized as the least example of this pattern. This result precisely expresses the sense in which the algorithms may pull down witnesses from the accidentally writable realm into the eventually writable or writable realms. At the heart of the proof and the result is the fact that every computation repeats the stage ζ configuration at stage Σ.

Many of the fundamental constructions of classical finite time computability theory carry over to the infinite time context. For example, one can prove the infinite time analogues of the *smn*-theorem, the Recursion theorem and the undecidability of the infinite time halting problem, by essentially the classical arguments. Some other classical facts, however, do not directly generalize. For example, it is not true in the infinite time context that if the graph of a function f is semi-decidable, then the function is computable. This is a consequence of the following:

Theorem 1.1 (Lost Melody Theorem). *There is a real c such that $\{c\}$ is infinite time decidable, but c is not writable.*

The real c, a lost melody that you cannot sing on your own, although you can recognize it yes-or-no when someone sings it to you, exhibits sufficient internal structure that $\{c\}$ is decidable, but is too complicated itself to be writable. That is, we can recognize whether a given real y is c or not, but we cannot produce c from nothing. The function $f(x) = c$ with constant value c, therefore, is not computable, because c is not writable, but the graph is decidable, because we can recognize whether a pair has the form (x, c).

The infinite time analogue of the halting problem breaks into lightface and boldface versions, $h = \{ p \mid \varphi_p(p)\downarrow \}$ and $H = \{ (p, x) \mid \varphi_p(x)\downarrow \}$, respectively. These are both semi-decidable and not decidable, but in the infinitary context, they are not computably equivalent.

The notion of oracle computation lifts to the infinitary context and gives rise to a theory of relative computability and a rich structure of degrees. In contrast to the classical theory on $\mathbb{N}$, however, in the infinite time context we have two natural sorts of oracles to be used in oracle computations, corresponding to the second order nature of the theory. First, one can use an individual real as an oracle in exactly the classical manner, by adjoining an oracle tape on which the values of that real are written out. This amounts to fixing a supplemental input parameter and can be viewed as giving rise to a boldface theory of infinitary computability, just as one allows arbitrary real parameters in the descriptive set-theoretic treatment of boldface $\utilde{\Delta}^1_1$ and $\utilde{\Pi}^1_1$. (We shall explicitly adopt such a boldface perspective in our application to the theory of equivalence relations under infinite time reducibility.) Second, however, one naturally wants somehow to use a *set* of reals as an oracle, although we cannot expect in general to write such a set out on the tape (perhaps it is even uncountable). Instead, the oracle tape is empty at the start of computation, and during the computation the machine may freely write on this tape; whenever the algorithm calls for it, the machine may make a membership query about whether the real currently written on the oracle tape is a member of the oracle or not. Thus, the machine is able to know of any real that it can produce, whether the real is in the oracle set or not.

Such oracle computations give rise to a notion of relative computability $\varphi_p^A(x)$ and therefore a notion of infinite time omputable reduction $A \leq_\infty B$ and the accompanying infinite time degree relation $A \equiv_\infty B$. For any set A, we have the lightface jump $A^\triangledown$ and the boldface jump $A^\blacktriangledown$, corresponding to the two halting problems, relativized to A. The boldface jump jumps much higher than the lightface jump, as [9] establishes that $A <_\infty A^\triangledown <_\infty A^\blacktriangledown$, as well as $A^{\triangledown\blacktriangledown} \equiv_\infty A^\blacktriangledown$ and a great number of other interesting interactions. The infinite time analogue of Post's problem, the question of whether there are intermediate semi-decidable degrees between 0 and the jump $0^\triangledown$, was settled by [10] in an answer that cuts both ways:

THEOREM 1.2. *The infinite time analogue of Post's problem has both affirmative and negative solutions.*

(1) *There are no reals z with $0 <_\infty z <_\infty 0^\triangledown$.*
(2) *There are sets of reals A with $0 <_\infty A <_\infty 0^\triangledown$. Indeed, there are incomparable semi-decidable sets of reals $A \perp_\infty B$.*

The degrees of the accidentally writable reals are linearly ordered and in fact form a well-ordered hierarchy of order type $\zeta + 1$, which corresponds also to their order of earliest appearance on any computation. In other work, Welch [28] found minimality in the infinite time Turing degrees. Hamkins and Seabold [15] analyzed one-tape versus multi-tape infinite time Turing machines, and Benedikt Löwe [23] observed the connection between infinite time Turing machines and revision theories of truth.

§2. Some applications and extensions. Let us briefly describe a few of the recent developments and extensions of infinite time Turing machines, such as the rise of infinite time complexity theory and the introduction of infinite time computable model theory. After this, in the following section we shall go into greater detail concerning the application of infinite time Turing machines to an analogue of the theory of Borel equivalence relations.

Ralf Schindler [26] initiated the study of infinite time complexity theory by solving the infinite time Turing machine analogue of the P versus NP question. To define the polynomial class P in the infinite time context, Schindler observed simply that all reals have length ω and the polynomial functions of ω are bounded by those of the form ω^n. Thus, he defined that a set $A \subseteq \mathbb{R}$ is in P if there is a program p and a natural number n such that p decides A and halts on all inputs in time before ω^n. The set A is in NP if there is a program p and a natural number n such that $x \in A$ if and only if there is y such that p accepts (x, y), and p halts on all inputs in time less than ω^n. Schindler proved $\mathrm{P} \neq \mathrm{NP}$ for infinite time Turing machines in [26], using methods from descriptive set theory to analyze the complexity of the classes P and NP. This has now been generalized in joint work [3] to the following, where the class co-NP consists of the complements of sets in NP.

THEOREM 2.1. $\mathrm{P} \neq \mathrm{NP} \cap \text{co-NP}$ *for infinite time Turing machines.*

The proof of Theorem 2.1 appears in [3]. It follows that $\mathrm{P} \neq \mathrm{NP}$ for infinite time Turing machines. (This result has no bearing whatsoever on the finitary classical $\mathrm{P} \neq \mathrm{NP}$ question.) Some of the structural reasons behind $\mathrm{P} \neq \mathrm{NP} \cap \text{co-NP}$ are revealed by placing the classes P and NP within a larger hierarchy of complexity classes P_α and NP_α using computations of size bounded below α. Results in [3] showed, for example, that the classes NP_α are identical for $\omega + 2 \leq \alpha \leq \omega_1^{\mathrm{CK}}$, but nevertheless, $\mathrm{P}_{\alpha+1} \subsetneq P_{\alpha+2}$ for any clockable limit ordinal α. It follows, since the P_α are steadily increasing while the classes $\mathrm{NP}_\alpha \cap \text{co-NP}_\alpha$ remain the same, that $\mathrm{P}_\alpha \subsetneq \mathrm{NP}_\alpha \cap \text{co-NP}_\alpha$ for any ordinal α with $\omega+2 \leq \alpha < \omega_1^{\mathrm{CK}}$. Thus, $\mathrm{P} \neq \mathrm{NP} \cap \text{co-NP}$. Nevertheless, we attain equality at the supremum ω_1^{CK} with

$$\mathrm{P}_{\omega_1^{\mathrm{CK}}} = \mathrm{NP}_{\omega_1^{\mathrm{CK}}} \cap \text{co-NP}_{\omega_1^{\mathrm{CK}}}.$$

In fact, this is an instance of the equality $\Delta_1^1 = \Sigma_1^1 \cap \Pi_1^1$, and one can thereby begin to see how the theory of infinite time Turing machines grows naturally into descriptive set theory.

This same pattern of inequality $\mathrm{P}_\alpha \subsetneq \mathrm{NP}_\alpha \cap \text{co-NP}_\alpha$ is mirrored higher in the hierarchy, whenever α lies strictly within a contiguous block of clockable ordinals, with the corresponding $\mathrm{P}_\beta = \mathrm{NP}_\beta \cap \text{co-NP}_\beta$ for any β that begins a gap in the clockable ordinals. In addition, the question is settled in [3] for the other complexity classes P^+, P^{++} and P^f. We remark that Benedikt Löwe has also introduced analogues of PSPACE [24].

The subject of infinite time computable model theory was introduced in [14]. Computable model theory is model theory with a view to the computability of the structures and theories that arise. Infinite time computable model theory carries out this program with the notion of infinite time computability provided by infinite time Turing machines. The classical theory began decades ago with such topics as computable completeness (Does every decidable theory have a decidable model?) and computable categoricity (Does every isomorphic pair of computable models have a computable isomorphism?), and the field has now matured into a sophisticated analysis of the complexity spectrum of countable models and theories.

The motivation for a broader context is that, while classical computable model theory is necessarily limited to countable models and theories, the infinitary computability context allows for uncountable models and theories, built on the reals. Many of the computational constructions in computable model theory generalize from structures built on $\mathbb{N}$, using finite time computability, to structures built on $\mathbb{R}$, using infinite time computability. The uncountable context opens up new questions, such as the infinitary computable Löwenheim-Skolem Theorem, which have no finite time analogue. Several of the most natural questions turn out to be independent of ZFC.

In joint work [14], we defined that a model $\mathcal{A} = \langle A, \dots \rangle$ is infinite time *computable* if $A \subseteq \mathbb{R}$ is decidable and all functions, relations and constants are uniformly infinite time computable from their Gödel codes and input. The structure $\mathcal{A}$ is *decidable* if one can compute whether $\mathcal{A} \models \varphi[\bar{a}]$ given $\ulcorner\varphi\urcorner$ and $\bar{a}$. A theory T is infinite time *decidable* if the relation $T \vdash \varphi$ is computable in $\ulcorner\varphi\urcorner$. Because we want to treat uncountable languages, the natural context for Gödel codes is $\mathbb{R}$ rather than $\mathbb{N}$.

The initial question, of course, is the infinite time computable analogue of the Completeness Theorem: Does every consistent decidable theory have a decidable model? The answer turns out to be independent of ZFC.

THEOREM 2.2 ([14]). *The infinite time computable analogue of the Completeness Theorem is independent of ZFC. Specifically:*

(1) *If $V = L$, then every consistent infinite time decidable theory has an infinite time decidable model, in a computable translation of the language.*
(2) *It is relatively consistent with ZFC that there is an infinite time decidable theory, in a computably presented language, having no infinite time computable or decidable model in any translation of the language.*

The proof of (1) uses the concept of a *well-presented* language $\mathcal{L}$, for which there is an enumeration of the symbols $\langle s_\alpha \mid \alpha < \delta \rangle$ such that from any $\ulcorner s_\alpha \urcorner$ one can uniformly compute a code for the prior symbols $\langle \ulcorner s_\beta \urcorner \mid \beta \le \alpha \rangle$. One can show that every consistent decidable theory in a well-presented language has a decidable model, and if $V = L$, then every computable language has a well presented computable translation. For (2), one uses the theory T extending the atomic diagram of $\langle \mathrm{WO}, \equiv \rangle$ while asserting that f is a choice function on the $\equiv$

classes. This is a decidable theory, but for any computable model $\mathcal{A} = \langle A, \equiv, f\rangle$ of T, the set $\{ f(c_u) \mid u \in \mathrm{WO} \}$ is Σ^1_2 and has cardinality ω_1. It is known to be consistent with ZFC that no Σ^1_2 set has size ω_1.

For the infinite time analogues of the Löwenheim-Skolem Theorem, we proved for the upward version that every well presented infinite time decidable model has a proper elementary extension with a decidable presentation, and for the downward version, every well presented uncountable decidable model has a countable decidable elementary substructure. There are strong counterexamples to a full direct generalization of the Löwenheim-Skolem theorem, however, because [14] provides a computable structure $\langle \mathbb{R}, U\rangle$ on the entire set of reals, which has no proper computable elementary substructure.

Some of the most interesting work involves computable quotients. A structure has an infinite time computable *presentation* if it is isomorphic to a computable structure, and has a computable *quotient presentation* if it is isomorphic to the quotient of a computable structure by a computable equivalence relation (a congruence). For structures on $\mathbb{N}$, in either the finite or infinite time context, these notions are equivalent, because one can computably find the least element of any equivalence class. For structures on $\mathbb{R}$, however, computing such distinguished elements of every equivalence class is not always possible.

QUESTION 2.3. *Does every structure with an infinite time computable quotient presentation have an infinite time computable presentation?*

In the finite time theory, or for structures on $\mathbb{N}$, the answer of course is Yes. But in the full infinite time context for structures on $\mathbb{R}$, the answer depends on the set theoretic background.

THEOREM 2.4. *The answer to Question 2.3 is independent of ZFC. Specifically,*

(1) *It is relatively consistent with ZFC that every structure with an infinite time computable quotient presentation has an infinite time computable presentation.*
(2) *It is relatively consistent with ZFC that there is a structure having an infinite time computable quotient presentation, but no infinite time computable presentation.*

Let us briefly sketch some of the ideas appearing in the proof. In order to construct an infinite time computable presentation of a structure, given a computable quotient presentation, we'd like somehow to select a representative from each equivalence class, in a computably effective manner, and build a structure on these representatives. Under the set theoretic assumption $V = L$, we can attach to the L-least member of each equivalence class an escort real that is powerful enough to reveal that it is the L-least member of its class, and build a computable presentation out of these escorted pairs of reals. (In particular, the new presentation is not built out of mere representatives from the original class, since these reals may be too weak; they need the help of their escorts.) Thus, if $V = L$, then every structure with a computable quotient presentation has a computable presentation. The other side of the independence is similar to statement (2) of Theorem 2.2. Specifically, the structure $\langle\omega_1, <\rangle$ always has a computable quotient presentation built from reals coding well orders, but there are forcing extensions in which no infinite time computable set has size ω_1, on descriptive set theoretic grounds. In these extensions, therefore, $\langle\omega_1, <\rangle$ has a computable quotient presentation, but no computable presentation.

Let us also briefly discuss some of the alternative models of ordinal computation to which infinite time Turing machines have given rise. Peter Koepke [20] introduced the *Ordinal Turing Machines*, which generalize the infinite time Turing machines by extending the tape to transfinite ordinal length. The limit rules are accordingly adjusted so that the machine can make use of this extra space. Specifically, rather than using a special *limit* state, the ordinal Turing machines simply have a fixed order on their (finitely many) states, and at any limit stage, the state is defined to be the lim inf of the prior states. The head position is then defined to be the lim inf of the head positions when the machine was previously in that resulting limit state. For uniformity, then, Koepke defines that the cells of the tape use the lim inf of the prior cell values (rather than lim sup as with the infinite time Turing machines). If the head moves left from a cell at a limit position, then it appears all the way to the left on the first cell.

These machines therefore provide a model of computation for functions on the ordinals, and notions of decidability for classes of ordinals. The main theorem is that the power of these machines is essentially the same as that of Gödel's constructible universe.

THEOREM 2.5 (Koepke). *The sets of ordinals that are ordinal Turing machine decidable, with finitely many ordinal parameters, are exactly the sets of ordinals in Gödel's constructible universe L.*

Several other infinitary models of ordinal computation are based on a concept of ordinal registers, and have given rise to a rich theory. See [20], [21], [19], [22], [1], [12], [13], and [11].

§3. Infinite time computable equivalence relation theory. Recently, we have introduced the natural analogue of Borel equivalence relation theory in which infinite time decidable relations are compared with respect to infinite time computable reduction functions. This is motivated in part by the occasional need in the study of Borel equivalence relations to go beyond Borel. Indeed, a more powerful notion of reducibility may be able to accurately compare more complex relations. In particular, we shall be able to consider the new relations which arise out of the infinite time complexity classes.

We begin with a quick introduction to the study of Borel equivalence relations. The name of the subject is somewhat of a misnomer – in fact the principle objects of study are arbitrary equivalence relations on *standard Borel spaces*, that is, sets equipped with the Borel structure of a complete separable metric space. In applications, we think of an equivalence relation as representing a classification problem from some other area of mathematics. For instance, since any group with domain $\mathbb{N}$ is determined by its multiplication function, studying the classification problem for countable groups amounts to studying the isomorphism equivalence relation on a suitable subspace of $2^{\mathbb{N}\times\mathbb{N}\times\mathbb{N}}$. For many more examples, see Section 1.2 of [27].

The theory of Borel equivalence relations revolves around the following key notion of complexity. If E, F are equivalence relations on standard Borel spaces X, Y, then following [4] and [18] we say that E is *Borel reducible* to F, written $E \leq_B F$, iff there exists a Borel function $f\colon X \to Y$ such that

$$(1) \qquad x\, E\, x' \iff f(x)\, F\, f(x')\,.$$

Borel reducibility measures the complexity of equivalence relations not as sets of pairs, but as *classification problems*. That is, if E is Borel reducible to F, then the classification of elements of X up to E is no harder than the classification of elements of Y up to F. The by now classical and highly successful study of Borel equivalence relations consists in part of two major endeavors. First, one wishes to map out the relationships between numerous well-understood and naturally occurring equivalence relations. Second, given a real-life classification problem one should measure its complexity by comparing it against the mapped-out benchmark relations.

Some definability condition on the reduction functions (in this case that they be Borel) is necessary. Indeed, without any such restriction reducibility would always be determined by cardinalities alone. However, there are cases of natural classifications by invariants which cannot be computed by a Borel reduction function. For instance, it is $\undertilde{\Delta}^1_2$ and not Borel to compute the classical Ulm invariants for a countable torsion abelian group. One might be tempted to form a theory of $\undertilde{\Delta}^1_2$ reducibility, but it turns out this notion is too generous. Indeed, as we shall see below in Theorem 3.4, it may lump most equivalence relations together into one trivial complexity class.

We will consider here reduction functions which are computable by an infinite time Turing machine (see [2] for a more complete exposition). Thus, for any two equivalence relations E, F on $\mathbb{R}$, we say that E is *infinite time computably reducible* to F, written $E \leq_c F$, if there is an infinite time computable function f (freely allowing real parameters) satisfying Equation (1). Similarly, we say that E is *eventually reducible* to F, written $E \leq_e F$, if there is an eventually computable function f satisfying Equation (1). Note here that since all uncountable standard Borel spaces are Borel isomorphic, we lose no generality by restricting ourselves to equivalence relations with domain $\mathbb{R}$.

Of course, by the remarks in Section 1 (and again emphasizing that we have allowed parameters) the infinite time computable reductions include all of the Borel reductions. Thus, our theory will extend the classical theory. Conversely, many classical proofs of *non*-reducibility $E \not\leq_B F$ rely on methods such as measure, category, or forcing. Hence, they frequently "overshoot" and show that there does not exist a reduction from E to F which is Lebesgue measurable, Baire measurable, or absolutely $\underset{\sim}{\Delta}^1_2$ (discussed below), respectively. Since the infinite time computable and eventually computable functions enjoy all three of these properties, it follows in each of these cases that $E \not\leq_c F$ and even $E \not\leq_e F$, and hence not too much is "collapsed" when we pass from the $\leq_B$ hierarchy to the $\leq_c$ and $\leq_e$ hierarchies.

The infinite time notions of reducibility are very closely related to that of absolutely $\underset{\sim}{\Delta}^1_2$ reducibility, which has been treated in the literature by Hjorth and others. Recall that a subset $A \subseteq \mathbb{R}$ is said to be *absolutely* $\underset{\sim}{\Delta}^1_2$ if it is defined by equivalent $\underset{\sim}{\Sigma}^1_2$ and $\underset{\sim}{\Pi}^1_2$ formulas which remain equivalent in every forcing extension. A function $f\colon \mathbb{R} \to \mathbb{R}$ is said to be absolutely $\underset{\sim}{\Delta}^1_2$ if its diagram $\{\,(x,n) \mid f(x) \in B_n\,\}$ is absolutely $\underset{\sim}{\Delta}^1_2$ (here, B_n runs through the basic open subsets of $\mathbb{R}$). We suspect that there are very few naturally occurring cases in which there is an absolutely $\underset{\sim}{\Delta}^1_2$ reduction between two equivalence relations but not an infinite time computable reduction. And when there is an infinite time computable reduction, one can demonstrate that this is the case by simply "coding up" an algorithm which implements the witnessing reduction function. This computational approach may be more satisfying than abstractly defining a reduction function and verifying that it is $\underset{\sim}{\Delta}^1_2$ in all forcing extensions. On the other hand, we do not have any general tools for establishing non-reducibility by infinite time computable functions beyond the already established tools mentioned above, all of which establish non-reducibility by absolutely $\underset{\sim}{\Delta}^1_2$ functions already. A brief summary of results due to Hjorth and Kanovei which establish non-reducibility for absolutely $\underset{\sim}{\Delta}^1_2$ functions can be found in Section 5 of [2]. Some deeper results on this notion of reducibility can be found Hjorth in Chapter 9 of [17].

For an example of "coding up" a new (non-Borel) reduction function, consider the E_{ck} relation defined by $x \; E_{ck} \; y$ if x and y compute (in the ordinary sense) the same ordinals. We will compare it against the relation $\cong_{\mathrm{WO}}$, which is just the isomorphism relation restricted to the set of codes for well-orders. These two relations are not comparable by Borel reductions; nevertheless they are closely related and this is made precise by the following result.

Theorem 3.1. *E_{ck} and $\cong_{\mathrm{WO}}$ are infinite time computably bireducible.*

For instance, there is an intuitive reduction from E_{ck} to $\cong_{\mathrm{WO}}$ – namely, map x to a code for the supremum of the ordinals which are computable (in the ordinary sense) from x. And indeed, this intuition easily translates into a program for an infinite time Turing machine. Briefly, the program simply simulates all ordinary Turing computations, and inspects the real enumerated by each. Whenever one of these reals is seen to be code for a well-order, this code is added to a list. Finally, the program computes and outputs a code for the supremum of the ordinals in its list.

Another obvious benefit to using infinite time computable and eventually computable reductions is that they are tailor-made to handle equivalence relations which arise in the study of infinite time complexity classes. As a very simple example, consider two of the most important such equivalence relations: the infinite time degree relation $\equiv_\infty$ which was introduced in Section 1, and the (light face) jump equivalence relation defined by $x\ J\ y$ if and only if $x^\triangledown \equiv_\infty y^\triangledown$. We have the following (somewhat trivial) relationship between the two.

THEOREM 3.2. *J is eventually reducible to $\equiv_\infty$ by the function which computes the infinite time jump of a real.*

The program which witnesses this simply simulates all infinite time programs on input x, and whenever one of them halts adds its index to a list on its output tape. Since all programs which will halt do so by stage λ, the output tape will eventually show $x^\triangledown$.

Meanwhile, the next result gives a sampling of non-reducibility results which can be obtained using the methods of Hjorth and Kanovei discussed above. Here $=$ of course denotes the equality relation on $\mathbb{R}$, and E_0 the almost equality relation defined by $x\ E_0\ y$ if and only if $x(n) = y(n)$ for almost all n. Next, $\cong_{\mathrm{HC}}$ denotes the isomorphism relation restricted to the set of codes for hereditarily countable sets. Finally, E_{set} denotes the relation defined by $x\ E_{\mathrm{set}}\ y$ if x and y, thought of as codes for countable sequences of reals, enumerate the same set.

THEOREM 3.3.

(1) *E_0 does not infinite time computably reduce to $=$.*
(2) *E_{set} does not infinite time computably reduce to E_0.*
(3) *$\cong_{\mathrm{HC}}$ and E_{set} do not infinite time computably reduce to $\cong_{\mathrm{WO}}$.*

Without strong set-theoretic hypotheses, such results cannot be obtained for reduction functions which are much more general than the absolutely $\utilde{\Delta}^1_2$ functions. For instance, the infinite time semi-computable reduction functions are still well inside the class $\utilde{\Delta}^1_2$, but if we were to allow reduction functions in this class, then all of the equivalence relations in Theorem 3.3 would be reducible to the equality relation.

THEOREM 3.4. *If $V = L$, then every infinite time computable equivalence relation on $\mathbb{R}$ is reducible to the equality relation by an infinite time semi-computable function.*

The proof of Theorem 3.4 uses the same ideas as in the proof of Theorem 2.4, and as in that argument, the reduction functions are not selectors for the relation. On the other hand, under suitable determinacy hypotheses, every infinite time semi-computable function is Lebesgue measurable. In this situation, infinite time semi-computable reducibility again resembles the more concrete reducibility notions.

We have seen that by expanding the class of reduction functions available, we are sometimes able to bring a wider class of equivalence relations under consideration. A major example of this is the following generalization of the class of countable Borel equivalence relations. Here, a Borel equivalence relation is said to be *countable* iff every equivalence class is countable. The countable relations have become one of the most important collections studied in the classical theory, since many natural relations lie at this level and some basic progress has been made in uncovering their structure under $\leq_B$. For instance, by a classical result of Silver, the equality relation $=$ is the $\leq_B$-least countable Borel equivalence relation. Moreover, by a deep result of Kechris-Harrington-Louveau, E_0 is the $\leq_B$-least Borel equivalence relation which is not reducible to $=$. Thirdly, we have that there is a $\leq_B$-greatest countable Borel equivalence relation, denoted E_∞. The remaining countable Borel equivalence relations lie in the interval (E_0, E_∞), and a result of Adams-Kechris implies that there are continuum many distinct relations up to Borel bireducibility.

This last result holds also in the context of $\leq_c$ and $\leq_e$ reducibility, since the arguments that Adams and Kechris use to establish non-reducibility are measure-theoretic. We presently define a class of infinite time computable relations which we propose is the correct analogue of the countable Borel equivalence relations, and investigate the corresponding generalizations of the remaining results. The idea comes from the classical proof of the maximality of E_∞, which hinges on the following characterization of the countable Borel equivalence relations. Namely, E is a countable Borel equivalence relation if and only if it admits a *Borel enumeration*, that is, a Borel function f such that $f(x)$ codes an enumeration of $[x]_E$, for all x. (This characterization is an immediate consequence of the Lusin-Novikov theorem from descriptive set theory.) Generalizing this, we say that the equivalence relation E is (infinite time) *enumerable* if there exists an infinite time computable function f such that $f(x)$ codes an enumeration of $[x]_E$, for all x. The *eventually enumerable* equivalence relations are defined analogously. This is a worthwhile generalization; for instance the relation defined by $x \equiv_{\mathsf{hyp}} y$ iff x and y are hyperarithmetic in one another is enumerable but not Borel.

Since we have said that the maximality of E_∞ depends on the above characterization of the countable Borel equivalence relations, and since we have defined the enumerable and eventually enumerable equivalence relations in the analogous way, the proof of maximality of E_∞ in the Borel context yields the same in our context.

THEOREM 3.5. *E_∞ is $\leq_c$-greatest among the enumerable relations, and $\leq_e$-greatest among the eventually enumerable relations.*

Perhaps surprisingly, one can establish the minimality of $=$ as well.

THEOREM 3.6. *$=$ is reducible to every eventually enumerable equivalence relation by a continuous function.*

This result is an immediate consequence of the fact (due originally to Welch) that there exists a perfect set of $\equiv_{e\infty}$-classes. (Here, $\equiv_{e\infty}$ denotes the *eventual degree* relation, which is defined analogously to $\equiv_{\infty}$.) The idea of Welch's proof (which is based on the arguments of [28]) is to use the theory of forcing over L_{Σ} to obtain a perfect set of mutually generic Cohen reals, and then argue that this set does the job. To see that Theorem 3.6 follows, observe that every eventually enumerable relation E is contained (as a set of pairs) in the relation $\equiv_{e\infty}$. Hence there exists a perfect set of E-classes, and it follows that there is a continuous reduction from $=$ to E.

Finally, we have been unable to establish the minimality of E_0 over the equality relation, and we leave this as a question. It is hoped that methods similar to the proof of Theorem 3.6 will provide an answer.

Question 3.7. *Is it true of every enumerable equivalence relation E that either E is reducible to $=$ or else E_0 is reducible to E?*

REFERENCES.

[1] Merlin Carl, Tim Fischbach, Peter Koepke, Russell Miller, Miriam Nasfi, and Gregor Weckbecker. The basic theory of infinite time register machines. *Archive for Mathematical Logic*, 49(2): 249–273, 2010.

[2] Sam Coskey and Joel David Hamkins. Infinite time computable equivalence relations. *Notre Dame Journal of Formal Logic*, 52(2): 203–228, 2011.

[3] Vinay Deolalikar, Joel David Hamkins, and Ralf-Dieter Schindler. $P \neq NP \cap co-NP$ for infinite time Turing machines. *Journal of Logic and Computation*, 15(5): 577–592, 2005.

[4] Harvey Friedman and Lee Stanley. A Borel reducibility theory for classes of countable structures. *J. Symbolic Logic*, 54(3): 894–914, 1989.

[5] Joel David Hamkins. Infinite time Turing machines. *Minds and Machines*, 12(4): 521–539, 2002. (special issue devoted to hypercomputation).

[6] Joel David Hamkins. Supertask computation. In Boris Piwinger Benedikt Löwe and Thoralf Räsch, editors, *Classical and New Paradigms of Computation and their Complexity Hierarchies*, Papers of the conference "Foundations of the Formal Sciences III" held in Vienna, September 21–24, 2001. Volume 23 of Trends in Logic, pages 141–158. Kluwer Academic Publishers, 2004.

[7] Joel David Hamkins. Infinitary computability with infinite time Turing machines. In Barry S. Cooper and Benedikt Löwe, editors, *New Computational Paradigms*, CiE, Amsterdam, June 8–12 2005. Volume 3526 of Lecture Notes in Computer Science, Springer, 2005.

[8] Joel David Hamkins. A survey of infinite time Turing machines. In Jérôme Durand-Lose and Maurice Margenstern, editors, *Machines, Computations, and Universality – 5th International Conference MCU 2007,*

Orleans, France. Volume 4664 of Lecture Notes in Computer Science, Springer, 2007.

[9] Joel David Hamkins and Andy Lewis. Infinite time Turing machines. *J. Symbolic Logic*, 65(2): 567–604, 2000.

[10] Joel David Hamkins and Andy Lewis. Post's problem for supertasks has both positive and negative solutions. *Archive for Mathematical Logic*, 41(6): 507–523, 2002.

[11] Joel David Hamkins, David Linetsky, and Russell Miller. The complexity of quickly decidable ORM-decidable sets. In Barry Cooper, Benedikt Löwe, and Andrea Sorbi, editors, *Computation and Logic in the Real World – Third Conference of Computability in Europe CiE 2007*, Siena, Italy. Volume 4497 of Lecture Notes in Computer Science, Springer, 2007.

[12] Joel David Hamkins and Russell Miller. Post's problem for ordinal register machines. In Barry Cooper, Benedikt Löwe, and Andrea Sorbi, editors, *Computation and Logic in the Real World – Third Conference of Computability in Europe CiE 2007*, Siena, Italy. Volume 4497 of Lecture Notes in Computer Science, Springer, 2007.

[13] Joel David Hamkins and Russell Miller. Post's problem for ordinal register machines: an explicit approach. *Annals of Pure and Applied Logic*, 160(3): 302–309, 2009.

[14] J. D. Hamkins, R. Miller, D. Seabold, and S. Warner. Infinite time computable model theory. In S.B. Cooper, Benedikt Löwe, and Andrea Sorbi, editors, *New Computational Paradigms: Changing Conceptions of What is Computable*, pages 521–557. Springer, 2007.

[15] Joel David Hamkins and Daniel Seabold. Infinite time Turing machines with only one tape. *Mathematical Logic Quarterly*, 47(2): 271–287, 2001.

[16] Joel David Hamkins and Philip Welch. $P^f \neq NP^f$ for almost all f. *Mathematical Logic Quarterly*, 49(5): 536–540, 2003.

[17] Greg Hjorth. *Classification and Orbit Equivalence Relations*, volume 75 of Mathematical Surveys and Monographs. American Mathematical Society, Providence, 2000.

[18] Greg Hjorth and Alexander S. Kechris. Borel equivalence relations and classifications of countable models. *Ann. Pure Appl. Logic*, 82(3): 221–272, 1996.

[19] Peter Koepke and Martin Koerwien. Ordinal computations. *Math. Structures Comput. Sci.*, 16(5): 867–884, 2006.

[20] Peter Koepke. Turing computations on ordinals. *Bulletin of Symbolic Logic*, 11(3): 377–397, 2005.

[21] Peter Koepke and Ryan Siders. Register computations on ordinals. *Archive for Mathematical Logic*, 47(6): 529–548, 2008.

[22] Peter Koepke and Benjamin Seyfferth. Ordinal machines and admissible recursion theory. *Ann. Pure Appl. Logic*, 160(3): 310–318, 2009.

[23] Benedikt Löwe. Revision sequences and computers with an infinite amount of time. *Logic Comput.*, 11(1): 25–40, 2001.

[24] Benedikt Löwe. Space bounds for infinitary computation. In Arnold Beckmann, Ulrich Berger, Benedikt Löwe, and John V. Tucker, editors, *Logical Approaches to Computations Barriers: Second Conference on Computability in Europe*. Volume 3988 of Lecture Notes in Computer Science, pages 319–329. Springer, 2006.

[25] Giacomo Lenzi and Erich Monteleone. On fixpoint arithmetic and infinite time Turing machines. *Information Processing Letters*, 91(3): 121–128, 2004.

[26] Ralf-Dieter Schindler. P $\neq$ NP for infinite time Turing machines. *Monatshefte für Mathematik*, 139(4): 335–340, 2003.

[27] Scott Schneider and Simon Thomas. Countable Borel equivalence relations. In: James Cummings and Ernest Schimmerling, editors, *Appalachian Set Theory 2006–2012*. Volume 402 of LMS Lecture Note Series, pages 25–62, Cambridge University Press, 2012.

[28] Philip Welch. Friedman's trick: Minimality arguments in the infinite time Turing degrees. In S.B. Cooper and J.K. Truss, editors, *Sets and Proofs, Proceedings ASL Logic Colloquium*, Volume 258 of LMS Lecture Note Series, pages 425–436, Cambridge University Press, 1999.

[29] Philip Welch. Eventually infinite time Turing machine degrees: Infinite time decidable reals. *Journal of Symbolic Logic*, 65(3): 1193–1203, 2000.

[30] Philip Welch. The lengths of infinite time Turing machine computations. *Bulletin of the London Mathematical Society*, 32(2): 129–136, 2000.

[31] Philip Welch. The transfinite action of 1 tape Turing machines. In Barry S. Cooper and Benedikt Löwe, editors, *New Computational Paradigms*, CiE, Amsterdam, June 8–12 2005. Volume 3526 of Lecture Notes in Computer Science, Springer, 2005.

[32] Philip D. Welch. Bounding lemmata for non-deterministic halting times of transfinite Turing machines. *Theoret. Comput. Sci.*, 394(3): 223–228, 2008.

The Fields Institute, Toronto, Ontario M5T 3J1, Canada; *and*
The York University,
Department of Mathematics and Statistics,
Toronto, Ontario M3J 1P3, Canada.
E-mail: scoskey@nylogic.org

The Graduate Center of CUNY, NY 10016, USA; *and*
College of Staten Island of CUNY,
Mathematics Department,
Staten Island, NY 10314, USA.
E-mail: jhamkins@gc.cuny.edu

COMPUTABLE STRUCTURE THEORY ON ω_1 USING ADMISSIBILITY

NOAM GREENBERG AND JULIA F. KNIGHT

Abstract We use the theory of recursion on admissible ordinals to develop an analogue of classical computable model theory and effective algebra for structures of size $\aleph_1$, which, under our assumptions, is equal to the continuum. We discuss both general concepts, such as computable categoricity, and particular classes of examples, such as fields and linear orderings.

§1. Introduction. Our aim is to develop computable structure theory for uncountable structures. In this paper we focus on structures of size $\aleph_1$. The fundamental decision to be made, when trying to formulate such a theory, is the choice of computability tools that we intend to use. To discover which structures are computable, we need to first describe which subsets of the domain are computable, and which functions are computable. In this paper, we use admissible recursion theory (also known as α-recursion theory) over the domain ω_1. We believe that this choice yields an interesting computable structure theory. It also illuminates the concepts and techniques of classical computable structure theory by observing similarities and differences between the countable and uncountable settings. In particular, it seems that as is the case for degree theory and for the study of the lattice of c.e. sets, the difference between true finiteness and its analogue in the generalised case, namely countability in our case, is fundamental to some constructions and reveals a deep gap between classical computability and attempts to generalise it to the realm of the uncountable.

In this paper, we make a sweeping simplifying assumption about our set-theoretic universe. We suppose Every real is constructible; in other words, L_{ω_1} is the collection of all hereditarily countable sets. The reason for this assumption is to make every subset of ω_1 amenable for L_{ω_1}. Recall that L_{ω_1} is the domain for our model of computation; it will be important for us that for all $A \subseteq L_{\omega_1}$, L_{ω_1} is closed under the function $x \mapsto A \cap x$. Admissible computability would be far more complicated if we had to deal with non-amenable sets. In the non-amenable setting, we would have to make more decisions about what it means for an uncountable structure to be computable. This may be a subject of future research.

[1]The first author was partially supported by the Marsden Fund of New Zealand.
[2]Math Subject Codes: 03D30, 03D60, 03E15. Keywords: infinite time Turing machines, infinitary computability, ordinal computation.

We choose to focus on the least uncountable cardinal $\aleph_1$, rather than work with arbitrary uncountable cardinals κ, mostly for simplicity. Most of the theory carries over without changes to any *successor* cardinal. The limit case, in particular the singular case, is more difficult. We remark, however, that some results regarding linear orderings do not seem to easily generalise from ω_1 to higher successor cardinals. Another reason to concentrate on ω_1 is that under our assumptions (which imply the continuum hypothesis), it coincides with the size of the continuum, which has natural mathematical interest.

§2. Admissible computability on ω_1. In this section we develop admissible computability over ω_1 from the very beginning. Since ω_1 is a regular cardinal, this development is significantly easier than the general treatment over an arbitrary admissible ordinal. For the general theory and historical notes, see [31].

We remark here that various older and more recent results show the robustness of the notion of computability that we are about to define. This notion can be defined not only using definitions by existential formulas, as we do below, but also using inductive schemes such as E-recursion or ω_1-calculability (again see Sacks [31]) or other schemes (Takeuti, Kripke, Machover, and others, see [32]), and generalised Turing machines that have a tape of length ω_1 and run for countably many steps (see [21]).

2.1. Computable sets and functions. As mentioned above, the "universal domain" for our theory of computability is L_{ω_1}. Computability in this setting is the result of investigating the early layers of definability on this domain.

We work in the language of set theory, which contains equality and one binary relation symbol $\in$. A formula in this language is *bounded*, or Δ_0, if it is built from the atomic formulas using Boolean combinations and the bounded quantifiers $\exists\, x \in y$ and $\forall\, x \in y$. A formula is Σ_1, if it is of the form $\exists\, x_1, \dots, x_n\, \varphi$, where φ is bounded. We remark that in this section, unlike later ones, all of the formulas are finitary.

We sometimes enrich our language by adding one constant for each element of L_{ω_1}. As is often done in model theory, we identify the elements of L_{ω_1} with the constants that they name. We use $\Delta_0(L_{\omega_1})$ to denote the collection of bounded formulas of this enriched language. Of course a formula of this enriched language is bounded if and only if it is of the form $\varphi(\bar{a}, \bar{x})$, where $\varphi(\bar{y}, \bar{x})$ is a bounded formula in the language of pure set theory (with no added constants), and $\bar{a}$ is a tuple of (constants naming) elements of L_{ω_1}. Similarly, we define the collection $\Sigma_1(L_{\omega_1})$ of Σ_1 formulas of the augmented language.

The syntactic notions have semantic counterparts on the intended structure $(L_{\omega_1}, \in)$. An n-ary relation $R \subseteq (L_{\omega_1})^n$ is $\Sigma_1(L_{\omega_1})$ if it is defined by some $\Sigma_1(L_{\omega_1})$ formula. These relations have a familiar name.

DEFINITION 2.1.

- A relation $R \subseteq (L_{\omega_1})^n$ is *computably enumerable*, or *c.e.*, if it is $\Sigma_1(L_{\omega_1})$.

- A relation $R \subseteq (L_{\omega_1})^n$ is *computable* if both R and $(L_{\omega_1})^n \setminus R$ are computably enumerable.
- A partial function $f\colon (L_{\omega_1})^n \to L_{\omega_1}$ is *partial computable* if its graph

$$\{(\bar{a}, f(\bar{a})) \ : \ \bar{a} \in \operatorname{dom} f\}$$

  is a computably enumerable relation.
- A *computable* function $f\colon (L_{\omega_1})^n \to L_{\omega_1}$ is a partial computable function whose domain is computable.

Traditionally, the terminology for the notions above is "ω_1-recursively enumerable", "ω_1-recursive", etc. We drop the prefix ω_1 right from the start (and shift to "computable" rather than "recursive"). When we wish to refer to classical computability, we say that we are in the "setting of ω", or the "standard" or "classical" setting.

In the definition above, we referred to relations of any finite arity. This could lead to confusion when we recall that L_{ω_1} is closed under taking ordered n-tuples. There is no confusion because the "tupling" functions are computable: for example, the function $x, y \mapsto \langle x, y\rangle$ and the functions $\langle x, y\rangle \mapsto x$, $\langle x, y\rangle \mapsto y$ are all computable (the set of ordered pairs is computable). Hence a relation $R \subseteq (L_{\omega_1})^2$ is computable (or computably enumerable) if and only if the set

$$\{\langle a, b\rangle \ : \ R(a, b)\}$$

is a computable (computably enumerable) subset of L_{ω_1}. Henceforth, we ignore such subtleties.

Indeed, this can also be applied to countable arities, which we shall discuss later.

Proposition 2.2. *A set $A \subseteq L_{\omega_1}$ is computable iff its characteristic function χ_A is computable.*

Proof. If χ_A is computable, then A is $\Sigma_1(L_{\omega_1})$-definable by the formula $\chi_A(x) = 1$; the complement of A is definable by a similar formula.

For the other direction, suppose that $\psi(x)$ defines A and that $\theta(x)$ defines $L_{\omega_1} \setminus A$, and that both formulas are $\Sigma_1(L_{\omega_1})$. Then $\chi_A(x) = y$ is $\Sigma_1(L_{\omega_1})$-definable by the formula

$$[y = 1 \ \& \ \psi(x)] \vee [y = 0 \ \& \ \theta(x)]\,.$$

⊣

Lemma 2.3.

(1) *If $R(\bar{y})$ is a computably enumerable relation, then R is definable by a formula of the form $\exists x\, \varphi(x, \bar{y})$, where φ is $\Delta_0(L_{\omega_1})$.*

(2) *If $R(x, \bar{y})$ is a computably enumerable relation, and Q is a quantifier ($\exists$ or $\forall$), then the relation $Q\, x \in z\, R(x, \bar{y})$ is also computably enumerable.*

Proof. (1). We replace $\exists x_1, \ldots, x_n\, \psi(\bar{x}, \bar{y})$ by

$$\exists t\ [\exists x_1, \ldots, x_n \in t\, \psi(\bar{x}, \bar{y})],$$

using the fact that L_{ω_1} is closed under taking finite subsets.

(2). This is immediate for the existential quantifier. For the universal quantifier, $\forall x \in z\, \exists w\, \psi(x, w, \bar{y})$ is equivalent to $\exists t \exists w\, \forall x \in z\, \exists w \in t\, \psi(x, w, \bar{y})$, using the fact that each $z \in L_{\omega_1}$ is countable, so picking witnesses for any relation can be done within a countable set. ⊣

The main method of constructing computable functions is by recursion. This is possible by the following.

Proposition 2.4. *Let* $I\colon L_{\omega_1} \to L_{\omega_1}$ *be a computable function. Then there is a unique computable function* $f\colon \omega_1 \to L_{\omega_1}$ *such that for all* $\alpha < \omega_1$,

$$f(\alpha) = I(f\restriction_\alpha).$$

Proof. Define f according to the inductive formula. The point is that if $\alpha < \omega_1$ and $f\restriction_\alpha$ is already defined, then $f\restriction_\alpha \in L_{\omega_1}$, and so $f(\alpha) = I(f\restriction_\alpha)$ is defined. Uniqueness follows by induction as well. The main point is that f is computable: $f(\alpha) = a$ iff there is some $g \in L_{\omega_1}$ whose domain is an ordinal greater than α and that satisfies the inductive formula defining f and such that $g(\alpha) = a$. Unravelling, this is a $\Sigma_1(L_{\omega_1})$ formula that defines the graph of f. ⊣

Defining computable functions by recursion allows us to view them dynamically, as is common in classical computability. Processes of computation, described by $\Delta_0(L_{\omega_1})$ formulas, for instance, and taking only countably many steps, can be used to define computable functions. An intuition for informal definitions of such computable objects develops with experience.

Recall that the constructible universe is globally well-ordered: there is a well-ordering $<_L$ of all constructible sets, such that whenever $\alpha < \beta$, $<_L\restriction_{L\alpha}$ is an initial segment of $<_L\restriction_{L\beta}$.

Lemma 2.5.

(1) *The function* $\alpha \mapsto L_\alpha$ *(for* $\alpha < \omega_1$*) is computable.*
(2) *The function* $\alpha \mapsto <_L\restriction_{L_\alpha}$ *(for* $\alpha < \omega_1$*) is computable.*

Proof. Both are constructed by recursion on α (Proposition 2.4). The corresponding recursive rules I can be written to be Σ_1.

For (1), $I(f)$ is the collection of definable subsets of $f(\alpha)$, if $\operatorname{dom} f = \alpha + 1$ is a successor ordinal; otherwise, $I(f)$ is the union of the range of f. A similar rule defines $<_L\restriction_{L\alpha}$ given the sequence $\left\langle <_L\restriction_{L\beta}\right\rangle_{\beta<\alpha}$. ⊣

Corollary 2.6.

(1) $<_L\restriction_{L_{\omega_1}}$ *is a computable relation.*
(2) *The map*

$$a \mapsto \{b \in L_{\omega_1} : b < a\}$$

is computable.

(3) *The map that takes* $\alpha < \omega_1$ *to the* α^{th} *element of* $<_L$ *is computable.*

Proof. (1). Let $a, b \in L_{\omega_1}$. Then $a <_L b$ iff there is some $\alpha < \omega_1$ such that $a, b \in L_\alpha$ and a precedes b in the ordering $<_L \upharpoonright L_\alpha$. This shows that $<_L \upharpoonright L_{\omega_1}$ is a computably enumerable relation. However, it is also a total ordering, so it must be computable. Alternatively, $a <_L b$ iff for every α such that $a, b \in L_\alpha$, a precedes b in the ordering $<_L \upharpoonright L_\alpha$; this shows that $<_L \upharpoonright L_{\omega_1}$ is co-c.e.

(2). x is the set of $<_L$-predecessors of a iff there is (for all) α such that $a \in L_\alpha$ and such that x is the set of $<_L$-predecessors of a which are in L_α. The last part involves only quantification over L_α, i.e., bounded quantification. By Lemma 2.3(2), this defines a Σ_1 relation.

(3). a is the αth element of $<_L$ iff in L_{ω_1}, there is some order-preserving map from α to the set of $<_L$-predecessors of a. ⊣

The map of Corollary 2.6(3) is a computable bijection between ω_1 and L_{ω_1}. Using this map, we may pass without mention between investigating computable and c.e. subsets of ω_1 and subsets of L_{ω_1}. This can be done in general:

Proposition 2.7. *The following are equivalent for a non-empty set* $A \subseteq L_{\omega_1}$:

(1) *A is computably enumerable.*

(2) *A is the range of a computable function* $f\colon \omega_1 \to L_{\omega_1}$.

(3) *A is the domain of a partial computable function.*

Moreover, an uncountable set $A \subseteq L_{\omega_1}$ *is c.e. iff it is the range of a total, 1-1 function* $f\colon \omega_1 \to L_{\omega_1}$.

Proof. The implications (2)→(1) and (3)→(1) are immediate.

For (1)→(3), let $\exists y\, \varphi(x, y)$ be a $\Sigma_1(L_{\omega_1})$ formula that defines A, where φ is bounded. Let $f(x) = y$ if y is the least, according to $<_L$, such that $\varphi(x, y)$ holds. Then f is partial computable and $\operatorname{dom} f = A$.

For (1)→(2), again let $\exists y\, \varphi(x, y)$ define A, with φ bounded. Suppose first that A is uncountable. By recursion, define a function $f\colon \omega_1 \to L_{\omega_1}$ by letting $f(\alpha) = x$ if there is some y such that:

- $\langle x, y \rangle$ is the $<_L$-least pair such that $\varphi(x, y)$ holds; and
- $x \notin \operatorname{range} f\upharpoonright_\alpha$.

Then f is total and the range of f is A.

If A is countable, but nonempty, we let $g\colon \omega \to A$ be a function whose range is A; $g \in L_{\omega_1}$. Let a be any element of A. We then let $f\colon \omega_1 \to L_{\omega_1}$ be defined as follows:

$$f(\alpha) = \begin{cases} g(\alpha), & \text{if } \alpha < \omega; \\ a, & \text{if } \alpha \geqslant \omega. \end{cases}$$

⊣

2.2. The universal c.e. set. In what follows, we regard formulas as mathematical objects, elements of L_{ω_1}.

LEMMA 2.8. *The relation*

$$\{(\alpha, \varphi) \,:\, \varphi \text{ is a sentence over } L_\alpha \;\&\; L_\alpha \models \varphi\}$$

is computable.

PROOF. This is proved by induction on the complexity of φ. The point is that all quantifiers are bounded, as they range only over L_α. The induction is turned into a computable definition in the style of the proof of Proposition 2.4: $L_\alpha \models \psi(\bar{a})$ if there is a sequence of relations on L_α, defined by the subformulas of ψ, and $\bar{a}$ belongs to the last relation on the list (the one defined by $\psi(\bar{x})$). ⊣

COROLLARY 2.9. *The collection of $\Sigma_1(L_{\omega_1})$ sentences that are true in L_{ω_1} is computably enumerable.*

PROOF. Let $\varphi(\bar{x})$ be a Σ_1 formula and let $\bar{a} \in L_{\omega_1}$. Then $L_{\omega_1} \models \varphi(\bar{a})$ if and only if there is some $\alpha < \omega_1$ such that $\bar{a} \in L_\alpha$ and $L_\alpha \models \varphi(\bar{a})$. The point is that since L_α is transitive, bounded formulas are absolute between L_α and L_{ω_1}. ⊣

Corollary 2.9 allows us to define a universal c.e. set. The collection $\mathcal{W}$ of all $\Sigma_1(L_{\omega_1})$ formulas with one free variable is computable. Let $\alpha \mapsto \psi_\alpha$ be a computable bijection between ω_1 and $\mathcal{W}$. For $\alpha < \omega_1$, let W_α be the subset of L_{ω_1} defined by ψ_α. The collection $\{W_\alpha \,:\, \alpha < \omega_1\}$ is the collection of all c.e. sets. Corollary 2.9 ensures that

$$\bigoplus_{\alpha<\omega_1} W_\alpha = \{(\alpha, x) \,:\, x \in W_\alpha\}$$

is a c.e. set. We take this to be the *universal* c.e. set. Of course, it depends on the enumeration $\langle \psi_\alpha \rangle$. Diagonalisation holds:

PROPOSITION 2.10. *The universal c.e. set is not computable.*

PROOF. Suppose, for contradiction, that the universal c.e. set $\bigoplus_\alpha W_\alpha$ is computable. Then the diagonal set

$$K = \{\alpha < \omega_1 \,:\, \alpha \in W_\alpha\}$$

is computable, and so its complement $\bar{K}$ is computable as well. Hence, there is some α such that $\bar{K} = W_\alpha$. We get a contradiction by examining whether $\alpha \in K$. ⊣

Similarly, there is an effective enumeration of all partial computable functions, and, hence, there is a universal partial computable function. Let $R(\alpha, x, z)$ be a computable relation such that for all x, $x \in W_\alpha$ iff there is some z such that $R(\alpha, x, z)$. Let $\varphi_\alpha(x) = y$ if there is some z such that $R(\alpha, \langle x, y\rangle, z)$ and such that for no $\langle y', z'\rangle <_L \langle y, z\rangle$ do we have $R(\alpha, \langle x, y'\rangle, z')$. Then $\langle \varphi_\alpha \rangle_{\alpha<\omega_1}$ is an effective enumeration of all partial computable functions: the set $\{(\alpha, x, y) : \varphi_\alpha(x) = y\}$ is computably enumerable.

From the universal c.e. set, we also obtain a uniform enumeration of all c.e. sets. Let f be a computable, injective function from ω_1 onto the universal c.e. set. For $\alpha, s < \omega_1$, we let $W_{\alpha,s}$ be the collection of $a \in L_{\omega_1}$ such that there is some $t < s$ such that $f(t) = \langle \alpha, a \rangle$. Each $W_{\alpha,s}$ is countable, the function $\alpha, s \mapsto W_{\alpha,s}$ is computable, and for every $\alpha < \omega_1$, the sequence $\langle W_{\alpha,s} \rangle_s$ is increasing and $W_\alpha = \bigcup_{s<\omega_1} W_{\alpha,s}$.

2.3. Some classical theorems. The proofs of the following two theorems are again analogous to the classical ones.

Proposition 2.11 (s-m-n theorem). *If $f(x, y)$ is a partial computable function, then there is a (total) computable function g such that for all x, $\varphi_{g(x)} = f(x, -)$.*

Proof. Effectively in x, we find a $\Sigma_1(L_{\omega_1})$ formula θ_x that defines the graph of $f(x, -)$; since $\alpha \mapsto \psi_\alpha$ is a bijection, from θ_x we can effectively find some $\alpha = g(x)$ such that $\theta_x = \psi_\alpha$ and so $\varphi_\alpha = f(x, -)$ ⊣

Theorem 2.12 (Recursion theorem). *If $f : \omega_1 \to \omega_1$ is a total computable function, then there is some $\alpha < \omega_1$ such that $\varphi_\alpha = \varphi_{f(\alpha)}$.*

Proof. We solve the equation

$$\varphi_{\varphi_\alpha(\alpha)} = \varphi_{f(\varphi_\alpha(\alpha))}. \tag{2}$$

By Proposition 2.11, there is a computable function g such that for all α,

$$\varphi_{g(\alpha)} = \varphi_{f(\varphi_\alpha(\alpha))}.$$

There is some α^* such that $g = \varphi_{\alpha^*}$. So, α^* solves Equation 2. ⊣

2.4. Relative computability. The basic notion of computable enumerability can be relativised to an oracle; equivalently, it can be used to define this relativisation, using functionals.

Definition 2.13. An *enumeration functional* is a c.e. set of pairs (σ, a) where $\sigma \in 2^{<\omega_1}$ and $a \in L_{\omega_1}$.

Here $2^{<\omega_1}$ is the collection of all binary strings of countable length, that is, functions from some countable ordinal to $\{0, 1\}$. As is standard in classical computability, we identify sets with their characteristic functions. We are considering subsets of ω_1. So, for $\sigma \in 2^{<\omega_1}$ and $B \subseteq \omega_1$ we write $\sigma \subset B$ if $\sigma \subset \chi_B$; that is, if for all $\alpha < \operatorname{dom} \sigma$, $x \in B \Leftrightarrow \sigma(x) = 1$.

If Φ is an enumeration functional and $B \subseteq \omega_1$, we let

$$\Phi^B = \{a : \exists \sigma \subset B\ [(\sigma, a) \in \Phi]\}.$$

In another direction, we consider an enrichment of the language of set theory with constants for the elements of L_{ω_1} by one unary predicate symbol. In this language, too, we define the collection of bounded and existential formulas. For a given $B \subseteq \omega_1$, we say that a subset of L_{ω_1} is $\Sigma_1(L_{\omega_1}, B)$ if it is defined over the structure $(L_{\omega_1}, \in, B)$ (again with the constants interpreted by themselves) by a Σ_1 formula.

PROPOSITION 2.14. *Let $B \subseteq \omega_1$. The following are equivalent for $A \subseteq L_{\omega_1}$:*

(1) *There is some enumeration functional Φ such that $A = \Phi^B$.*
(2) *A is $\Sigma_1(L_{\omega_1}, B)$.*

Such a set A is called *c.e. in* (or *relative to*) B.

PROOF. The collection of countable initial segments of B is $\Sigma_1(L_{\omega_1}, B)$ (in fact it is $\Delta_0(L_{\omega_1}, B)$), as it is defined by bounded universal quantification on the domain of the binary string. Hence if Φ is an enumeration functional, then the relation

$$\exists \sigma \subset B \ (\sigma, a) \in \Phi$$

is $\Sigma_1(L_{\omega_1}, B)$, which shows that Φ^B is $\Sigma_1(L_{\omega_1}, B)$.

In the other direction, let $\varphi(x, \bar{b})$ be an existential formula in the enriched language with a unary predicate. Let Φ be the collection of pairs (σ, a) such that for $\alpha = \operatorname{dom} \sigma$, $a, \bar{b} \in L_\alpha$ and $(L_\alpha, \in, \sigma) \models \varphi(a, \bar{b})$ (where by σ we mean the subset of α whose characteristic function is σ). Then Φ is an enumeration functional, and Φ^B is the subset of L_{ω_1} defined by $\varphi(x, \bar{b})$ over $(L_{\omega_1}, \in, B)$. ⊣

A set is B-computable if it is both c.e. and co-c.e. relative to B. A partial function is *B-partial computable* if its graph is c.e. in B, and B-computable if also its domain is B-computable. The relativisation of Proposition 2.7 holds: for any set $B \subseteq \omega_1$, a set $A \subseteq L_{\omega_1}$ is c.e. in B if and only if it is the domain of a B-partial computable function if and only if it is empty or the range of a B-computable function.

The results of Subsection 2.2 also relativise: for any $B \subseteq \omega_1$, there is a universal B-c.e. set and universal B-partial computable function. This is because there is an effective enumeration $\langle \Phi_\alpha \rangle$ of enumeration functionals, and letting $W^B_\alpha = \Phi^B_\alpha$, the universal set $\bigoplus_\alpha W^B_\alpha$ is c.e. in B. Equivalently, there is an effective enumeration of all existential formulas in the enriched language with a unary predicate, and the satisfaction relation for these formulas is c.e. in the oracle B. The universal B-c.e. set is often denoted B'.

The universal B-c.e. set also gives us a uniform B-computable enumeration of all B-c.e. sets. This, in fact, can be also done uniformly in B. For a given enumeration functional Φ and a binary string σ, let

$$\Phi^\sigma = \{a : \exists \sigma' \subseteq \sigma \ [(\sigma', a) \in \Phi_{\operatorname{dom} \sigma}]\},$$

where $\langle \Phi_s \rangle_{s<\omega_1}$ is an effective enumeration of Φ. For all σ, Φ^σ is countable, and the function $\sigma \mapsto \Phi^\sigma$ is computable (indeed, the function $\sigma, \alpha \mapsto W^\sigma_\alpha$ is computable). If $\sigma \subset s'$ then $\Phi^\sigma \subseteq \Phi^{\sigma'}$, and for all $B \subseteq \omega_1$, $\Phi^B = \bigcup_{\sigma \subset B} \Phi^\sigma$.

Relative computability can also be given by functionals.

Definition 2.15. A *Turing functional* is a c.e. set Φ of pairs of binary strings of countable length, which is *consistent* in the sense that if $(\sigma, \tau), (\sigma', \tau') \in \Phi$ and $\sigma \subseteq \sigma'$ then τ and τ' are compatible, that is, $\tau \subseteq \tau'$ or $\tau' \subseteq \tau$.

If Φ is a Turing functional and $B \subseteq \omega_1$, then we let

$$\Phi^B = \bigcup \{\tau : \exists \sigma \subset B\ [(\sigma, \tau) \in \Phi]\}.$$

For all B, $\Phi^B \in 2^{\leqslant \omega_1}$; we say that Φ^B is *total* if $\Phi^B \in 2^{\omega_1}$.

Proposition 2.16. *Let $A, B \subseteq \omega_1$. Then A is computable in B if and only if there is a Turing functional Φ such that $A = \Phi^B$.*

Proof. Suppose that $A = \Phi^B$ for some Turing functional Φ. Then $a \in A$ if and only if there is some $\sigma \subset B$ and some τ such that $\tau(a) = 1$ and $(\sigma, \tau) \in \Phi$. This is a $\Sigma_1(L_{\omega_1}, B)$ definition of A. By replacing 1 by 0, we get a definition of $\bar{A}$.

For the other direction, we use the regularity of ω_1. Suppose that A is B-computable. Let φ and ψ be existential formulas that define A and $\bar{A}$, respectively, over $(L_{\omega_1}, \in, B)$. Let Φ be the collection of pairs (σ, τ) of countable binary strings such that for $\alpha = \operatorname{dom} \sigma$, L_α contains the parameters of φ and ψ, $\alpha > \operatorname{dom} \tau$, and for all $a < \operatorname{dom} \tau$,

- If $\tau(a) = 1$, then $(L_\alpha, \in, \sigma) \models \varphi(a)\ \&\ \neg\psi(a)$;
- If $\tau(a) = 0$, then $(L_\alpha, \in, \sigma) \models \neg\varphi(a)\ \&\ \psi(a)$.

Then Φ is a Turing functional. The main point is that $\Phi^B = A$ because countably much information about A is decided by a countable level of $(L_{\omega_1}, \in, B)$. ⊣

We write $A \leqslant_{\mathrm{T}} B$ if A is computable from B.

Proposition 2.17. *The relation $\leqslant_T$ is reflexive and transitive.*

Proof. Reflexivity is immediate. Suppose that $A \leqslant_{\mathrm{T}} B \leqslant_{\mathrm{T}} C$; let Φ and Ψ be Turing functionals such that $A = \Phi^B$ and $B = \Psi^C$. Let Θ be the collection of pairs (σ, ρ) of countable binary strings such that there are strings σ', ρ', τ and τ' with $\sigma \subseteq \sigma', \rho' \subseteq \rho, \tau' \subseteq \tau$ and $(\rho', \tau) \in \Psi$, $(\tau', \sigma') \in \Phi$. Then Θ is a Turing functional, and $\Theta^C = A$. ⊣

We write $A \equiv_{\mathrm{T}} B$ if $A \leqslant_{\mathrm{T}} B$ and $B \leqslant_{\mathrm{T}} A$. By Proposition 2.17, this is an equivalence relation on the subsets of ω_1. The equivalence classes are called the (ω_1)-Turing degrees. The relation $\leqslant_{\mathrm{T}}$ induces a partial ordering on the Turing degrees. The map $B \mapsto B'$ is degree-invariant and induces an order-preserving, increasing map on the degrees.

The jump function can be iterated, giving an analogue of the arithmetical hierarchy along all countable ordinals. Details can be found in [5].

§3. Computable structures. In classical computability, a structure with a computable domain is identified with its atomic diagram. We can use the same definition for ω_1. It turns out that for some of our applications, the correct generalisation of the concept of a (model theoretic) structure should allow countably infinite arities.

3.1. Definitions.

Definition 3.1. A *signature* (or *language*) is a collection of function and relation symbols, together with an arity function, which associates with every symbol a countable ordinal.

A constant symbol is a function symbol with arity 0; a proposition is a predicate symbol of arity 0. From now on, we assume that all signatures are computable. This means that the set of symbols is computable, and the function mapping a symbol to its arity is computable. We call a signature *finitary* if every relation and function symbol of the signature has finite arity.

Definition 3.2. A *structure* $\mathcal{M}$ for a signature $\mathcal{L}$ consists of a non-empty set M and an interpretation of the symbols of $\mathcal{L}$:

- for every α-ary relation symbol R, a set $R^{\mathcal{M}} \subseteq M^{\alpha}$;
- for every α-ary function symbol f, a function $f^{\mathcal{M}} \colon M^{\alpha} \to M$.

Here M^{α} is the collection of all sequences of element of M of length α. The key point regarding infinite arities, is that if $M \subseteq L_{\omega_1}$, then for all α, $M^{\alpha} \subset L_{\omega_1}$, and in fact, if M is computable, then M^{α} is computable, uniformly in α.

3.1.1. *The degree of a structure.* Let $\mathcal{M}$ be a structure for a computable signature $\mathcal{L}$, and suppose that the universe M of $\mathcal{M}$ is a subset of L_{ω_1}. To give $\mathcal{M}$ a Turing degree, we identify $\mathcal{M}$ with the 'infinite join" of the relations and functions of $\mathcal{M}$. Let $\tilde{\mathcal{M}}$ consist of:

- the pairs $(R, \bar{a})$ where R is an α-ary relation symbol of $\mathcal{L}$ and $\bar{a} \in R^{\mathcal{M}}$;
- the triples $(f, \bar{a}, b)$ where f is an α-ary relation symbol of $\mathcal{L}$, $\bar{a} \in M^{\alpha}$ and $f^{\mathcal{M}}(\bar{a}) = b$.

Then $\tilde{\mathcal{M}} \subset L_{\omega_1}$, and so has a Turing degree. We often identify $\mathcal{M}$ with $\tilde{\mathcal{M}}$, and so we write $\deg_T(\mathcal{M})$ to denote the Turing degree of $\tilde{\mathcal{M}}$. We remark that $\tilde{\mathcal{M}}$ is Turing equivalent to the atomic diagram $D(\mathcal{M})$ of $\mathcal{M}$, which we define in Section 5. If $\mathcal{L}$ is finitary, then our definition of $D(\mathcal{M})$ agrees with the standard definition of the atomic diagram in elementary first-order logic, and the Turing equivalence of $\tilde{M}$ and $D(\mathcal{M})$ is immediate.

We assume that every language contains the binary relation symbol =, which is always interpreted as true equality. It follows that the domain M of a structure $\mathcal{M}$ is computable from $\deg_T(\mathcal{M})$, as $a \in M$ if and only if $(=, (a, a)) \in \tilde{\mathcal{M}}$, if and only if the sentence $a = a$ is in $D(\mathcal{M})$.

3.2. Examples. We give some natural examples of computable structures, all of finitary signature.

Proposition 3.3. *There is a computable copy of the ordered real field* $(\mathbb{R}; +, \cdot, <, 0, 1)$.

Proof. Any standard set-theoretic construction of the real numbers will do. We use, for example, Dedekind cuts. We fix a copy of the rational numbers $(\mathbb{Q}; +, \cdot, <, 0, 1)$ (this is a countable object, hence an element of L_{ω_1}). Recall that a Dedekind cut is a nonempty initial segment of $\mathbb{Q}$ with no greatest element. Writing the definition in the language of set theory, we see that all quantification ranges over $\mathbb{Q}$ and the potential cut, and so is bounded: an element A of L_{ω_1} is a Dedekind cut if

$$\begin{aligned} \forall x \in A\,(x \in \mathbb{Q}) \ \ \& \ \ \exists x \in \mathbb{Q}\,(x \in A) \ \ \& \ \ \exists x \in \mathbb{Q}\,(x \notin A) \\ \& \ \ \forall x \in A\, \forall y \in \mathbb{Q}\,(y < x \rightarrow y \in A) \\ \& \ \ \forall x \in A\, \exists y \in A\,(x < y). \end{aligned}$$

It follows that the collection of all Dedekind cuts is computable. We let $\mathbb{R}$ be the collection of all Dedekind cuts. The ordering on $\mathbb{R}$ is set containment: $A \leqslant B$ if $\forall x \in A\,(x \in B)$, which is a computable relation. Addition is also computable: $C = A + B$ if

$$\forall x \in \mathbb{Q}\,(x \in C \ \Leftrightarrow \ \exists y \in A\, \exists z \in B\,(x = y + z)).$$

Similarly, multiplication on $\mathbb{R}$ (defined in the standard way on cuts) is computable. ⊣

Note that the standard embedding of the rationals into the reals is also computable.

Proposition 3.4. *There is a computable copy of the complex field* $(\mathbb{C}; +, \cdot, 0, 1)$.

Proof. The algebraic construction works. Let $(\mathbb{R}; +, \cdot, 0, 1)$ be a computable copy of the real field. Let $\mathbb{C} = \mathbb{R}^2$; this is a computable subset of L_{ω_1}. The standard formulas for the operations of addition and multiplication in terms of the real and imaginary parts show that these operations are computable. ⊣

Fix a computable copy $(\mathbb{R}; +, \cdot, <, 0, 1)$ of the ordered real field.

Lemma 3.5. *Every continuous function* $f\colon \mathbb{R} \to \mathbb{R}$ *is computable.*

By cardinality considerations, most functions from $\mathbb{R}$ to $\mathbb{R}$ are not computable.

Proof. Let $f\colon \mathbb{R} \to \mathbb{R}$ be continuous. $f{\upharpoonright}_{\mathbb{Q}}$ is countable, and so an element of L_{ω_1}; so we can use it as a parameter in the following computable definition of f: for $a, b \in \mathbb{R}$, $f(a) = b$ if and only if

$$\begin{aligned} \forall \epsilon \in \mathbb{Q}\, \exists \delta \in \mathbb{Q}\, [\epsilon \leqslant 0 \ \vee (\delta > 0 \ \& \ \forall x \in \mathbb{Q} \\ (a - \delta < x < a + \delta) \rightarrow (b - \epsilon < f{\upharpoonright}_{\mathbb{Q}}(x) < b + \epsilon))]. \end{aligned}$$

⊣

It follows that every total analytic function on $\mathbb{R}$ is computable. This fact has a uniform proof.

Lemma 3.6.

(1) *The collection of triples* $(\langle a_0, a_1, \ldots \rangle, b, c)$ *such that* $\sum a_n b^n = c$ *is computable.*

(2) *The set of sequences* $\langle a_0, a_1, \ldots \rangle$ *such that* $\sum_n a_n x^n$ *defines a total analytic function on* $\mathbb{R}$ *is computable.*

Proof. For (1), we have $\sum a_n b^n = c$ if and only if for all rational $\epsilon > 0$ there is some natural number N such that for all natural numbers $m > N$ we have $|c - \sum_{n \leqslant m} a_n b^n| < \epsilon$. As this quantification ranges over the countable sets $\mathbb{N}$ and $\mathbb{Q}$, this is a computable definition.

Similarly, the collection of pairs $(\langle a_0, a_1, \ldots \rangle, b)$ such that $\sum a_n b^n$ converges to a finite value is also computable; we say that $\langle \sum_{n \leqslant m} a_n b^n \rangle_{m<\omega}$ is a Cauchy sequence.

(2) follows from the root test, or directly from the previous sentence: a power series converges everywhere if and only if it converges on all integers. ⊣

Corollary 3.7. *There is a computable expansion of the real field consisting of all total analytic functions.*

3.3. A basic result of computable model theory. Several basic results of classical computable model theory transfer to ω_1, often with similar or simplified proofs. For example, Morley and Millar ([27], [24]) showed that a countable, complete, decidable theory T has a decidable saturated model if and only if there is a computable enumeration of the complete types consistent with T.

In this subsection, we restrict ourselves to elementary (first-order) theories in countable languages. That is, the languages we deal with are finitary, and all formulas are finitary. We call a structure for a finitary language *decidable* if its elementary diagram is computable.

Proposition 3.8. *Every countable elementary theory with infinite models has a decidable saturated model of size* $\aleph_1$.

Proof. There are two main points:

(1) There is a computable function, which, given an elementary theory in a countable language, produces a complete extension of the theory in the same language.
(2) For any countable elementary theory T, there is an effective listing of all complete types consistent with T (uniformly in T).

To see (1), we first note that the function that takes a countable signature $\mathcal{L}$ and produces the collection of all $\mathcal{L}$-sentences is computable; we state the existence of a countable sequence that gives the recursive construction of all $\mathcal{L}$-formulas. Hence the set of pairs $(T, \mathcal{L})$ such that T is a complete $\mathcal{L}$-theory is computable. Now to get a completion for a given theory, we merely output the $<_L$-least complete extension. The argument for (2) is similar.

Let T_0 be a countable elementary theory with infinite models. Without loss of generality, we suppose that T_0 is complete, so every model of T_0 is infinite. Let $\langle c_\alpha \rangle_{\alpha<\omega_1}$ be a computable enumeration of an uncountable collection of new constants. For $\beta \leqslant \omega_1$, let $C_\beta = \{c_\alpha : \alpha < \beta\}$. By effective recursion, we define an increasing sequence $\langle T_\alpha \rangle_{\alpha<\omega_1}$ of countable elementary first order theories, each complete for the language $\mathcal{L} \cup C_\alpha$, where $\mathcal{L}$ is the language of T_0.

At stage $\alpha + 1$, we are given T_α. We let $\left\langle p^\alpha_\beta \right\rangle_{\beta<\omega_1}$ be a computable enumeration of all complete 1-types consistent with T_α. Let (β, γ) be the least pair of ordinals $\beta, \gamma \leqslant \alpha$ which has not been dealt with in a previous stage. Since T_γ is the restriction of T_α to its language $\mathcal{L} \cup C_\gamma$, the type p^γ_β is consistent with T_α. Since c_α is not a symbol of the language $\mathcal{L} \cap C_\alpha$ of T_α, the set

$$T_\alpha \cup p^\gamma_\beta(c_\alpha)$$

is consistent. We let $T_{\alpha+1}$ be a completion of $T_\alpha \cup p^\gamma_\beta(c_\alpha)$, in the language $\mathcal{L} \cup C_{\alpha+1}$.

At limit stages α, we let $T_\alpha = \bigcup_{\gamma<\alpha} T_\alpha$.

Now $T_{\omega_1} = \bigcup_{\alpha<\omega_1} T_\alpha$ is a computable and complete $\mathcal{L} \cup C_{\omega_1}$ theory extending T_0. As in the standard Henkin construction, we derive a structure $\mathcal{M}$ from T_{ω_1}. The universe M of $\mathcal{M}$ is the collection of $<_L$-least elements of the equivalence classes of C_{ω_1} under the equivalence relation $T_{\omega_1} \vdash c = d$; M is computable since T_{ω_1} is computable. The interpretation of the symbols of $\mathcal{L}$ which defines $\mathcal{M}$ is straightforward; for example, for any relation symbol R of $\mathcal{L}$, for any $\bar{c}$ in M, $\bar{c} \in R^{\mathcal{M}}$ if and only if $T_{\omega_1} \vdash R(\bar{c})$. Since T_{ω_1} is computable, so is $\mathcal{M}$.

By construction, for any $\alpha < \omega_1$, for any 1-type p of $\mathcal{L} \cup C_\alpha$ consistent with T_α, there is some $c \in M$ such that $T_{\omega_1} \vdash p(c)$. Let $\varphi(x)$ be an $\mathcal{L} \cup C_{\omega_1}$-formula such that $T_{\omega_1} \vdash \exists x \varphi$. Since φ is finite, there is some $\alpha < \omega_1$ such that φ is an $\mathcal{L} \cup C_\alpha$-formula. Since T_α is the restriction of T_{ω_1} to $\mathcal{L} \cup C_\alpha$, we have $T_\alpha \vdash \exists x \varphi$. It follows that there is some $\beta < \omega_1$ such that $\varphi(x) \in p^\alpha_\beta$. There is some $\delta > \alpha$ such that $\varphi(c_\delta) \in T_{\delta+1}$, and so there is some $c \in C_{\omega_1}$ such that $\varphi(c) \in T_{\omega_1}$. Thus, T_{ω_1} has the witness property, which implies that for all $\bar{c} \in M$ and all $\mathcal{L}$-formulas φ, $\mathcal{M} \models \varphi(\bar{c})$ if and only if $T_{\omega_1} \vdash \varphi(\bar{c})$. It follows that $\mathcal{M} \models T_0$. Hence, $\mathcal{M}$ is infinite. It also follows that $\mathcal{M}$ is decidable, again because T_{ω_1} is computable.

Now a similar argument shows that $\mathcal{M}$ is $\aleph_1$-saturated (and so that $\mathcal{M}$ is uncountable). Let A be a countable subset of M, and let p be a 1-type over A consistent with the theory of $(\mathcal{M}, A)$. There is some α such that $A \subset C_\alpha$. There is some 1-type q of $\mathcal{L} \cup C_\alpha$, consistent with T_α, extending p. The construction of T_{ω_1} and the argument above show that q, and so p, is realised in $\mathcal{M}$. ⊣

COROLLARY 3.9. *Let T be a countable, uncountably categorical elementary theory. The unique model of T of size $\aleph_1$ has a computable copy.*

§4. Elementary effective algebra. We give examples of analogues of some results from classical effective algebra, some of which require new approaches.

4.1. Vector spaces. In the standard setting, Metakides and Nerode [23] showed that there is a computable infinite-dimensional $\mathbb{Q}$-vector space with no infinite c.e. linearly independent set. By contrast, for any finite field F, any computable F-vector space has a computable basis. In the setting of ω_1, we can generalise these results and obtain the following.

Proposition 4.1. *For any countable field F, every computable F-vector space has a computable basis.*

Theorem 4.2. *There is computable $\mathbb{R}$-vector space, of uncountable dimension, with no uncountable c.e. independent set.*

Proof of Proposition 4.1. Let F be a countable field, and let V be a computable F-vector space. The collection of countable subsets of V that are linearly independent is computable: formally, if we write down the property of being independent, we see that we only need quantify over the countable field F, and so we get a (finitary) $\Delta_1(L_{\omega_1})$ property in the language of set theory defining the collection of countable independent subsets of V. Informally, a countable tuple is independent if satisfies a countable conjunction of computable statements saying that the non-trivial linear combinations do not result in 0_V, and the countable conjunction of computable statements is also computable. We can therefore build a computable basis for V by effective recursion. ⊣

In the rest of this Subsection, we sketch a proof of Theorem 4.2. The proof of Metakides and Nerode's can be adjusted to the setting of ω_1. We present a generalisation of Ash's proof of the same result. In the uncountable, the algebraic aspects of the proof are cleaner. This is because the rationals have no finite subfields, but the reals have countable subfields.

To construct a computable $\mathbb{R}$-vector space of uncountable dimension, with no uncountable c.e. independent set, we start with a canonical presentation V of $\mathbb{R}^{\omega_1}$, by specifying a computable uncountable set B to be the basis of V, and letting V be the collection of formal linear combinations of the elements of B over $\mathbb{R}$. By design, the collection of countable, linearly independent subsets of V is computable.

We construct an isomorphic copy $\hat{V}$ of V by "twisting" V to avoid uncountable independent c.e. subsets. We show that this twisting can be performed without affecting atomic statements about $\hat{V}$ to which we are already committed, thus making $\hat{V}$ computable.

We recursively define an increasing sequence $\langle F_s \rangle_{s<\omega_1}$ of countable subfields of $\mathbb{R}$ (at limit stages we take unions) whose union is $\mathbb{R}$. If F is a subfield of a field G, and U is a vector space over G, then we denote by $U{\restriction}_F$ the reduct of U to an F-vector space. We construct a (not necessarily increasing) sequence $\langle U_s \rangle$ of countable subsets of V such that for all s, U_s is an F_s-vector space, a subspace of $V{\restriction}_{F_s}$.

To define $\hat{V}$, we define an increasing sequence of countable sets $\left\langle \hat{U}_s \right\rangle$ and bijections $p_s \colon U_s \to \hat{U}_s$. The map p_s endows $\hat{U}_s$ with an F_s-vector space structure, by making p_s an isomorphism between U_s and $\hat{U}_s$. The point now is that even though the sequence $\langle p_s \rangle$ is not increasing, we will ensure that for all $s < t$, U_s is a subspace of $U_t{\restriction}_{F_s}$. This would ensure that $\hat{V} = \bigcup_s \hat{U}_s$ is a computable $\mathbb{R}$-vector space.

To ensure that $\hat{V}$ does not contain uncountable c.e. independent sets, we meet the following requirements:

R_α: If W_α is uncountable, then it is not a linearly independent subset of $\hat{V}$.

That $\hat{V}$'s dimension is $\aleph_1$ is ensured by the fact that it is isomorphic to V by the Δ^0_2 map $\lim_s p_s$. To ensure that this limit exists and is an isomorphism, we meet the following requirements:

N_α: There is some stage s such that for all $t \geqslant s$, $B \cap \alpha \subset U_t$, and $p_t\restriction_{B\cap\alpha} = p_s\restriction_{B\cap\alpha}$.

The construction is a finite injury priority argument, where the requirements are ordered in order-type ω_1. The algebraic content of the proof is the following.

Lemma 4.3. *Let F be a proper subfield of $\mathbb{R}$, and let U be a subspace of $V\restriction_F$. Let $u_0, u_1 \in U$ be F-independent. Then there is an F-linear map $f\colon U \to V\restriction_F$ such that $f(u_0)$ and $f(u_1)$ are not $\mathbb{R}$-independent in V.*

Proof. Let $X \subseteq U$ be the F-span of $\{u_0, u_1\}$, and let Y be a linear complement of X in U, so $U = X \oplus Y$. Let $v \in V$ be any vector which is not in Y, and let $Z \subset V$ be the $\mathbb{R}$-span of $\{v\}$. Since F is a proper subfield of $\mathbb{R}$, the F-vector space $Z\restriction_F$ has dimension greater than 1, so we can find $v_0, v_1 \in Z$ which are F-independent; so the F-span of $\{v_0, v_1\}$ is isomorphic to X. We then let f be the F-linear map determined by $f\restriction_Y = \mathrm{id}_Y$ and $f(u_i) = v_i$ for $i = 0, 1$. ⊣

At stage s of the construction, every requirement R_α is provided with a restraint $p_s(\alpha)$ which is the restriction of p_s to some F_s-linear subspace $U_s(\alpha)$ of U_s. A requirement P_α requires attention at stage s if $W_{\alpha,s} \cap \hat{U}_s$ is F_s-independent, and moreover contains a pair $\{u_0, u_1\}$ which is independent over $\hat{U}_s(\alpha) = \operatorname{range} p_s(\alpha)$. Suppose that R_α receives attention at stage s (which happens if it is the strongest requirement that requires attention at stage s). Since $\{u_0, u_1\}$ is independent over $\hat{U}_s(\alpha)$, there is a complement X of $U_s(\alpha)$ in U_s that contains $p_s^{-1}(u_0)$ and $p_s^{-1}(u_1)$. By Lemma 4.3, we find some Y, a subspace of $V\restriction_{F_s}$, disjoint from $U_s(\alpha)$, and an F_s-isomorphism $f\colon Y \to X$ such that $(f \circ p_s)^{-1}(u_0)$ and $(f \circ p_s)^{-1}(u_1)$ are not $\mathbb{R}$-independent in V. We then let $q = f \oplus \mathrm{id}_{U_s(\alpha)}$; q is an isomorphism from $U_s(\alpha) \oplus Y$ to $\hat{U}_s$ such that $q^{-1}(u_0)$ and $q^{-1}(u_1)$ are not $\mathbb{R}$-independent in V. Let F_{s+1} be a countable subfield of $\mathbb{R}$ containing F_s such that $q^{-1}(u_0)$ and $q^{-1}(u_1)$ are not F_{s+1}-independent. Let U_{s+1} be the F_{s+1}-span of $U_s(\alpha) \oplus Y$ in V, let $\hat{U}_{s+1}$ be a countable set containing $\hat{U}_s$, and let p_{s+1} be a bijection from U_{s+1} to $\hat{U}_{s+1}$ extending q; as described above, $\hat{U}_{s+1}$ is now equipped with an F_{s+1}-vector space structure that makes p_{s+1} an isomorphism. Since q is an F_s-isomorphism, U_s is a subspace of $U_{s+1}\restriction_{F_s}$. Note that $p_s(\alpha) \subset p_{s+1}$, so the restraint on R_α was respected; and that u_0 and u_1 are not independent in $\hat{U}_{s+1}$, so R_α is met.

A requirement N_α requires attention at stage s if $B \cap \alpha$ is not a subset of U_s. If N_α receives attention, then we let $F_{s+1} = F_s$ and let U_{s+1} be the F_s-span of $U_s \cup (B \cap \alpha)$, and let p_{s+1} be an extension of p_s; the requirement N_α then imposes the restraint $p_{s+1} \subset p_t(\beta)$ for all $\beta > \alpha$.

Special care is required at limit stages s. Since the sequences $\langle F_t \rangle$ and $\left\langle \hat{U}_t \right\rangle$ are increasing, we can take unions at s and obtain F_s and $\hat{U}_s$. We then need to define p_s and U_s; we limit the damage to requirements N_α to a minimum. Let A_s be the collection of all ordinals $\alpha < s$ such that there is some $t_\alpha < s$ such that requirement N_α is not injured at any stage $t \in [t_\alpha, s)$. For any $\alpha \in A_s$, for all $t \in [t_\alpha, s)$, we have $p_t \upharpoonright_{B\cap\alpha} = p_{t_\alpha} \upharpoonright_{B\cap\alpha}$, which is a map from $B \cap \alpha$ to an independent subset of $\hat{U}_t$. It follows that the function

$$\bigcup_{\alpha\in A_s} p_{t_\alpha} \upharpoonright_{B\cap\alpha}$$

is a well-defined function from $B \cap (\sup A_s)$ to an independent subset of $\hat{U}_s$. We let p_s be any linear map extending this function. This definition ensures that no N_α for $\alpha \in A_s$ is injured at stage s. This completes our sketch of the proof of Theorem 4.2.

4.2. Fields. In the classical setting, any finite extension of a computable field has a computable copy. We have the following analogous result.

Proposition 4.4. *Every countable extension of a computable field has a computable copy.*

Proof. We start by noting that if F is a computable field and $G = F(a)$ is a one-element extension of F, then G has a computable copy; in fact, if the universe of F is co-uncountable, then there is some computable $G' \supset F$ and $a' \in G'$ such that $\mathrm{id}_F \cup \{a' \mapsto a\}$ determines an isomorphism from G' to G (so $G' = F(a')$). This is done in cases: if a is transcendental over F, then we take $G' = F(x)$ be the field of rational functions over F; if a is algebraic over F, with f being its minimal polynomial, then we let $G' = F(x)/(f)$, which has a computable copy into which F canonically embeds.

Now suppose that F is a computable field and that G is a countable extension of F; that is, $G = F(A)$ for some countable set $A \subset G$. Let $\langle a_n \rangle_{n<\omega}$ be an enumeration of A, and for $n < \omega$, let $G_n = F(a_0, a_1, \ldots, a_{n-1})$. We may assume that the universe of F is co-uncountable. By recursion on $n < \omega$ we define computable fields G'_n, whose universes are co-uncountable, and isomorphisms $\varphi_n \colon G'_n \to G_n$. We start with $G'_0 = G_0 = F$, and $\varphi_0 = \mathrm{id}_F$. Given G'_n and φ_n, we know that $G_{n+1} = G_n(a_n)$ and so by the previous paragraph, we may find a computable $G'_{n+1} = G'_n(a'_n)$ and an extension φ_{n+1} of φ_n which is an isomorphism from G'_{n+1} to G_{n+1}, determined by mapping a'_n to a_n.

Since $\omega < \omega_1$, not only is each G'_n computable, but in fact they are uniformly computable. It follows that $G' = \bigcup_n G'_n$ is computable. The map $\bigcup_n \varphi_n$ shows that G' is isomorphic to G. ⊣

Van der Waerden [34] gave the idea for producing a computable field without a splitting algorithm; this was fleshed out by Fröhlich and Shepherdson [12]: to the field of rationals we add a (p_n)th root of unity for $n \in \emptyset'$, where $\langle p_n \rangle$ is an enumeration of all prime numbers. This field has the feature that no computable copy has a splitting algorithm. For more on the early results of effective algebra and computable model theory, see Miller [26]. The idea of van der Waerden, and Fröhlich and Shepherdson, cannot be immediately applied in the setting of ω_1; there are no new prime numbers. Nevertheless, the analogous results hold.

PROPOSITION 4.5. *There is a computable field F with no splitting algorithm: the collection of irreducible polynomials in $F[x]$ is not computable.*

We need the following:

LEMMA 4.6. *Let $(\mathbb{C}; +, \cdot, 0, 1)$ be a computable copy of the complex field. There is an uncountable computable subset of $\mathbb{C}$ which is algebraically independent.*

In fact, we can obtain an algebraic basis for $\mathbb{C}$ over $\mathbb{Q}$.

PROOF. The relation of algebraic independence in $\mathbb{C}$, namely the collection of pairs (A, b) such that $A \subset \mathbb{C}$ is computable and b belongs to the algebraic closure of A in $\mathbb{C}$, is computable, since it requires quantification over only countably many polynomials over A. A computable algebraic basis for $\mathbb{C}$ over $\mathbb{Q}$ can be then defined recursively, at each stage choosing, for the next element, the $<_L$-least element independent from the elements chosen so far. $\dashv$

LEMMA 4.7. *Let F be a field. The collection of elements $a \in F$ such that F contains a square root of a is computable from the set of irreducible polynomials in $F[x]$.*

PROOF. F contains a square root of a if and only if the polynomial $x^2 - a$ is reducible in $F[x]$. $\dashv$

PROOF OF PROPOSITION 4.5. By Lemma 4.6, let $\{a_\alpha : \alpha < \omega_1\}$ be a computable, algebraically independent subset of $\mathbb{C}$; let

$$F_0 = \mathbb{Q}(a_\alpha)_{\alpha<\omega_1}.$$

It is not difficult to see that F_0 is computable. Now we let

$$F = F_0(\sqrt{a_\alpha})_{\alpha\in\emptyset'}.$$

It is not hard to show that a_α has a square root in F if and only if $\alpha \in \emptyset'$; by Lemma 4.7, the collection of irreducible polynomials in $F[x]$ computes $\emptyset'$.

The field F is not computable (as a subfield of $\mathbb{C}$); the technique of the proof of Proposition 4.4 shows that F has a computable copy which extends F_0, so that the mapping $\alpha \mapsto a_\alpha$ remains computable as a function from ω_1 to F. $\dashv$

Unlike the Fröhlich-Shepherdson field, the field F of the previous proof has a computable copy with a computable splitting algorithm. Using more robust coding, we get the full result.

THEOREM 4.8 (with Hirschfeldt and Montalbán). *There is a computable field F (of characteristic 0) such that for any computable field K isomorphic to F, the collection of irreducible polynomials in $K[x]$ computes $\emptyset'$.*

The reduction of $\emptyset'$ to the collection of irreducible polynomials in $K[x]$ is uniform in K.

The proof of Theorem 4.8 uses a technique of coding of graphs into fields which is due to Friedman and Stanley [10], extending results of [3]. In this context, a (simple, undirected) *graph* is a structure for a language with a single binary relation symbol which is interpreted by a symmetric, irreflexive relation (rather than being a two-sorted structure containing vertices and edges).

DEFINITION 4.9. A graph $G = (V, E)$ is *c.e.* if V is a c.e. set and E is a c.e. set of pairs.

Theorem 4.8 follows from the following two propositions:

PROPOSITION 4.10. *There is a c.e. graph, every copy of which computes $\emptyset'$.*

PROPOSITION 4.11. *For any c.e. graph G there is a field F with a computable copy such that if K is isomorphic to F, then the collection of irreducible polynomials in $K[x]$ computes a copy of G.*

PROOF OF PROPOSITION 4.10. By taking a computable bijection between ω_1 and the power set of ω, we may assume that $\emptyset'$ is a set of subsets of ω. The graph G is a disjoint union of "daisy graphs". Fix $A \subseteq \omega$. The graph G_A consists of a vertex c_A; for every $n \in A$, a cycle of length $2n + 4$ starting and ending in c_A; for every $n \notin A$, a cycle of length $2n + 5$ starting and ending in c_A; two extra vertices a_A and b_A, both connected by an edge to c_A; and if $A \in \emptyset'$, we connect a_A and b_A by an edge. The graph G is the disjoint union of G_A for all $A \subseteq \omega$.

It is clear that G has a c.e. copy. Suppose that $H \cong G$; let $f : G \to H$ be an isomorphism. Let V_A be the collection of vertices of G_A; it is uniformly computable in A. The map $A \mapsto f\restriction_{V_A}$ is H-computable. To see this, let $\tilde{G}_A$ be G_A, except that there is no edge between a_A and b_A. Thus $\tilde{G}_A$ is uniformly computable in A. For any function $g \in L_{\omega_1}$, $g = f\restriction_{V_A}$ if and only if g is an isomorphism between $\tilde{G}_A$ and the restriction of H to range g, except that we ignore any H-edge between $g(a_A)$ and $g(b_A)$. It follows that the map $A \mapsto \{f(a_A), f(b_A)\}$ is H-computable, and $A \in \emptyset'$ if and only if $f(a_A)$ and $f(b_A)$ are connected by an edge in H, which is an H-computable predicate. ⊣

In the rest of the section, we prove Proposition 4.11. Let $G = (V, E)$ be a c.e. graph. By pulling back by an effective enumeration, we may assume that the set V of vertices of G is computable. By Lemma 4.6, we may in fact assume that V is an algebraically independent subset of $\mathbb{C}$. Let F_0 be the compositum of the algebraic closures of $\mathbb{Q}(v)$ for $v \in V$, and let

$$F = F_0(\sqrt{v + w})_{(v,w)\in E}.$$

As in the proof of Proposition 4.5, F is not computable, but has a computable copy; when we discover that $(v, w) \in E$, we add a new element and declare that its square is $v + w$.

For $a, b \in \mathbb{C}$, let $a \leqslant_{\mathbb{C}} b$ if a is algebraic over $\mathbb{Q}(b)$; since $a \leqslant_{\mathbb{C}} b$ if and only if the index $[\mathbb{Q}(a,b) : \mathbb{Q}(a)]$ is finite, and the field extension index is multiplicative, the relation $\leqslant_{\mathbb{C}}$ is transitive; hence algebraic equivalence $a \sim_{\mathbb{C}} b$, defined by $a \leqslant_{\mathbb{C}} b$ & $b \leqslant_{\mathbb{C}} a$, is an equivalence relation. We have $a \sim_{\mathbb{C}} b$ if and only if the algebraic closure $\mathrm{acl}^{\mathbb{C}}(\mathbb{Q}(a))$ of $\mathbb{Q}(a)$ in $\mathbb{C}$ equals the algebraic closure $\mathrm{acl}^{\mathbb{C}}(\mathbb{Q}(b))$ of $\mathbb{Q}(b)$ in $\mathbb{C}$. We use the following important fact.

Fact 4.12 (Friedman, Stanley [10]).

(1) *If $a \in F$ is transcendental and $\mathrm{acl}^{\mathbb{C}}(\mathbb{Q}(a)) \subset F$ then there is some $v \in V$ such that $a \sim_{\mathbb{C}} v$.*
(2) *If $v, w \in V$, $a, b \in F$, $a \sim_{\mathbb{C}} v$, $b \sim_{\mathbb{C}} w$ and F contains a square root of $a + b$, then $(v, w) \in E$.*

Lemma 4.13. *Let K be a field. The collection of countable subfields of K which are algebraically closed is computable from K.*

Proof. $A \subset K$ is a subfield if it is closed under the field operations; this involves quantifying over A. A is algebraically closed if every polynomial over A has a root in A; again this involves only quantifying over A. ⊣

Let $K \cong F$; let $f : K \to F$ be an isomorphism. We build a copy $H = (V_H, E_H)$ of G, computable from K and the collection of irreducible polynomials in $K[x]$. We let V_H be the set of countable subfields of K which are algebraically closed and distinct from the algebraic closure of $\mathbb{Q}$ in K. By Lemma 4.13, V_H is K-computable. For $A, B \in V_H$, we let $(A, B) \in E_H$ if there are $a \in A$ and $b \in B$ such that K contains a square root of $a + b$. By Lemma 4.7, E_H is computable from the collection of irreducible polynomials in $K[x]$ (as A and B are countable).

Finally, fact 4.12 implies that $g : V \to V_H$ defined by $g(v) = \mathrm{acl}^K(\mathbb{Q}(f(v)))$ is an isomorphism from G to H. This completes the proof of Proposition 4.11.

§5. Computable categoricity and intrinsically c.e. relations. In the classical setting, much research has gone into two related notions, computable categoricity and intrinsically c.e. relations. There is also work on relative versions of these notions. We recall the definitions, beginning with intrinsically c.e. and relatively intrinsically c.e. relations.

Definition 5.1. Let $\mathcal{A}$ be a computable structure, and let R be a relation on $\mathcal{A}$.

- R is *intrinsically c.e. on* $\mathcal{A}$ if for all computable $\mathcal{B} \cong \mathcal{A}$, for all isomorphisms $f : \mathcal{A} \to \mathcal{B}$, $f[R]$ is c.e.
- R is *relatively intrinsically c.e.* on $\mathcal{A}$ if for all $\mathcal{B} \cong \mathcal{A}$, for all isomorphisms $f : \mathcal{A} \to \mathcal{B}$, $f[R]$ is c.e. relative to $\mathcal{B}$.

The notion of an intrinsically c.e. relation seems natural, particularly if computable structures are of primary interest. It turns out, however, that the relative version is better behaved. This is true in the standard setting (see [4] and [6]). It is true also in our setting, as we shall soon see.

In a computable graph, the collection of vertices that are contained in an infinite complete subgraph is relatively intrinsically c.e., as is the collection of vertices that are contained in an infinite, totally disconnected subgraph. In a computable partial ordering, the collection of elements that are contained in an infinite chain is relatively intrinsically c.e., as is the collection of elements that are contained in an infinite antichain. In an Abelian group, the set of divisible elements is relatively intrinsically c.e.

Next, we give the definitions of computable categoricity, and relative computable categoricity.

DEFINITION 5.2. Let $\mathcal{A}$ be a computable structure.

- $\mathcal{A}$ is *computably categorical* if for all computable $\mathcal{B} \cong \mathcal{A}$, there is a computable isomorphism from $\mathcal{A}$ to $\mathcal{B}$.
- $\mathcal{A}$ is *relatively computably categorical* if for all $\mathcal{B} \cong \mathcal{A}$, there is an isomorphism from $\mathcal{A}$ to $\mathcal{B}$ computable from $\mathcal{B}$.

For example, the linear ordering $(\mathbb{R}, <)$ is relatively computably categorical, whereas the linear orderings $(\omega_1, <)$ and $\mathbb{Q}\times\omega_1$ and $\mathbb{R}\times\omega_1$ (the long line) are not. Again, it turns out that the relative notion is better behaved, just as in the standard setting ([4], [6]).

5.1. The appropriate infinitary logic. What is intrinsic to a structure should be definable. Similarly, we expect that if a structure is computably categorical, then there should be an easy way to describe it which would enable the standard back-and-forth construction to produce the promised computable isomorphism. This is indeed the case for the relative notions.

In the standard setting, we use formulas of $\mathcal{L}_{\omega_1\omega}$, of a special kind, to define the intrinsically c.e. relations and the orbits in the computably categorical structures. In particular, we use "computable Σ_1 formulas", where these are c.e. disjunctions of finitary existential formulas, with a fixed finite tuple of variables. Similar results hold in the uncountable case. Below, we say what, in the setting of ω_1, are the computable infinitary Σ_1 formulas. We start by describing the analogue of $\mathcal{L}_{\omega_1\omega}$ in our setting, which is a generalisation of $\mathcal{L}_{\omega_2\omega_1}$, in that it allows quantification over countably many variables, and conjunctions and disjunctions of families of formulas of size $\aleph_1$.

5.1.1. *The infinitary logic.* Let $\mathcal{L}$ be a computable signature (see Subsection 3.1). The collection of $\mathcal{L}$-*terms* is defined recursively, starting with an uncountable computable collection of variables, by applying function symbols of $\mathcal{L}$ with arity α to α-tuples of terms of lower rank. This recursion is effective, and so the collection of all $\mathcal{L}$-terms is computable.

An *atomic $\mathcal{L}$-formula* is an expression of the form $R(\bar{t})$, where R is a relation symbol of $\mathcal{L}$ of arity α, and $\bar{t}$ is a tuple of $\mathcal{L}$-terms of length α.

A *(countable) quantifier-free* $\mathcal{L}$-formula (sometimes called a Σ_0 or a Π_0 formula) is recursively obtained from atomic formulas by applying negation and countable conjunctions and disjunctions. Each countable quantifier-free $\mathcal{L}$-formula is an element of L_{ω_1}, and as with terms, the collection of such formulas is computable. These formulas are the analogue of the elementary ($\mathcal{L}_{\omega\omega}$) quantifier-free formulas.

Semantics are defined as expected. This gives us the notion of the *atomic diagram* $D(\mathcal{M})$ of an $\mathcal{L}$-structure, which is the collection of all atomic sentences, and negations of atomic sentences true in $\mathcal{M}$. Really, the sentences are in the language $\mathcal{L}_{\mathcal{M}}$ that is obtained from $\mathcal{L}$ by adding constants naming the elements of $\mathcal{M}$. As mentioned above, the atomic diagram (which is a subset of L_{ω_1}) is Turing equivalent to $\mathcal{M}$, and to the collection of all countable quantifier-free $\mathcal{L}_{\mathcal{M}}$-sentences tnat hold in $\mathcal{M}$.

Infinitary (uncountable) formulas are in general obtained from countable quantifier-free formulas by allowing uncountable conjunctions and disjunctions, negation, and quantification over countably many variables. We restrict to formulaswith only countably many free variables.

5.1.2. *The effective infinitary formulas.* A (countable) *existential formula* is a formula of the form $\exists \bar{u}\, \varphi(\bar{x}, \bar{u})$, where φ is a countable quantifier-free formula, and $\bar{x}$ and $\bar{u}$ are countable tuples of variables. Again, every countable existential formula is an element of L_{ω_1}, and the collection of all countable existential formulas is a computable subset of L_{ω_1}.

A *computable (infinitary)* Σ_1 *formula* is the disjunction of a c.e. set (possibly uncountable) of existential formulas, with a fixed countable tuple $\bar{x}$ of free variables.

5.1.3. *Defining relatively intrinsically c.e. relations.* In the standard setting, every relatively intrinsically c.e. relation is definable by a computable Σ_1 formula with a finite tuple of parameters (see [4] and [6]). The analogous result holds in our setting.

THEOREM 5.3. *Let $\mathcal{A}$ be a computable structure, and let R be a relation on $\mathcal{A}$. The following are equivalent:*

(1) *R is relatively intrinsically c.e. on $\mathcal{A}$.*
(2) *There is an expansion of $\mathcal{A}$ by countably many constants, in which R is definable by a computable Σ_1 formula.*

The proof of Theorem 5.3 is given in [15]. Theorem 5.3 was also extended to the arithmetical hierarchy; see [5].

We remark that in the classical setting, Manasse [22] and Goncharov [13] gave examples of relations that are intrinsically c.e. but not relatively intrinsically c.e. It seems likely that such examples can be given in the setting of ω_1.

5.2. Continuous Scott sets. A new ingredient enters the picture when we consider computable categoricity. In the classical setting, a computable structure $\mathcal{A}$

is relatively computably categorical if and only if some expansion $(\mathcal{A}, \bar{c})$ of $\mathcal{A}$ by finitely many constants has a formally Σ_1^0 Scott family, where this is a c.e. set Φ of computable Σ_1 formulas that includes definitions of the orbits of all finite tuples in $\mathcal{A}$ under the action of the group of automorphisms of $(\mathcal{A}, \bar{c})$. The formally Σ_1^0 Scott family enables us to effectively carry out a back-and-forth construction between any two copies of $\mathcal{A}$. In fact, taking the disjuncts of the members of Φ, we obtain a c.e. Scott family consisting of finitary existential formulas.

There are similar results for relative Δ_α^0 categoricity where α is an arbitrary computable ordinal, where the isomorphism is Δ_α^0 relative to the copy, and the formulas in the c.e. Scott family are "computable Σ_α".

In the setting of ω_1, it is not sufficient to have a c.e. set of computable existential formulas that define the orbits of tuples in our structure. The key point is that the back-and-forth construction now needs to pass through limit stages, and we need to preserve the property of effectively describing the orbits of the tuples that we have already included in the domain and range of our partial isomorphism. Below, we add a condition to take care of this problem.

Definition 5.4. Let A be an uncountable set. Recall that $[A]^{\aleph_0}$ denotes the collection of all countable subsets of A. We say that a set $\mathcal{C} \subseteq [A]^{\aleph_0}$ is *unbounded* if for all countable $b \subset A$ there is some $c \in \mathcal{C}$ such that $b \subseteq c$. We say that $\mathcal{C} \subseteq [A]^{\aleph_0}$ is *closed* if whenever $a_0 \subset a_1 \subset a_2 \subset \dots$ is a sequence of elements of $\mathcal{C}$, $\bigcup_n a_n$ is also an element of $\mathcal{C}$.

Definition 5.5. Let $\mathcal{A}$ be a structure. A *continuous formally c.e. Scott family* for $\mathcal{A}$ consists of a computable closed unbounded subset $\mathcal{C}$ of $[A]^{\aleph_0}$ and a c.e. collection Φ of (c.e. indices for) computable Σ_1 formulas such that:

(1) for every $\bar{a} \in \mathcal{C}$ there is a formula $\varphi_{\bar{a}} \in \Phi$ that defines the orbit of $\bar{a}$ in $\mathcal{A}$; and
(2) if $\bar{a}_0 \subset \bar{a}_1 \subset \bar{a}_2 \subset \dots$ is an increasing sequence of elements of $\mathcal{C}$, then

$$\varphi_{\bigcup_n \bar{a}_n} = \bigwedge_n \varphi_{\bar{a}_n}.$$

Theorem 5.6. *The following are equivalent for a computable structure $\mathcal{A}$:*

(1) *$\mathcal{A}$ is relatively computably categorical.*
(2) *There is an expansion of $\mathcal{A}$ by countably many constants, which has a continuous formally c.e. Scott family.*

The restriction to a computable closed unbounded set is necessary. Theorem 5.6 is proved in [15].

As for intrinsically c.e. relations, in the classical setting there are structures that are computably categorical but not relatively so; again we expect the same to hold for ω_1.[3]

[3]*Note added in Proof.* In his thesis, Johnson, [18], gives categoricity and non-categoricity results for members of "quasi-minimal excellent" classes. In particular, the Zil'ber field of size K^+ is not computably categorical, while the Zil'ber cover is relatively computably categorical.

§6. Linear orderings. A significant body of work concerns the effective properties of countable linear orderings. In the uncountable case, we get both analogues of classical results that require new proofs, and results that are completely opposite to classical ones.

6.1. A saturated linear ordering. We may apply Proposition 3.8 to the theory DLO of dense linear orderings without endpoints. The resulting saturated model of DLO of size $\aleph_1$, which we denote by η_1, is the ω_1-dense linear ordering: it is characterised by the property that whenever A and B are countable subsets of η_1 with the property that $A < B$ (which means that for all $a \in A$ and $b \in B$, $a < b$), there is some $c \in \eta_1$ such that $A < c < B$ (which means that for all $a \in A$ and $b \in B$, $a < c < b$). Since DLO has quantifier elimination, η_1 is saturated with respect to simple embeddings: if $\mathcal{L}$ and $\mathcal{L}'$ are countable linear orderings, and $f \colon \mathcal{L} \to \mathcal{L}'$, $g \colon \mathcal{L} \to \eta_1$ are order-preserving functions, then there is an order-preserving function $h \colon \mathcal{L}' \to \eta_1$ such that $g = h \circ f$.

In the classical setting, Nurtazin [28] showed that there is a "computable numbering" of the computable linear orderings; that is, a uniformly computable sequence of linear orderings, that includes a copy of every computable linear ordering. We can transfer Nurtazin's proof to the uncountable case.

Proposition 6.1. *There is a uniformly computable sequence $\langle \mathcal{L}_\alpha \rangle_{\alpha<\omega_1}$ of linear orderings such that for every computable linear ordering $\mathcal{L}$ there is some α such that $\mathcal{L} \cong \mathcal{L}_\alpha$.*

Proof. Let $\mathcal{M} = (M, <^{\mathcal{M}})$ be a computable copy of η_1. Since M is computable, by taking a computable bijection we may suppose that $M = L_{\omega_1}$. First, we note that if $\mathcal{L}$ is a computable linear ordering, then we can recursively embed $\mathcal{L}$ into $\mathcal{M}$ such that the image of the embedding is c.e. Next, we show that each c.e. set W_α, the substructure of $\mathcal{M}$ with universe W_α is isomorphic to a computable ordering. For all $a \in W_\alpha$, let $g_\alpha(a) = (a, s)$, where s is the (ordinal) stage at which a enters W_α. We let $\mathcal{L}_\alpha$ be the linear ordering whose universe is range g_α, where the ordering is defined so that g_α is an isomorphism from $(W_\alpha, <^{\mathcal{M}} \restriction_{W_\alpha})$ to $\mathcal{L}_\alpha$. This is the desired uniformly computable sequence representing all computable order types. ⊣

6.2. Computable well orderings. Much of hyperarithmetic theory is based on the theory of computable well-orderings. In the uncountable case, there is no analogue for hyperarithmetic theory, and the situation is distorted by the fact that well-foundedness is now a co-c.e. phenomenon rather than Π^1_1. Nevertheless, it is natural to study the computable well-orderings of ω_1 in their own right.

As in the classical setting, the initial segment of any computable well-ordering has a computable copy. Hence the collection of ordinals that are isomorphic to computable linear orderings is an initial segment of the class of ordinals (of size $\aleph_1$, the number of computable linear orderings). As in the classical case, no ordinal beyond ω_1 with a computable copy can be admissible; if L_α is admissible and $\alpha > \omega_1$, then L_α contains every computable well-ordering of ω_1, and given a computable well-ordering $\mathcal{L}$ of ω_1, by effective recursion, L_α must also contain the order-type of $\mathcal{L}$. [By effective recursion in L_α, define a function f from the ordinals onto initial segments of $\mathcal{L}$. At stage β, if $f\restriction_\beta$ is not onto $\mathcal{L}$, we let $f(\beta)$ be the $\mathcal{L}$-least-upper-bound of the range of $f\restriction_\beta$; this definition is effective since it involves only quantification over the elements of $\mathcal{L}$. If the construction doesn't halt before stage α, we get an effective injective map from α onto an initial segment of $\mathcal{L}$; but since $\mathcal{L}$ is a well-ordering, every initial segment of $\mathcal{L}$ is an element of L_α, for a contradiction.]

The following diverges markedly from the classical case, in which the least noncomputable ordinal ω_1^{CK} is admissible. It follows from results of S. Friedman [11]. We include a proof for completeness and for illustration of our methods, that will be used for Proposition 6.3.

Proposition 6.2 (Friedman). *The least ordinal that is not isomorphic to any computable well-ordering of ω_1 is not admissible.*

Proof. Let β be this least ordinal; and let α be the least admissible ordinal greater than ω_1. Every computable linear ordering of ω_1 is an element of L_{ω_1+1}, and the collection $\mathcal{W}$ of computable well-orderings is definable over L_{ω_1+1} (since the countable descending sequences are elements of L_{ω_1}). Then $\mathcal{W}$ is an element of L_α. The function that takes every element of $\mathcal{W}$ to its ordertype is L_α-effective, and so by admissibility, the range β of this function is bounded below α. ⊣

Even though by definition there is no (classically) computable linear ordering isomorphic to ω_1^{CK}, there is a (classically) computable linear ordering with an initial segment isomorphic to ω_1^{CK}. This is Harrison's linear ordering, which is isomorphic to $\omega_1^{\mathrm{CK}}(1+\mathbb{Q})$. It is obtained by taking a non-standard extension of ω_1^{CK}, one with no infinite descending hyperarithmetic sequences (equivalently, it is an initial segment of the Kleene-Brouwer ordering of a computable tree that is ill-founded but has no hyperarithmetical paths). We obtain a similar result in the setting of ω_1 using completely different tools.

Proposition 6.3 (with Shore). *There is a computable ordering with an initial segment of order-type β, where β is the least ordinal that is not isomorphic to any computable well-ordering of ω_1.*

Proof. By Proposition 6.1, let $\langle \mathcal{L}_\alpha \rangle$ be a uniformly computable list of all computable linear orderings. We note that the set

$$\{\alpha < \omega_1 \ : \ \mathcal{L}_\alpha \text{ is not a well-ordering}\}$$

is c.e.

We construct a computable linear ordering $\mathcal{L}$ by recursion with priorities. At stage $s < \omega_1$ we have constructed a countable part of $\mathcal{L}$, and identified our guess A_s for the initial segment A that should be isomorphic to β. As long as $\mathcal{L}_\alpha$ apears to be well-founded, we add a copy of $\mathcal{L}_\alpha$ to A. If at stage s we discover that

$\mathcal{L}_\alpha$ is not well-founded, we declare that all the elements that were put in order to copy $\mathcal{L}_\alpha$ lie to the right of A, and we injure $\mathcal{L}_\gamma$ for $\gamma > \alpha$, so that we code such well-founded $\mathcal{L}_\gamma$ in A all over again. ⊣

6.3. Least degrees and jump-degrees.

Definition 6.4. The *degree spectrum* degSpec($\mathcal{M}$) of a structure $\mathcal{M}$ is the collection of Turing degrees that compute a structure isomorphic to $\mathcal{M}$.

Much work has gone into the study of degree spectra of structures in general and of linear orderings in particular. A fundamental theorem of the second author states that if $\mathbf{b} \in$ degSpec($\mathcal{M}$) then (except for a trivial case) there is a copy of $\mathcal{M}$ of degree $\mathbf{b}$. This holds for ω_1 as well.

The degree spectrum degSpec($\mathcal{M}$) is of particular interest in case it is the cone above some degree $\mathbf{b}$; in this case, we say that $\mathbf{b}$ is *the* degree of (the isomorphism type of) $\mathcal{M}$. In the countable context, Richter [30] showed that no nonzero degree can be the degree of a linear ordering. While it is possible to code an arbitrary subset of ω in the isomorphism type of a countable linear ordering, it is not possible to do this so that the set is computable relative to all copies of the ordering. Richter showed that for any countable linear ordering $\mathcal{L}$, there is some $\mathcal{L}'$, isomorphic to $\mathcal{L}$, which forms a minimal pair with $\mathcal{L}$. This result strongly uses the true finiteness of the finite cardinals, and cannot be replicated in the case of ω_1. In fact, we have the opposite.

Theorem 6.5. *Every Turing degree is the degree of some linear ordering.*

Proof. We show that for any uncountable $A \subseteq \mathbb{R} \setminus \mathbb{Q}$ there is some linear ordering $\mathcal{L}_A$ such that for any $X \subseteq \omega_1$, X computes a copy of $\mathcal{L}_A$ if and only if A is c.e. relative to X. Then, given any Turing degree $\mathbf{b}$, we fix some $A \subseteq \mathbb{R} \setminus \mathbb{Q}$ of degree $\mathbf{b}$; the degree spectrum of the linear ordering $\mathcal{L}_{A \oplus \bar{A}}$ is then precisely the cone above $\mathbf{b}$.

So, fix some uncountable $A \subseteq \mathbb{R}$; let $\mathcal{L}_A = A \cup \mathbb{Q}$, with the ordering inherited from the real line.

Let $X \subseteq \omega_1$. If A is c.e. relative to X, let $f\colon \omega_1 \to A$ be injective and extend f to a function $g\colon \omega_1 \cup \mathbb{Q} \to A \cup \mathbb{Q}$ by letting $g\restriction_{\mathbb{Q}} = \mathrm{id}_{\mathbb{Q}}$. Let $\mathcal{L}$ be the linear ordering defined on $\omega_1 \cup \mathbb{Q}$ that makes g into an isomorphism from $\mathcal{L}$ to $\mathcal{L}_A$; $\mathcal{L}$ is computable in X.

In the other direction, suppose that $\mathcal{L} \cong \mathcal{L}_A$; we show that A is c.e. in $\mathcal{L}$. Let $f\colon \mathcal{L} \to \mathcal{L}_A$ be an isomorphism. Let $Q = f^{-1}\mathbb{Q}$; the point is that $f\restriction_Q$ is countable. For all a in the domain of $\mathcal{L}$, and which is not in Q, let

$$C_a = \left\{q \in Q : q <^{\mathcal{L}} a\right\}$$

and

$$D_a = \left\{q \in Q : q >^{\mathcal{L}} a\right\}.$$

The sets C_a and D_a (which are countable) are $\mathcal{L}$-computable given a; since $f\restriction_Q \in L_{\omega_1}$, so are $f[C_a]$ and $f[D_a]$. Now $f(a)$ is the unique irrational real number such that

$$f[C_a] < f(a) < f[D_a],$$

This definition involves quantifying only over $f[C_a]$ and $f[D_a]$; so f is computable from $\mathcal{L}$. Hence A, which is the range of f (minus the rationals), is c.e. relative to $\mathcal{L}$. ⊣

In the absence of a degree for a linear ordering $\mathcal{L}$, one can ask about a jump-degree of $\mathcal{L}$: a least degree for the Turing jumps of all isomorphic copies of $\mathcal{L}$, and similarly the double-jump-degree and so on. In the countable case, the second author [19] showed that $\mathbf{0}'$ is the only jump-degree of a countable linear ordering; but every degree above $\mathbf{0}''$ is the double-jump-degree of a linear ordering ([1], [9]).

In recent work, the first author, with Kach, Lempp and Turetsky, showed that in the uncountable setting, every degree above $\mathbf{0}'$ is the proper jump-degree of a linear ordering.

6.4. Techniques of Jockusch and Soare. Theorem 6.5 implies the analogue of a result of the second author for the countable case, that for any nonzero Turing degree $\mathbf{b}$, there is a $\mathbf{b}$-computable linear ordering that has no computable copy; in other words, for all $\mathbf{b} > \mathbf{0}$ there is some linear ordering $\mathcal{L}$ such that $\mathbf{b} \in \text{degSpec}(\mathcal{L})$ but $\mathbf{0} \notin \text{degSpec}(\mathcal{L})$. The proof of the result in the countable setting is quite elaborate, relying on double-jump inversion and arguments of Seetapun's. Seetapun's proofs are in turn generalisations of a technique of Jockusch's and Soare's [17], who first proved the result for c.e. degrees $\mathbf{b}$. These techniques can be simplified in the uncountable setting to get analogues of other classical results.

PROPOSITION 6.6. *There is a computable function f such that for all $e < \omega_1$, if e is an index for an infinite computable linear ordering $\mathcal{L}$, $f(e)$ is a computable index for an infinite linear ordering $\mathcal{L}'$ (of universe ω_1) such that $\mathcal{L} \not\cong \mathcal{L}'$.*

Thus the fixed-point theorem fails for infinite linear orderings.

PROOF. Suppose that we are given an enumeration $(\mathcal{L}_s)_{s<\omega_1}$ of an infinite computable linear ordering $\mathcal{L}$. We know that $\mathcal{L}$ contains either an infinite ascending chain or an infinite descending chain, so we can wait for a stage $s < \omega_1$ at which we see such a chain. If we see an infinite descending chain in $\mathcal{L}$, we output a computable copy of ω_1; if we see an infinite ascending chain in $\mathcal{L}$, we output a computable copy of ω_1^*. ⊣

In the countable setting, Miller [25] extended the Jockusch-Soare and Seetapun technique and showed that there is a linear ordering $\mathcal{L}$ with no computable copy whose degree spectrum contains every hyperimmune degree, in particular every nonzero Δ^0_2 degree. The same holds in the uncountable case.

THEOREM 6.7 (Greenberg, Kach, Lempp, Turetsky). *There is a linear ordering of ω_1 that has no computable copy, but whose degree spectrum contains every nonzero Δ^0_2 degree.*

SKETCH OF PROOF. We give an axiomatic approach, which originates from understanding the algebraic aspects of Miller's proof for the countable case. A *computable directed system of linear orderings* is a sequence $\left\langle C_\beta, h_\beta \right\rangle_{\beta<\omega_1}$ such that:

- for every $\beta < \omega_1$, C_β is a countable linear ordering, and h_β is an embedding of C_β into $C_{\beta+1}$;
- $\beta \mapsto (C_\beta, h_\beta)$ is a computable function;
- C_0 is empty;
- for limit β, C_β is the direct limit of $\langle C_\alpha, h_\alpha \rangle_{\alpha<\beta}$.

What we require is a pair of computable directed systems $\left\langle A_\beta, f_\beta \right\rangle_{\beta<\omega_1}$ and $\left\langle B_\beta, g_\beta \right\rangle_{\beta<\omega_1}$ of linear orderings with the following properties:

(1) For all β, A_β does not embed into any proper initial segment of A_β.
(2) If C is a nonempty initial segment of B_{ω_1} (the direct limit of the system $\left\langle B_\beta, g_\beta \right\rangle$), then for all $\beta < \omega_1$, $A_{\beta+1}$ embeds into $A_\beta + C$.
(3) For all $\beta < \omega_1$ there is some $\alpha > \beta$ such that $A_\beta + B_\beta$ embeds into A_α, extending the embedding of A_β into A_α induced by $\left\langle f_\gamma \right\rangle_{\gamma\in[\beta,\alpha)}$.
(4) If β is a limit ordinal, then no proper initial segment of A_β contains copies of A_γ for all $\gamma < \beta$.

The existence of such systems is exactly what is required for Miller's argument to work, with Δ^0_2 (and in fact hyperimmune) permitting. Miller's argument (with ω_1 replaced by ω) used $A_\omega = \omega$ and $B_\omega = \omega^*$, with $A_n = B_n = n$ with the obvious embeddings; we use $A_\beta = \sum_{\gamma<\beta} \mathbb{Z}^\gamma$ and $B_\beta = A^*_\beta$. ⊣

We note that the Jockusch–Soare techniques fail for classes close to linear orderings, such as Boolean algebras: in the countable context, Downey and Jockusch [8] showed that every low Boolean algebra has a computable copy; this was extended by Thurber [33] to low$_2$ and by Stob and the second author [20] for low$_3$ and low$_4$. So far as we know, no work has been yet done on an analogue of these results in the uncountable setting.

6.5. Computable categoricity. Dzgoev and Goncharov [14], and, independently, Remmel [29], showed in the standard setting that a computable linear ordering is computably categorical (equivalently, relatively computably categorical) if and only if it contains only finitely many successor pairs (pairs $a < b$ such that b is the immediate successor of a in the linear ordering).

A naïve attempt to generalise this result would guess that a computable linear ordering of size $\aleph_1$ is computably categorical if and only if it contains only countably many successor pairs. This fails in both directions:

- The linear ordering $2 \cdot \mathbb{R}$ (replacing every real number by a successor pair) contains uncountably many successor pairs, but is computably categorical: after fixing a copy of the double rationals in two computable copies, there is a unique extension to an isomorphism, which is computable.
- The linear ordering $\mathbb{Q} \cdot \mathbb{R}$ does not contain any successor pairs, but is not computably categorical: we can build a "bad" computable copy of $\mathbb{Q} \cdot \mathbb{R}$ and defeat all computable attempts at an isomorphism by waiting for the eth computable function to be defined on the eth copy of $\mathbb{Q}$ in the standard copy, and then add a point to the eth copy of $\mathbb{Q}$ in our copy.

When we try to generalise the result of Dzgoev, Goncharov, and Remmel, we encounter the same difficulty as with Richter's result. The Dzgoev–Goncharov–Remmel theorem relies on the true finiteness of the finite cardinals. The correct generalisation involves an effectiveness condition that does not appear in the countable case. We believe that understanding the correct generalisation to the uncountable case sheds new light on the countable case as well.

To state the correct generalisation, we make use of the following notions. Let $\mathcal{L} = (L, <^{\mathcal{L}})$ be a linear ordering, and let C be a subset of the universe L of $\mathcal{L}$. A *C-cut* is a nonempty initial segment of C (by the inherited ordering from $\mathcal{L}$). If A is a C-cut, then the *$\mathcal{L}$-interval* determined by A, is the set

$$I^{\mathcal{L}}(A) = \left\{b \in L \setminus C \,:\, \forall c \in C \;\left[c <^{\mathcal{L}} b \Leftrightarrow c \in A\right]\right\}.$$

Theorem 6.8 (Greenberg, Kach, Lempp, Turetsky). *A computable linear ordering $\mathcal{L}$ is computably categorical if and only if there is a countable subset C of the domain of $\mathcal{L}$ and a uniformly c.e. partition $\langle S_n\rangle_{n\geqslant 1}$ of the collection of all C-cuts that define nonempty $\mathcal{L}$-intervals such that for all $n \geqslant 1$, for all $A \in S_n$, $I^{\mathcal{L}}(A)$ either has size n or is isomorphic to η_1.*

For the proof, and for further results on the degree spectrum of the successor relation on computable linear orderings, see [16]. We remark that the proof makes use of our understanding of countable linear orderings, particularly the scattered / nonscattered dichotomy, and does not seem to immediately generalise to higher cardinality.

Question 6.9. *Which ω_2-computable linear orderings of ω_2 are computably categorical?*

§7. Conclusion. The following are desirable features for a model of computability in an uncountable setting.

1. **Applications**. There should be interesting results about familiar mathematical objects such as the field of real numbers.
2. **Implementation**. There should be an implementation of the model, or at least a way of thinking about the computations that makes them "feel" effective.

3. **Comprehensibility**. A working mathematician should be able to understand the model.
4. **Insight**. Studying the new model should deepen our understanding of the standard model.

We believe that our model meets all criteria except possibly the third one. We have given some examples of computable structures, and some results in computable structure theory. We find it pleasing that our model lets us think of the real numbers, with the analytic functions, as a computable structure. The model also provides insight into standard notions and constructions, as was shown when discussing the generalisations and failure thereof of Richter's result and the Dzgoev-Goncharov-Remmel theorem.

As in the countable setting, after gaining experience with admissible recursion, one develops a solid intuition to what constitutes a computable construction and what does not; it becomes natural to describe such constructions in an informal way, and rely on an analogue of Church's thesis. In Sacks's terminology, one gets a dynamic feeling which lifts beyond the static nature of existential formulas in the language of set theory.

As for comprehensibility, the apparatus of constructible sets and the Levy hierarchy of formulas are certainly not widely known. As indicated earlier, though, there are alternative approaches to defining the notions of c.e. and computable subsets of ω_1. The two main kinds of development rely on either inductive schemes for partial computable functions, or by transfinite runs of Turing machines. These definitions may seem more immediately effective and more easily comprehensible. What they do assume – what cannot be omitted by any development of the subject – are the countable ordinals and ω_1. Mathematicians who are familiar with the application of Zorn's Lemma may find this less of a barrier.

REFERENCES.

[1] Chris J. Ash, Carl G. Jockusch, Jr., and Julia F. Knight. Jumps of orderings. *Trans. Amer. Math. Soc.*, 319(2): 573–599, 1990.

[2] Chris J. Ash and Julia F. Knight. *Computable Structures and the Hyperarithmetical Hierarchy*, volume 144 of Studies in Logic and the Foundations of Mathematics. North-Holland Publishing Co., Amsterdam, 2000.

[3] Shreeram Abhyankar. On the compositum of algebraically closed subfields. *Proc. Amer. Math. Soc.*, **7**: 905–907, 1956.

[4] Chris J. Ash, Julia F. Knight, Mark Manasse, and Theodore A. Slaman. Generic copies of countable structures. *Ann. Pure Appl. Logic*, **42**(3): 195–205, 1989.

[5] J. Carson, Julia F. Knight, Karen Lange, Charles McCoy, and John Wallbaum. The arithmetical hierarchy in the setting of ω_1. In preparation.

[6] John Chisholm. Effective model theory vs. recursive model theory. *J. Symbolic Logic*, **55**(3): 1168–1191, 1990.

[7] Rod Downey, Denis R. Hirschfeldt, Asher M. Kach, Steffen Lempp, Joseph R. Mileti, and Antonio Montalbán. Subspaces of computable vector spaces. *J. Algebra*, **314**(2): 888–894, 2007.

[8] Rod Downey and Carl G. Jockusch, Jr. Every low Boolean algebra is isomorphic to a recursive one. *Proc. Amer. Math. Soc.*, **122**(3): 871–880, 1994.

[9] Rod Downey and Julia F. Knight. Orderings with αth jump degree $\mathbf{0}^{(\alpha)}$. *Proc. Amer. Math. Soc.*, **114**(2): 545–552, 1992.

[10] Harvey Friedman and Lee Stanley. A Borel reducibility theory for classes of countable structures. *J. Symbolic Logic*, **54**(3): 894–914, 1989.

[11] Sy D. Friedman. Uncountable admissibles I: forcing. *Trans. Amer. Math. Soc.*, **270**(1): 61–73, 1982.

[12] A. Fröhlich and J.C. Shepherdson. Effective procedures in field theory. *Philos. Trans. Roy. Soc. London. Ser. A.*, **248**: 407–432, 1956.

[13] Sergei S. Gončarov. The number of nonautoequivalent constructivizations. *Algebra i Logika*, **16**(3): 257–282, 377, 1977.

[14] Sergei S. Gončarov and V.D. Dzgoev. Autostability of models. *Algebra i Logika*, **19**(1): 45–58, 132, 1980.

[15] Noam Greenberg and Julia F. Knight. Relative computable categoricity and Scott families in uncountable computable model theory. In preparation.

[16] Noam Greenberg, Asher M. Kach, Steffen Lempp, and Dan Turetsky. Computable properties of uncountable linear orderings. In preparation.

[17] Carl G. Jockusch, Jr. and Robert I. Soare. Degrees of orderings not isomorphic to recursive linear orderings. *Ann. Pure Appl. Logic*, **52**(1-2): 39–64, 1991. International Symposium on Mathematical Logic and its Applications (Nagoya, 1988).

[18] Jesse Johnson. *Computable Model Theory for Uncountable Structures*. PhD dissertation, University of Notre Dame, 2013.

[19] Julia F. Knight. Degrees coded in jumps of orderings. *J. Symbolic Logic*, **51**(4): 1034–1042, 1986.

[20] Julia F. Knight and Michael Stob. Computable Boolean algebras. *J. Symbolic Logic*, **65**(4): 1605–1623, 2000.

[21] Peter Koepke and Benjamin Seyfferth. Ordinal machines and admissible recursion theory. *Ann. Pure Appl. Logic*, **160**(3): 310–318, 2009.

[22] Mark Manasse. *Techniques and Counterexamples in Almost Categorical Recursive Model Theory*. PhD thesis, University of Wisconsin – Madison, 1982.

[23] George Metakides and Anil Nerode. Recursively enumerable vector spaces. *Ann. Math. Logic*, **11**(2): 147–171, 1977.

[24] Terrence S. Millar. Foundations of recursive model theory. *Ann. Math. Logic*, **13**(1): 45–72, 1978.

[25] Russell G. Miller. The Δ^0_2-spectrum of a linear order. *J. Symbolic Logic*, **66**(2): 470–486, 2001.

[26] Russell G. Miller. Computable fields and Galois theory. *Notices Amer. Math. Soc.*, **55**(7): 798–807, 2008.

[27] Michael Morley. Decidable models. *Israel J. Math.*, **25**(3–4): 233–240, 1976.

[28] A.T. Nurtazin. Strong and weak constructivizations, and enumerable families. *Algebra i Logika*, **13**: 311–323, 364, 1974.

[29] Jeffrey B. Remmel. Recursively categorical linear orderings. *Proc. Amer. Math. Soc.*, **83**(2): 387–391, 1981.

[30] Linda Jean Richter. Degrees of structures. *J. Symbolic Logic*, **46**(4): 723–731, 1981.

[31] Gerald E. Sacks. *Higher Recursion Theory*. Perspectives in Mathematical Logic. Springer-Verlag, Berlin, 1990.

[32] Richard A. Shore, α-recursion theory. In *Handbook of Mathematical Logic, Part C*, pages 525–815. Studies in Logic and the Foundations of Math., Vol. 90. North-Holland, Amsterdam, 1977.

[33] John J. Thurber. Every low_2 Boolean algebra has a recursive copy. *Proc. Amer. Math. Soc.*, **123**(12): 3859–3866, 1995.

[34] Bartel L. van der Waerden. Eine Bemerkung über die Unzerlegbarkeit von Polynomen. *Math. Ann.*, **102**(1): 738–739, 1930.

School of Mathematics, Statistics and Operations Research,
Victoria University of Wellington,
Wellington, New Zealand.
E-mail: greenberg@msor.vuw.ac.nz

Department of Mathematics,
University of Notre Dame,
Notre Dame, Indiana 46556, USA.
E-Mail: Julia.F.Knight.1@nd.edu

LOCAL COMPUTABILITY AND UNCOUNTABLE STRUCTURES

RUSSELL MILLER

§1. Introduction. Turing computability has always been restricted to maps on countable sets. This restriction is inherent in the nature of a Turing machine: a computation is performed in a finite length of time, so that even if the available input was a countable binary sequence, only a finite initial segment of that sequence was actually used in the computation. The *Use Principle* then says that an input of any other infinite sequence with that same initial segment will result in the same computation and the same output. Thus, while the domain might have been viewed as the (uncountable) set of infinite binary sequences, the countable domain containing all finite initial segments would have sufficed.

To be sure, there are approaches that have defined natural notions of computable functions on uncountable sets. The bitmap model, detailed in [3] and widely used in computable analysis, is an excellent model for computability on Cantor space 2^ω. On the real numbers $\mathbb{R}$, however, it fails to compute even the simplest discontinuous functions, which somewhat limits its utility. The Blum–Shub–Smale model (see [2]) expands the set of functions which we presuppose to be computable. Having done so, it gives an elegant account of computable functions on the reals, with nice analogies to computability on ω, but the initial assumption immediately distances it from Turing's original concept of computability.

Nevertheless, mathematicians are hardly daunted by the prospect of doing actual computations on $\mathbb{R}$. When faced with a real number whose binary expansion is not immediately accessible, they do not flinch; they simply call that real "x." All field operations can then be performed with ease within the subfield of $\mathbb{R}$ generated by x; the mathematician only needs to know whether x is algebraic or transcendental, and, in the former case, what its minimal polynomial over $\mathbb{Q}$ is. Similar devices handle the situation of several unknown reals at once. The binary expansions of these reals are not required for the algebraic operations.

In this chapter we formalize this process. Starting with the notion of a *computable model*, which is entirely in keeping with Turing's notion of computability,

[1]The author was partially supported by by grant #13397 from the Templeton Foundation, by award #DMS-1001306 from the National Science Foundation, and by grants numbered 67182–00–36, 68470–00 37, 69723–00 38, 61467–00 39, 62632–00 40, 63286–00 41, and 80209–04–12 from the Research Foundation of the City University of New York.

[2]Math Subject Codes: 03D30, 03D60, 03E15. Keywords: infinite time Turing machines, infinitary computability, ordinal computation.

we will view the real numbers and other fields as locally computable structures. No claim is made that the real numbers can be presented globally, as a single structure with programs for the arithmetic operations on its entire domain, but we develop a definition in which a countable collection of countable objects is used to describe all finitely generated substructures of a (potentially uncountable) structure $\mathcal{S}$. Then the *local computability* of $\mathcal{S}$ is determined by the computability of the countable objects. In cases such as the field $\mathbb{R}$, where every finitely generated substructure is computably presentable, we will say that we have a *computable cover* of the structure. Indeed, for $\mathbb{R}$, a single algorithm can list out all elements of this cover.

The term "cover" is borrowed from the definition of a manifold, and the analogy, while imprecise, can be useful for intuitions about our definitions. For instance, for a topological space M, being a manifold does not just require the existence of a cover by open subsets of $\mathbb{R}^n$, but also that the charts within M given by the cover should fit together in a nice way: the transition functions between open subsets of $\mathbb{R}^n$, defined whenever two charts in M intersect, should be continuous (or differentiable, or C^∞, depending on how nicely we wish the manifold to behave). In short, it is not sufficient just to describe the local behavior of M; one must ensure that where the descriptions overlap, they agree with one another in a reasonable way.

For us, it will certainly be true that finitely generated substructures of a structure $\mathcal{S}$ can overlap. Therefore, our description of finitely generated substructures of $\mathcal{S}$ will include an account of which such substructures extend to others. Since any two finitely generated substructures of $\mathcal{S}$ lie within a single larger finitely generated substructure, it is sufficient for our purposes to consider the question of extensions among them. Topological notions do not fit our setting very well, but embeddings among finitely generated computable structures are themselves inherently computable, since they are determined by their values on the generators of the domain. (This is our main reason for considering only finitely generated substructures of $\mathcal{S}$, in fact, rather than all countable substructures.) In order for a structure to be called *locally computable*, we will require not only that the finitely generated substructures be computably presentable in a uniform way, but also that there be a computable enumeration of the embeddings among them corresponding to extensions in the structure $\mathcal{S}$. Various strengthenings of this requirement, mostly in Section 4, will allow us to prove stronger theorems about certain of the structures.

The technical content of this chapter is not especially high, but when computability-theoretic notions arise, we refer the reader to [13], the standard source, for notation and definitions. A good overview of the field of computable model theory is given in [6]. Certain examples arise in each section to illustrate the concepts discussed, but several other examples are grouped together in Section 8, and it may be useful for the reader to work back and forth between this section and the others.

§2. Local Computability. Let T be a $\forall$-axiomatizable theory in a finite language. We first consider simple covers of a model $\mathcal{S}$ of T. These describe only the finitely generated substructures of $\mathcal{S}$, with no attention paid to any relations between those substructures.

DEFINITION 2.1. A *simple cover of* $\mathcal{S}$ is a (finite or countable) collection $\mathfrak{A}$ of finitely generated models $\mathcal{A}_0, \mathcal{A}_1, \ldots$ of T, such that:

- every finitely generated substructure of $\mathcal{S}$ is isomorphic to some $\mathcal{A}_i \in \mathfrak{A}$; and
- every $\mathcal{A}_i \in \mathfrak{A}$ embeds isomorphically into $\mathcal{S}$.

If T is finitely axiomatizable but not $\forall$-axiomatizable, we can Skolemize to give it a set of $\forall$-axioms, while keeping the language finite. The theory of fields is a natural example: one makes the axioms for inverses universal by adding a unary function symbol for negation and another for reciprocation, with 0 defined to be its own reciprocal. In this expanded language, every substructure of a model of T is also a model of T, since the axioms, being universal, hold in all substructures.

Definition 2.1 could allow uncountable covers, of course, but since we will mostly be interested in the possibility of presenting all $\mathcal{A}_i$ computably, the uncountable case is irrelevant for our purposes. We often write $(\mathcal{A}_i; \vec{a}_i)$ to denote that $\vec{a}_i = \langle a_i^1, \ldots, a_i^{k_i} \rangle$ is a finite tuple of generators for $\mathcal{A}_i$. The intention is that $\mathcal{S}$ itself should not be finitely generated, of course, although the definition is still valid in this case. Indeed, $\mathcal{S}$ is not at all required to be countable, since a single $\mathcal{A}_i$ may be isomorphic to many substructures of $\mathcal{S}$. For countable structures $\mathcal{S}$, a related notion is Fraïssé's concept of the *age* of $\mathcal{S}$, i.e. the set of all finitely generated substructures of $\mathcal{S}$. All elements of the cover $\mathfrak{A}$ must be models of T, by the $\forall$-axiomatizability of T. The existence of a cover does mean that in some sense only countably many different things can happen within $\mathcal{S}$. (Model theorists would say that the Σ_0-type space of $\mathcal{S}$ is countable.) Similarly, in our next definition, a computable simple cover suggests that all parts of $\mathcal{S}$ are computably presentable.

DEFINITION 2.2. A simple cover $\mathfrak{A}$ is *computable* if every $\mathcal{A}_i \in \mathfrak{A}$ is a computable structure whose domain is an initial segment of ω. $\mathfrak{A}$ is *uniformly computable* if the sequence $\langle(\mathcal{A}_i, \vec{a}_i)\rangle_{i\in\omega}$ can be given uniformly: there must exist a single computable function which, on input i, outputs a tuple of elements

$$\langle e_1, \ldots, e_n, \langle a_0, \ldots, a_{k_i}\rangle\rangle \in \omega^n \times \mathcal{A}_i^{<\omega}$$

such that $\mathcal{A}_i$ is generated by $\{a_0, \ldots, a_{k_i}\}$ and φ_{e_j} computes the jth function, relation, or constant of the language in $\mathcal{A}_i$. (Here n is the cardinality of the language, which we assumed to be finite.)

Notice that the definition requires that the generators of $\mathcal{A}_i$ be given as a tuple $\langle a_0, \ldots, a_{k_i} \rangle$, so that k_i is computable uniformly in i and we know how many values from $\mathcal{A}_j$ are needed to define an embedding of $\mathcal{A}_i$ into $\mathcal{A}_j$. (In the language of [13, II.2.4], the definition requires that we compute the *canonical index* for the set $\{a_0, \ldots, a_{k_i}\}$.)

As an example, we show that the best-known uncountable structure in mathematics is locally computable. The pieces of the proof have long been established, but for completeness we repeat the details.

PROPOSITION 2.3. *The field* $\mathcal{R} = (\mathbb{R}, +, \cdot, -, r, 0, 1)$ *of real numbers has a uniformly computable simple cover.*

Our proof here will be re-used in Proposition 2.10 to show that this structure $\mathcal{R}$ is locally computable, according to Definition 2.9. Notice that we have added the unary operations of negation and inversion (r, for *reciprocal*) to the usual language of fields, in order to get a Π_1 axiom set. For definiteness we set $r(0) = 0$.

Proof We construct a uniformly computable cover $\mathfrak{A}$ of the field $\mathcal{R}$. For this purpose we will use a computable list $\langle n_0, p_0 \rangle, \langle n_1, p_1 \rangle, \ldots$ of the set

$$\{\langle n, p \rangle : n \in \omega \ \&\ p \in \mathbb{Q}[X_1, \ldots, X_n, Y]\}.$$

SUBLEMMA 2.4. *There is an algorithm which decides, for an arbitrary* $\langle n, p \rangle$ *in this set, whether or not* p *is irreducible in* $\mathbb{Q}[X_1, \ldots, X_n, Y]$ *and has a solution* $(x_1, \ldots, x_n, y) \in \mathbb{R}^{n+1}$ *with* $\{x_1, \ldots, x_n\}$ *algebraically independent over* $\mathbb{Q}$.

Proof The algorithm for deciding the irreducibility of p in $\mathbb{Q}(X_1, \ldots, X_n)[Y]$, uniformly in n, was developed by Kronecker in [9]; and p is irreducible there iff it is irreducible in $\mathbb{Q}[\vec{X}, Y]$. (Details can be found in [4] and in Lemma 2 on p. 92 of [14].) We immediately rule out the reducible polynomials p. For the second part of the sublemma, we show that the set of those p which have such a solution (and are irreducible) is both Σ^0_1 and Π^0_1.

We claim first that there exists a solution as required iff there exist $q_1, \ldots, q_n, q'$, $q'' \in \mathbb{Q}$ such that $p(\vec{q}, q') > 0 > p(\vec{q}, q'')$. If such rational numbers exist, then there exist algebraically independent real numbers $x_1, \ldots x_n$, with each x_i sufficiently close to q_i that $p(\vec{x}, q') > 0 > p(\vec{x}, q'')$ still holds. But then the Intermediate Value Theorem yields the $y \in \mathbb{R}$ with $p(\vec{x}, y) = 0$. Conversely, if we have the solution $(\vec{x}, y)$ as required, then since the set $\{\vec{x}\}$ is algebraically independent, the real polynomial $p(\vec{x}, Y) \in \mathbb{R}[Y]$ is irreducible, and so its derivative $\frac{dp}{dY}$ is nonzero at $(\vec{x}, y)$. Therefore there exist y' and y'' with $p(\vec{x}, y') > 0 > p(\vec{x}, y'')$. But the density of $\mathbb{Q}$ in $\mathbb{R}$ then shows that the rationals $\vec{q}, q', q''$ exist.

Next, we claim that the required solution fails to exist iff there exist $k \in \omega$, polynomials $g_1, \ldots, g_k \in \mathbb{Q}[\vec{X}, Y]$, and rational numbers $c_1, \ldots, c_k \geq 0$ such that $p(\vec{X}, Y) = \pm \sum_{i=1}^{k} c_i \cdot (g_i(\vec{X}, Y))^2$. If these elements exist, then clearly $p(\vec{X}, Y)$ is either positive semidefinite (i.e. takes on only values ≥ 0 on $\mathbb{R}^{n+1}$) or negative semidefinite. But we saw above that the existence of a solution $(\vec{x}, y)$ (with $\vec{x}$ algebraically independent) implies that $p(\vec{x}, Y) \in \mathbb{R}[Y]$ is neither positive nor negative semidefinite. Therefore no solution exists in which $\vec{x}$ is algebraically independent.

For the converse, suppose that no solution with $\vec{x}$ algebraically independent exists. Sublemma 2.6 below, applied with $F = \mathbb{Q}$, shows that in this case $p(\vec{X}, Y)$ must be either positive or negative semidefinite, and the existence of the required k, g_i, and c_i then follows from Artin's Theorem. (For details, we refer the reader to [10, VIII.1.12].)

THEOREM 2.5 (Artin's Theorem.). *Let F be an ordered field. A polynomial $f \in F[Y_1, \dots, Y_m]$ is positive semidefinite iff there exist $g_1, \dots, g_k \in F[\vec{Y}]$ and $c_1, \dots c_k \geq 0$ in F such that $f = \sum_{i=1}^{k} c_i \cdot g_i^2$.*

SUBLEMMA 2.6. *Fix any $n \in \omega$, and let $F \subset \mathbb{R}$ be a finitely generated field extension of $\mathbb{Q}$. Let $p \in F[X_1, \dots, X_n, Y]$ be a polynomial, irreducible in this polynomial ring, which assumes both positive and negative values on $\mathbb{R}^{n+1}$. Then there exists a solution $(x_1, \dots, x_n, y) \in \mathbb{R}^{n+1}$ to the equation $p = 0$ such that the set $\{x_1, \dots x_n\}$ is algebraically independent over F.*

Proof of Sublemma 2.6. We induct on n, with the statement for $n = 0$ following from the Intermediate Value Theorem. Fix $n > 0$.

Write $p(\vec{X}, Y)$ as a polynomial in $X_2, \dots, X_n, Y$, with coefficients $q_i(X_1)$ in $F[X_1]$, and suppose for a contradiction that for every $x_1 \in \mathbb{R}$, the polynomial $p(x_1, X_2, \dots, X_n, Y)$ is either positive definite or negative semidefinite. By the assumption of the sublemma, each of these possibilities (positive and negative semidefinite) does hold for some value of x_1. By completeness of $\mathbb{R}$, there must be an x_1 such that $p(x_1, X_2, \dots, X_n, Y)$ is identically 0. In particular, choosing some $\{x_2, \dots, x_n, y\}$ algebraically independent over $F(x_1)$, we have $p(\vec{x}, y) = 0$, so $p(x_1, X_2, \dots, X_n, Y)$ must be the zero polynomial. Thus every $q_i(x_1) = 0$, yet since $p(\vec{X}, Y)$ was nonzero, some $q_i(X_1)$ is nonzero. So x_1 is algebraic over F, and its minimal polynomial in $F[X_1]$ is a factor of every $q_i(X_1)$, hence divides $p(\vec{X}, Y)$, contradicting irreducibility.

Therefore there exists an $x_1' \in \mathbb{R}$ such that $p(x_1', X_2, \dots, X_n, Y)$ assumes both positive and negative values as a function on $\mathbb{R}^n$. But then there exists an x_1 transcendental over F and sufficiently close to x_1' that the polynomial $q = p(x_1, X_2, \dots, X_n, Y)$ also assumes both positive and negative values. Moreover, since p is irreducible in F, we see from [14, Lemma 2, p. 92] that q is irreducible in the field $K = F(x_1)$. So we apply the inductive hypothesis with the field K and the polynomial q to get the required $x_2, \dots, x_n, y$.

This completes the proof of Sublemma 2.6, and also of Sublemma 2.4. ⊣

Hence we may enumerate finitely generated fields $\mathcal{A}_i$ uniformly in i, by considering pairs $\langle n, p \rangle$ as above until we find the least one which has not been used for any $\mathcal{A}_j$ with $j < i$ and which has a solution $(x_1, \dots, x_n, y) \in \mathbb{R}^{n+1}$ with $\vec{x}$ algebraically independent. When we find such a pair, it is straightforward to build a computable field $\mathcal{A}_i$, with domain ω, isomorphic to the quotient field of $\mathbb{Q}[X_1, \ldots, X_n, Y]/(p(\vec{X}, Y))$. These fields $\mathcal{A}_i$ will be the elements of our cover $\mathfrak{A}$. Clearly the uniform computability conditions on the $\mathcal{A}_i$ themselves are satisfied. (Below we consider the embeddings in $I^{\mathfrak{A}}$, once $I^{\mathfrak{A}}$ has appeared in Definition 2.9.) It is also clear that every such field is isomorphic to the subfield $\mathbb{Q}(\vec{x}, y)$ of $\mathcal{R}$. Conversely, we have the following.

Sublemma 2.7. *Every finitely generated subfield of $\mathcal{R}$ is isomorphic to some $\mathcal{A}_i \in \mathfrak{A}$ given by this process.*

Proof Let F be a finitely generated subfield of $\mathcal{R}$. By the Noether Normalization Lemma, F is an algebraic extension of a purely transcendental extension K of $\mathbb{Q}$. Since F is finitely generated, K must have a finite (possibly empty) transcendence basis $\{x_1, \ldots, x_n\}$ over $\mathbb{Q}$, so $K = \mathbb{Q}(\vec{x})$. Finite generation also implies that F is a finite algebraic extension of K. Since we are in characteristic 0, the Theorem of the Primitive Element applies, showing that $F = K(y)$ for some $y \in F$ algebraic over K. Choose $p \in \mathbb{Q}[X_1, \ldots, X_n, Y]$ such that $p(\vec{x}, Y)$, when divided by its lead coefficient, is the minimal polynomial of y over K. Thus p is irreducible in $K[Y]$, and by dividing by the content of p in the ring $\mathbb{Q}[\vec{X}]$, we may assume that p is also irreducible in $\mathbb{Q}[X_1, \ldots, X_n, Y]$. p still has the solution $(\vec{x}, y) \in \mathbb{R}^{n+1}$, with $\vec{x}$ algebraically independent over $\mathbb{Q}$, so F is isomorphic to that $\mathcal{A}_i$ which we enumerated into $\mathfrak{A}$ when we reached the pair $\langle n, p\rangle$. This completes the proofs of Sublemma 2.7 and Proposition 2.3. ⊣

It is also useful to see a negative example. Although the real numbers form a locally computable field, adding the usual $<$ relation to the structure destroys local computability.

Proposition 2.8. *The ordered field $(\mathcal{R}, <)$ of real numbers, with $\mathcal{R}$ as in Proposition 2.3, has no computable simple cover, uniform or otherwise.*

Proof Let b be any noncomputable real number. (That is, the Dedekind cut of b should be a noncomputable subset of $\mathbb{Q}$.) We claim that the ordered subfield $\mathcal{B}$ of $\mathcal{R}$ generated by b has no computable presentation. Clearly this implies the proposition.

Suppose $\mathcal{A}$ were a computable presentation of $\mathcal{B}$, with $a \in \mathcal{A}$ the image of b under the isomorphism from $\mathcal{B}$ onto $\mathcal{A}$. Then just from knowing the additive and multiplicative identity elements in $\mathcal{A}$, we could compute the representation in $\mathcal{A}$ of any rational number $\frac{p}{q}$. But then we could compute the Dedekind cut of b, just by using the computable relation $<$ in $\mathcal{A}$ to compare a to each rational. Therefore no such $\mathcal{A}$ can exist. ⊣

We will be concerned mainly with the full definition of a cover, in which we also describe how the substructures of $\mathcal{S}$ fit together.

DEFINITION 2.9. A *cover of* $\mathcal{S}$ consists of a simple cover $\mathfrak{A} = \{\mathcal{A}_0, \mathcal{A}_1, \dots\}$ of $\mathcal{S}$, along with sets $I_{ij}^{\mathfrak{A}}$ (for all $\mathcal{A}_i, \mathcal{A}_j \in \mathfrak{A}$) of injective homomorphisms $f : \mathcal{A}_i \hookrightarrow \mathcal{A}_j$, such that:

- for all substructures $\mathcal{B} \subseteq \mathcal{C}$ of $\mathcal{S}$, there exist $i, j \in \omega$ and $f \in I_{ij}^{\mathfrak{A}}$ and isomorphisms $\beta : \mathcal{A}_i \twoheadrightarrow \mathcal{B}$ and $\gamma : \mathcal{A}_j \twoheadrightarrow \mathcal{C}$ with $\beta = \gamma \circ f$; and
- for every k and m and every $g \in I_{km}^{\mathfrak{A}}$, there exist substructures $\mathcal{D} \subseteq \mathcal{E}$ of $\mathcal{S}$ and isomorphisms $\delta : \mathcal{A}_k \twoheadrightarrow \mathcal{D}$ and $\epsilon : \mathcal{A}_m \twoheadrightarrow \mathcal{E}$ with $\delta = \epsilon \circ g$.

This cover is *uniformly computable* if $\mathfrak{A}$ is a uniformly computable simple cover of $\mathcal{S}$ and there exists a c.e. set W such that for all $i, j \in \omega$,

$$I_{ij}^{\mathfrak{A}} = \{\varphi_e \restriction \mathcal{A}_i : \langle i, j, e \rangle \in W\}.$$

A structure $\mathcal{B}$ is *locally computable* if it has a uniformly computable cover.

Diagrams of the situation are often useful. Solid arrows represent given maps, and dotted arrows represent maps whose existence is required by the definitions. Definition 2.9 demands that the following diagrams both commute.

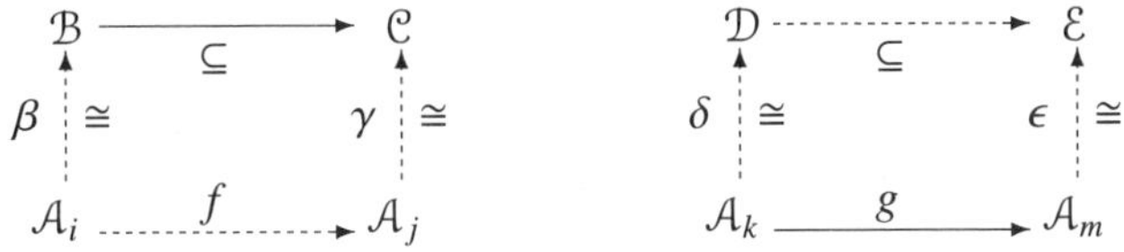

If $\mathfrak{A}$ is a computable simple cover, then every embedding of any $\mathcal{A}_i$ into any $\mathcal{A}_j$ is determined by its values on the generators of $\mathcal{A}_i$. Since $\mathcal{A}_i$ must be finitely generated, all such embeddings are computable, and therefore it is reasonable to call $\mathfrak{A}$ a computable cover without any further requirements on the sets $I_{ij}^{\mathfrak{A}}$. (Our main reason for considering only the finitely generated substructures of $\mathcal{S}$, rather than countable ones, is that embeddings among such structures are always computable.) For a uniformly computable cover, on the other hand, the sets $I_{ij}^{\mathfrak{A}}$ will play a key role in our development of the subject, and it should be kept in mind that $I_{ij}^{\mathfrak{A}}$ need not contain every possible embedding of $\mathcal{A}_i$ into $\mathcal{A}_j$.

It is an easy exercise to see that the second condition of Definition 2.9 follows trivially from the definition of a simple cover, for any embedding $f : \mathcal{A}_i \hookrightarrow \mathcal{A}_j$. We include this second condition here because it is the dual of the first, and in the rest of our study of local computability, this duality between inclusion maps within $\mathcal{S}$ and embeddings among structures in $\mathfrak{A}$ will appear repeatedly.

In the uniformly computable simple cover $\mathfrak{A}$ of the reals built in Proposition 2.3, we now consider the embeddings between the structures $\mathcal{A}_i$.

PROPOSITION 2.10. *There exists a uniformly computable cover of the field $\mathcal{R}$ of real numbers in which every set $I_{ij}^{\mathfrak{A}}$ is not only c.e., but actually computable uniformly in i and j.*

Proof The simple cover is the same $\mathfrak{A}$ built in Proposition 2.3. We enumerate into $I_{ij}^{\mathfrak{A}}$ all possible embeddings of the field $\mathcal{A}_i$ into $\mathcal{A}_j$. Fix i and j, and suppose that $\mathcal{A}_i$ was built from the pair $\langle n, p\rangle$ as above. Embeddings are given simply by naming the images of the elements $x_1, \dots, x_n, y$ in $\mathcal{A}_j$ and then extending the embedding to the rest of $\mathcal{A}_i$ using the function symbols of the language. Of course, though, not all choices of images extend to an embedding, so we need the following sublemma.

SUBLEMMA 2.11. *Let* $\mathcal{A}_i = \mathbb{Q}(z_1, \dots, z_m, w)$ *and* $\mathcal{A}_j = \mathbb{Q}(x_1, \dots, x_n, y)$ *with* $\vec{x}$ *and* $\vec{z}$ *each algebraically independent over* $\mathbb{Q}$, *and let* $q \in \mathbb{Q}(\vec{z})[W]$ *and* $p \in \mathbb{Q}(\vec{x})[Y]$ *be the minimal polynomials of* w *and* y *over* $\mathbb{Q}(\vec{z})$ *and* $\mathbb{Q}(\vec{x})$ *respectively. Then there exists an algorithm, uniform in* n, m, p *and* q, *which decides for any* $(a_1, \dots a_m, b) \in \mathcal{A}_j^{m+1}$ *whether the map* f *with* $f(z_i) = a_i$ *and* $f(w) = b$ *extends to an injective homomorphism of* $\mathcal{A}_i$ *into* $\mathcal{A}_j$.

Proof We immediately check whether $q(\vec{a}, b) = 0$ and whether $n \geq m$. If either of these fails, then of course f does not extend to an embedding, so assume that they do both hold. Then f extends to an embedding iff $\vec{a}$ is algebraically independent over $\mathbb{Q}$ in $\mathcal{A}_j$.

On one hand, we can search for a nonzero polynomial in $\mathbb{Q}[A_1, \dots, A_m]$ for which $\vec{a}$ is a solution in $\mathcal{A}_j$. Such a polynomial exists iff $\vec{a}$ is algebraically dependent over $\mathbb{Q}$, so clearly this outcome is Σ_1^0.

On the other hand, knowing that $n \geq m$, we search for $a_{m+1}, \dots a_n \in \mathcal{A}_j$ and polynomials $p_1, \dots p_n \in \mathbb{Q}[A_1, \dots, A_n, X]$ such that for every $i \leq n$ we have

$$p_i(a_1, \dots, a_n, x_i) = 0 \ \& \ p_i(a_1, \dots, a_n, X) \neq 0.$$

If we find such elements and polynomials, then $\{a_1, \dots, a_n\}$ is a transcendence basis for $\mathbb{Q}(x_1, \dots, x_n)$, since it spans $\mathbb{Q}(x_1, \dots, x_n)$ algebraically and has the minimum possible size for such a spanning set. In this case $\{a_1, \dots, a_m\}$ is an algebraically independent set. Conversely, if this set really is algebraically independent over $\mathbb{Q}$ in $\mathcal{A}_j$, then it does extend to a transcendence basis, and so such elements and polynomials must exist. Therefore algebraic independence is also a Σ_1^0 condition. This proves Sublemma 2.11. ⊣

Since $I_{ij}^{\mathfrak{A}}$ includes every possible embedding of $\mathcal{A}_i$ into $\mathcal{A}_j$, the first condition of Definition 2.9 is immediate, and we have already noted that the second condition is trivial. Hence we have proven Proposition 2.10. ⊣

The point of Proposition 2.10 is the computability of the sets $I_{ij}^{\mathfrak{A}}$. That they can be computably enumerated would have followed immediately from our next result.

LEMMA 2.12. *A structure* $\mathcal{S}$ *has a uniformly computable cover (i.e. is locally computable) iff* $\mathcal{S}$ *has a uniformly computable simple cover.*

Proof Assume that $\mathfrak{A} = \{\mathcal{A}_0, \mathcal{A}_1, \dots\}$ is a uniformly computable simple cover of $\mathcal{S}$. The domain $\{a_{i,0}, a_{i,1}, \dots\}$ of each $\mathcal{A}_i$ is enumerable uniformly in i, and so it is straightforward to enumerate the domain of the substructure

$$\mathcal{B}_{\langle i,j\rangle} = \langle\{a_{i,k} \in \mathcal{A}_i : k \in D_j\}\rangle \subseteq \mathcal{A}_i$$

where the generating set is defined using the finite set D_j with canonical index j, as defined in [13]. Now every A_i is equal to some $\mathcal{B}_{\langle i,j\rangle}$, so $\{\mathcal{B}_{\langle i,j\rangle}\}_{i,j}$ is another uniformly computable simple cover of $\mathcal{S}$. Next, define the set $I_{\langle i,j\rangle,\langle i',j'\rangle}$ to be empty if $i' \neq i$. For $i = i'$, we wait until all elements of D_j have appeared in $\mathcal{B}_{\langle i,j'\rangle}$. If this ever happens, we enumerate the identity map into $I_{\langle i,j\rangle,\langle i,j'\rangle}$; if not, then $I_{\langle i,j\rangle,\langle i,j'\rangle}$ is empty.

Clearly this uniformly enumerates the sets $I_{\langle i,j\rangle,\langle i',j'\rangle}$, and we claim that with these sets, $\mathfrak{B} = \{\mathcal{B}_{\langle i,j\rangle} : i, j \in \omega\}$ forms a cover of $\mathcal{S}$. First, if $f \in I_{\langle i,j\rangle,\langle i',j'\rangle}$, then $i = i'$ and $\mathcal{B}_{\langle i,j\rangle} \subseteq \mathcal{B}_{\langle i,j'\rangle} \subseteq A_i$. Since $\mathcal{A}_i$ is isomorphic to a substructure $\mathcal{B} \subseteq \mathcal{S}$, we can simply lift f to the identity map on the corresponding substructures of this $\mathcal{B}$. Conversely, if $\mathcal{B} \subseteq \mathcal{C}$ are finitely generated substructures of $\mathcal{S}$, then $\mathcal{C}$ is isomorphic to some $\mathcal{A}_i$, and there are some j and j' such that $\mathcal{B}_{\langle i,j'\rangle} = \mathcal{A}$ and $\mathcal{B}_{\langle i,j\rangle}$ is the substructure of $\mathcal{A}_i$ corresponding to $\mathcal{B}$ within $\mathcal{C}$. But then the identity map from $\mathcal{B}_{\langle i,j\rangle}$ into $\mathcal{B}_{\langle i,j'\rangle}$ must have appeared in $I_{\langle i,j\rangle,\langle i,j'\rangle}$, and matches the inclusion map from $\mathcal{B}$ into $\mathcal{C}$. ⊣

In light of this lemma, one naturally asks why we bothered to give Definition 2.9. The answer is that local computability will be the $\theta = 0$ case in the definition of *θ-extensionally computable structures*, which appears in the next section as Definition 3.6 and which uses the enumeration of the sets $I_{ij}^{\mathfrak{A}}$ extensively. Indeed, it is the enumeration of the embeddings, rather than that of the finitely generated substructures of $\mathcal{S}$, which will be the heart of our study of local computability.

§3. Theory of Locally Computable Structures.

Definition 3.1. Let $\mathfrak{A}$ be a cover of a structure $\mathcal{S}$. We say that an $\mathcal{A}_i \in \mathfrak{A}$ matches a substructure $\mathcal{B} \subseteq \mathcal{S}$ *extensionally* if there is an isomorphism $\beta : \mathcal{A}_i \twoheadrightarrow \mathcal{B}$ for which the following hold.

- for every finitely generated $\mathcal{C}$ with $\mathcal{B} \subseteq \mathcal{C} \subseteq \mathcal{S}$, there exists $j \in \omega$, $f \in I_{ij}^{\mathfrak{A}}$, and an isomorphism γ mapping $\mathcal{A}_j$ onto $\mathcal{C}$ such that $\beta = \gamma \circ f$; and
- for every $m \in \omega$ and every $g \in I_{im}^{\mathfrak{A}}$, there exists a $\mathcal{E} \subseteq \mathcal{S}$ and an isomorphism ϵ mapping $\mathcal{A}_m$ onto $\mathcal{E}$ such that $\mathcal{B} \subseteq \mathcal{E}$ and $\beta = \epsilon \circ g$.

This β is called an *extensional match* between $\mathcal{A}_i$ and $\mathcal{B}$.

Again, diagrams help explain this definition. The two conditions may be expressed as follows.

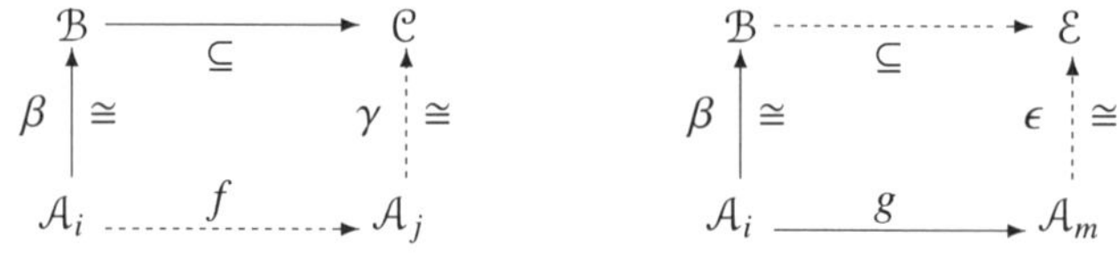

The difference from the diagrams of Definition 2.9 is that now $\mathcal{A}_i$, $\mathcal{B}$, and the isomorphism β are all fixed: the required j, f, and γ and the required $\mathcal{E}$ and ϵ must all work for this particular $\beta : \mathcal{A}_i \to \mathcal{B}$ on the left edge of the diagram. We refer to β as an *extensional match* between $\mathcal{A}_i$ and $\mathcal{B}$. The idea is that the embeddings in the sets $I^{\mathfrak{A}}_{ij}$ (for all j) correspond precisely to the finitely generated superstructures of $\mathcal{B}$ in $\mathcal{S}$, rather than just to possible extensions of various $\mathcal{B}' \cong \mathcal{B}$ within $\mathcal{S}$. This distinction will be illustrated in the examples below.

Definition 3.2. We say that a uniformly computable cover $\mathfrak{A}$ of $\mathcal{S}$ is *extensionally computable* (and we call $\mathfrak{A}$ an *extensional cover* of $\mathcal{S}$) if every $\mathcal{A}_i \in \mathfrak{A}$ extensionally matches some substructure $\mathcal{B} \subseteq \mathcal{S}$ and every finitely generated substructure $\mathcal{B} \subseteq \mathcal{S}$ extensionally matches some $\mathcal{A}_i \in \mathfrak{A}$.

If such a cover exists, we say that $\mathcal{S}$ is *extensionally locally computable*.

The point of this definition is that the extra conditions strengthen the idea that each finitely generated substructure of $\mathcal{S}$ is represented by some $\mathcal{A}_i \in \mathfrak{A}$: not only are they isomorphic, but the embeddings (enumerated effectively by $I^{\mathfrak{A}}$) of $\mathcal{A}_i$ into other structures in $\mathfrak{A}$ coincide exactly with the extensions of $\mathcal{B}$ to larger finitely generated substructures of $\mathcal{S}$. This point is best illustrated by the negative example of Proposition 3.5 below. However, we first show that the definition holds for the field of complex numbers.

Proposition 3.3. *Every algebraically closed field* $\mathcal{C} = (C, +, \cdot, -, r, 0, 1)$ *of characteristic* 0 *is extensionally locally computable. In particular, the field* $\mathbb{C}$ *of complex numbers is extensionally locally computable.*

Proof The construction of a uniformly computable cover $\mathfrak{A}$ of $\mathcal{C}$ is largely the same as that for the field of real numbers in Proposition 2.3. Of course, we no longer need worry whether a given polynomial of degree > 0 has a root in $\mathcal{C}$, and so Sublemma 2.4 in the construction of $\mathfrak{A}$ is not needed, except for the remark about Kronecker's algorithm. Also, in case the transcendence degree d of $\mathcal{C}$ over $\mathbb{Q}$ is finite, we make sure that $\mathfrak{A}$ only contains fields with transcendence degree $\leq d$ over $\mathbb{Q}$. Once again we define $I^{\mathfrak{A}}_{ij}$ to contain every embedding of $\mathcal{A}_i$ into $\mathcal{A}_j$, and the same analysis from the proof of Sublemma 2.11 shows that the sets $I^{\mathfrak{A}}_{ij}$ are computably enumerable (indeed computable) uniformly in i and j, and that $\mathfrak{A}$ is a uniformly computable cover of $\mathcal{C}$. (In fact, if two algebraically closed fields of characteristic 0 are isomorphic, or if they both have infinite transcendence degree, then we have built the same $\mathfrak{A}$ for both of them.)

It remains to show that $\mathfrak{A}$ is an extensionally computable cover of $\mathcal{C}$. Indeed, we will show more.

Lemma 3.4. *In this situation, every embedding of any $\mathcal{A}_i \in \mathfrak{A}$ into $\mathcal{C}$ is extensional.*

Proof Fix any $\mathcal{A}_i \in \mathfrak{A}$, with generators $x_1, \dots, x_m, y$ given in the construction of $\mathfrak{A}$ and with an irreducible polynomial $q \in \mathbb{Q}[\vec{X}, Y]$ such that $q(\vec{x}, y) = 0$, and let $\mathcal{B} \subseteq \mathcal{C}$ be any subfield isomorphic to $\mathcal{A}_i$, via any isomorphism β. We claim that β is an extensional match.

Now for any $g \in I_{ij}^{\mathfrak{A}}$, we start with the embedding $\gamma_0 = \beta \circ g^{-1}$ of $g(\mathcal{A}_i)$ into $\mathcal{C}$ and extend it one-by-one to the generators $z_1, \dots z_l, z_{l+1} = w$ of $\mathcal{A}_j$ given by the construction of $\mathfrak{A}$. For each $k \leq l + 1$, if z_k is transcendental over the domain of γ_{k-1} (i.e. the subfield of $\mathcal{A}_j$ generated by $g(\mathcal{A}_i)$ and $z_1, \dots, z_{k-1}$), then we choose $\gamma_k(z_k)$ to be any element of $\mathcal{C}$ transcendental over the image of γ_{k-1}. (Such an element of $\mathcal{C}$ must exist, since we ensured that $\mathcal{A}_j$ cannot have transcendence degree $> d$ over $\mathbb{Q}$.) Otherwise z_k is algebraic over $\text{dom}(\gamma_{k-1})$, so we let $p(Z)$ be its minimal polynomial over that subfeld. and choose $\gamma_k(z_k)$ to be any root in $\mathcal{C}$ of the polynomial $\overline{p}(Z) \in \mathcal{C}[Z]$ gotten by applying γ_{k-1} to the coefficients of p. In either case, γ_k then extends to an embedding into $\mathcal{C}$ of the subfield of $\mathcal{A}_j$ generated by $g(\mathcal{A}_i)$ and $z_1, \dots, z_k$. By induction, the map $\gamma = \gamma_{k+1}$ is an isomorphism from $\mathcal{A}_j$ onto a finitely generated substructure of $\mathcal{C}$, and since γ extends $\gamma_0 = \beta \circ g^{-1}$, it is clear that $\beta = \gamma \circ g$.

The converse is quicker. Fix any finitely generated subfield $\mathcal{B}$ of $\mathcal{C}$ with $g(\mathcal{A}_i) \subseteq \mathcal{B}$. Extend the transcendence basis $\{\beta(x_1), \dots, \beta(x_m)\}$ to a (finite) transcendence basis X for $\mathcal{B}$ over $\mathbb{Q}$, and pick a primitive element generating $\mathcal{B}$ over $\mathbb{Q}(X)$, using the Primitive Element Theorem. Then $\mathcal{B}$ is isomorphic to that field $\mathcal{A}_j \in \mathfrak{A}$ with the same transcendence degree over $\mathbb{Q}$ and the same minimal polynomial for its primitive element. Let γ be this isomorphism. Then $\gamma^{-1} \circ \beta$ is an embedding of $\mathcal{A}_i$ into $\mathcal{A}_j$, hence lies in $I_{ij}^{\mathfrak{A}}$, and we take this embedding as our g. Thus this $\mathcal{B}$ is an extensional match for $\mathcal{A}_i$. ⊣

Since $\mathfrak{A}$ was shown to be a cover of $\mathcal{C}$, every $\mathcal{A}_i \in \mathfrak{A}$ is isomorphic to some subfield $\mathcal{B} \subset \mathcal{C}$, and every finitely generated subfield of $\mathcal{C}$ is isomorphic to an element of $\mathfrak{A}$. By Lemma 3.4, the isomorphism in each case is an extensional match. Thus $\mathfrak{A}$ is an extensional cover of $\mathcal{C}$. ⊣

To make the meaning of Definition 3.2 more obvious, we now give an instance where it does not apply.

Proposition 3.5. *The field $\mathcal{R}$ of real numbers is not extensionally locally computable.*

Proof Suppose that $\mathfrak{A}$ were an extensionally computable cover of $\mathcal{R}$. Fix any noncomputable real number $t \in \mathbb{R}$. Definition 3.2 gives an $\mathcal{A}_i \in \mathfrak{A}$ which extensionally matches (via some isomorphism β) the subfield $\mathcal{B}$ of $\mathcal{R}$ generated by

t, and we may assume we know i and $\beta^{-1}(t)$, since they constitute finitely much information.

Now we can enumerate the lower cut $\{q \in \mathbb{Q} : q < t\}$ defined by t, knowing that extensions of $\mathcal{B}$ in $\mathcal{R}$ correspond to embeddings $f \in I_{ij}^{\mathfrak{A}}$ (for all j) in the extensionally computable cover. For any rational $q \in \mathbb{R}$:

$$\begin{aligned}
\models_{\mathcal{R}} q < t &\iff \models_{\mathcal{R}} (\exists x)x^2 = t - q \\
&\iff (\exists \text{ f.g. } \mathcal{C})[\mathcal{B} \subseteq \mathcal{C} \subset \mathbb{R} \;\&\; \models_{\mathcal{C}} (\exists x)x^2 = t - q] \\
&\iff (\exists j)(\exists f \in I_{ij}^{\mathfrak{A}}) \models_{\mathcal{A}_j} (\exists x)x^2 = f(\beta^{-1}(t - q)) \\
&\iff (\exists j)(\exists f \in I_{ij}^{\mathfrak{A}})(\exists a \in \mathcal{A}_j) \models_{\mathcal{A}_j} a^2 = f(\beta^{-1}(t)) - f(\beta^{-1}(q)).
\end{aligned}$$

A similar argument holds for the upper cut $\{q \in \mathbb{Q} : q > t\}$, using square roots of $(q-t)$ in $\mathcal{R}$. So the lower cut is both Σ_1^0 and Π_1^0, contradicting the noncomputability of t. (Of course, β and f fix the rationals, so $f(\beta^{-1}(q)) \in \mathcal{A}_j$ is just the element of $\mathcal{A}_j$ representing q. The domain element of $\mathcal{A}_j$ representing any particular rational q can easily be computed from the numerator and denominator of q, uniformly in j, by using the functions of $\mathcal{A}_j$.) ⊣

So the extensional local computability of $\mathcal{C}$ does not follow solely from the existential closure of the structure; after all, $\mathcal{R}$, viewed as a real closed field, is also existentially closed. The difficulty for $\mathcal{R}$ is that real closed fields have an implicit order on their elements, whether it is included in the language of the structure or not, and as we saw in Proposition 2.8, adding the order relation to $\mathcal{R}$ destroys local computability. $\mathcal{R}$ itself can still be locally computable, because the relation $<$ cannot be defined in $\mathcal{R}$ without quantifiers (even though it is both Σ_1-definable and Π_1-definable!) and existential questions about $\mathcal{R}$ can be left unanswered by a uniformly computable cover. An extensional cover, on the other hand, answers all such questions, as will be seen in Proposition 3.8 and Theorem 3.10.

Next we entend Definition 3.2. Isomorphisms that were called extensional matches will now be called 1-extensional, according to the following.

Definition 3.6. Let $\mathfrak{A}$ be a cover of a structure $\mathcal{S}$. Every isomorphism β between any $\mathcal{A}_i \in \mathfrak{A}$ and any substructure $\mathcal{B} \subseteq \mathcal{S}$ will be called *0-extensional*. For any ordinal $\theta > 0$, we say that such an isomorphism β is *θ-extensional* if:

- for every finitely generated $\mathcal{C}$ with $\mathcal{B} \subseteq \mathcal{C} \subseteq \mathcal{S}$, and every ordinal $\zeta < \theta$, there exists $j \in \omega$, $f \in I_{ij}^{\mathfrak{A}}$, and a ζ-extensional γ mapping $\mathcal{A}_j$ onto $\mathcal{C}$ such that $\beta = \gamma \circ f$; and
- for every $m \in \omega$ and every $g \in I_{im}^{\mathfrak{A}}$, and every ordinal $\zeta < \theta$, there exists an $\mathcal{E} \subseteq \mathcal{S}$ and a ζ-extensional ϵ mapping $\mathcal{A}_m$ onto $\mathcal{E}$ such that $\mathcal{B} \subseteq \mathcal{E}$ and $\beta = \epsilon \circ g$.

A uniformly computable cover $\mathfrak{A}$ of $\mathcal{S}$ is *θ-extensional* if every $\mathcal{A}_i \in \mathfrak{A}$ θ-extensionally matches some substructure of $\mathcal{S}$ (i.e. there exists a θ-extensional isomorphism between them) and every finitely generated substructure of $\mathcal{S}$ θ-extensionally matches some $\mathcal{A}_i \in \mathfrak{A}$. If such a cover exists, we say that $\mathcal{S}$ is *θ-extensionally locally computable*. Often we will abbreviate this and just call $\mathcal{S}$ itself θ-extensional.

The diagram here is exactly the same as that for Definition 3.1. The only difference is the stronger requirement about the isomorphisms γ and ϵ being ζ-extensional.

Notice that $\mathcal{S}$ is 0-extensional iff $\mathcal{S}$ is locally computable, iff $\mathcal{S}$ has a uniformly computable simple cover (by Lemma 2.12). In Section 4 we will add one more version of extensionality, even stronger than θ-extensional local computability. The rest of this section is devoted to generalizing results such as Proposition 3.5 and extending them to results about the complexity of the theory of $\mathcal{S}$ and of various fragments of its elementary diagram.

Knowing that a structure $\mathcal{S}$ is globally computable gives information about the decidability of the atomic diagram and the Σ_n-diagram of $\mathcal{S}$ for each $n > 0$: the former is computable, and each of the rest is 1-reducible to $\emptyset^{(n)}$. (Indeed, if we allow computable infinitary formulas from the hyperarithmetic hierarchy, then for any computable ordinal θ, the Σ_θ diagram is likewise Σ^0_θ, i.e. 1-reducible to $\emptyset^{(\theta)}$, under reasonable definitions.) When $\mathcal{S}$ is computably presentable, these observations still hold for the quantifier-free theory and the Σ_θ-theory of $\mathcal{S}$, although they may not hold for the actual diagrams. (The *Σ_θ-diagram of* $\mathcal{S}$ refers to the Σ_θ-theory of the structure $\mathcal{S}_S$ in the extended language with a constant for every $s \in \mathcal{S}$.) If $\mathcal{S}$ is uncountable, then it is pointless to talk about the Σ_θ-diagrams in terms of Turing computability, since the diagrams themselves are uncountable. However, we can still prove analogous results about the Σ_θ-theory for any θ-extensionally locally computable structure.

For these purposes, a first-order formula is Σ_n if it can be written in prenex normal form with n blocks of like quantifiers, beginning with an existential. For even n, this means:

$$(\exists x_1^1 \cdots \exists x_1^{k_1})(\forall x_2^1 \cdots \forall x_2^{k_2}) \cdots (\forall x_n^1 \cdots \forall x_n^{k_n})\varphi(\vec{x})$$

where φ is quantifier-free, and similarly for odd n, with $(\exists x_n^1 \cdots \exists x_n^{k_n})$. These notions generalize with computable ordinals $\theta \geq \omega$ in place of n; we refer the reader to [1] for details. We begin with simple results which do not require extensional local computability.

Proposition 3.7. *If $\mathcal{S}$ has a computable cover, then the quantifier-free theory of $\mathcal{S}$ is computable. If $\mathcal{S}$ is locally computable, then the Σ_1-theory of $\mathcal{S}$ is computably enumerable.*

Proof The truth of a quantifier-free sentence φ in $\mathcal{S}$ can be checked just by determining whether φ holds in $\mathcal{A}_i$, for any fixed $\mathcal{A}_i$ in a computable cover of $\mathcal{A}$. If $\mathfrak{A}$ is a uniformly computable cover of $\mathcal{S}$, then we can enumerate the Σ_1-theory of $\mathcal{S}$ by enumerating, for every i, every existential sentence which holds in the structure $\mathcal{A}_i$ in $\mathfrak{A}$. ⊣

This is as much as we can say in general about locally computable structures, but with extensional or perfect local computability we can develop results for more complex sentences. For simplicity we stick to finitary formulas in Proposition 3.8 and its proof. The subsequent Theorem 3.10 will generalize to computable infinitary formulas, as well as to parameters from $\mathcal{S}$.

PROPOSITION 3.8. *For $m \in \omega$, any m-extensionally locally computable structure $\mathcal{S}$, and any $n \leq m+1$, the Σ_n-theory of $\mathcal{S}$,*

$$\{\varphi \in Th\mathcal{S} : \varphi \text{ is a } \Sigma_n \text{ sentence}\},$$

is itself a Σ_n^0 set in the arithmetic hierarchy. (For $n > 0$, this means that the Σ_n-theory is 1-reducible to $\emptyset^{(n)}$, and for $n = 0$, the Σ_0-theory is computable.)

Proof Let $\langle \mathcal{A}_i \rangle_{i \in \omega}$ be an m-extensionally computable cover of $\mathcal{S}$. Proposition 3.7 already proved the result for the case $m \leq 1$. For arbitrary m, the key fact is simply that for any formula $\varphi(x_1, \dots, x_j)$,

$$\models_{\mathcal{S}} (\exists \vec{x})\varphi(\vec{x}) \text{ iff } (\exists \text{ f.g. } \mathcal{B} \subseteq \mathcal{S})(\exists \vec{b} \in \mathcal{B}^j) \models_{\mathcal{S}} \varphi(\vec{b}).$$

We restate this fact:

$$\models_{\mathcal{S}} (\forall \vec{x})\varphi(\vec{x}) \text{ iff } (\forall \text{ f.g. } \mathcal{B} \subseteq \mathcal{S})(\forall \vec{b} \in \mathcal{B}^j) \models_{\mathcal{S}} \varphi(\vec{b}).$$

When we have alternating quantifiers, we need to take superstructures at each step. For an arbitrary formula $\varphi(\vec{x}, \vec{y})$,

$$\begin{aligned}
&\models_{\mathcal{S}} (\exists \vec{x})(\forall \vec{y})\varphi(\vec{x}, \vec{y}) \\
&\text{iff } (\exists \text{ f.g. } \mathcal{B} \subseteq \mathcal{S})(\exists\, \vec{x} \in \mathcal{B}^k) \models_{\mathcal{S}} (\forall \vec{y})\varphi(\vec{x}, \vec{y}) \\
&\text{iff } (\exists \text{ f.g. } \mathcal{B} \subseteq \mathcal{S})(\exists\, \vec{x} \in \mathcal{B}^k)(\forall \text{ f.g. } \mathcal{C} \text{ s.t. } \mathcal{B} \subseteq \mathcal{C} \subseteq \mathcal{S})(\forall \vec{y} \in \mathcal{C}^p) \models_{\mathcal{S}} \varphi(\vec{x}, \vec{y})
\end{aligned}$$

If the original sentence was Σ_2, then the matrix (after all the quantifiers) will be the truth in $\mathcal{S}$ of the quantifier-free formula $\varphi(\vec{x}, \vec{y})$. In this case, $\varphi(\vec{x}, \vec{y})$ holds in $\mathcal{S}$ iff it holds in $\mathcal{C}$, so we can continue as follows:

$$\begin{aligned}
&\text{iff } (\exists \text{ f.g. } \mathcal{B} \subseteq \mathcal{S})(\exists\, \vec{x} \in \mathcal{B}^k)(\forall \text{ f.g. } \mathcal{C} \text{ s.t. } \mathcal{B} \subseteq \mathcal{C} \subseteq \mathcal{S})(\forall \vec{y} \in \mathcal{C}^p) \models_{\mathcal{C}} \varphi(\vec{x}, \vec{y}) \\
&\text{iff } (\exists i)(\exists\, \vec{b} \in \mathcal{A}_i^k)(\forall j)(\forall f \in I_{ij}^{\mathfrak{A}})(\forall \vec{c} \in \mathcal{A}_j^p) \models_{\mathcal{A}_j} \varphi(f(\vec{b}), \vec{c}).
\end{aligned}$$

The definition of 1-extensional cover shows these last two lines to be equivalent. Specifically, if the last line holds, then the witness $\mathcal{A}_i$ has a 1-extensional match α onto some $\mathcal{B} \subseteq \mathcal{S}$, and Definition 3.6, applied to any $\mathcal{A}_j$ and $f \in I_{ij}^{\mathfrak{A}}$, provides a 0-extensional match ϵ from $\mathcal{A}_j$ onto some $\mathcal{E} \supseteq \mathcal{B}$ such that $\epsilon \circ f = \alpha$. Then $\varphi(\epsilon(f(\vec{b})), \epsilon(\vec{c}))$ must hold in $\mathcal{C}$, since $\varphi(f(\vec{b}), \vec{c})$ holds in $\mathcal{A}_j$ and ϵ is an isomorphism. Conversely, if the next-to-last line holds, then there is some 1-extensional match β onto the witness $\mathcal{B}$ from some $\mathcal{A}_i \in \mathfrak{A}$, and a similar argument applies to any $\mathcal{C}$ extending $\mathcal{B}$, yielding the j, f, and γ required by the last line. This completes the proof of the result on 1-extensionally locally computable structures.

The obvious iteration of this process, applied to any Σ_n sentence about $\mathcal{S}$, yields a statement consisting of a Σ_n-sequence of quantifiers over structures in $\mathfrak{A}$, their elements, and the sets $I_{ij}^{\mathfrak{A}}$, followed by a quantifier-free statement about an $\mathcal{A}_j \in \mathfrak{A}$. The argument requires that each $\mathcal{A}_i$ correspond to some $\mathcal{B}$ via an $(n-1)$-extensional map, so that the extensions must then correspond via $(n-2)$-extensional maps, and so on down to 0-extensional maps once all the quantifiers have been moved outside the turnstile $\models$. Therefore, for an m-extensionally locally computable $\mathcal{S}$ with $m \geq (n-1)$, the Σ_n^0 statement yielded by iterating the process holds iff the original Σ_n sentence held in $\mathcal{S}$. Since the structures in $\mathfrak{A}$, the sets $I_{ij}^{\mathfrak{A}}$ and the atomic diagram of such an $\mathcal{A}_j$ are all computable uniformly in i and j, the truth of the original Σ_n-sentence in $\mathcal{S}$ is itself a Σ_n^0 fact. Moreover, this process is entirely uniform in n. ⊣

Notice that for 1-extensionally locally computable structures, the equivalence of a Σ_3 sentence in $\mathcal{S}$ with the corresponding Σ_3 statement about $\mathfrak{A}$ would not follow from this argument. Although the initial $\mathcal{A}_i$ would correspond to some $\mathcal{B}$ via a 1-extensional map, the isomorphism between the $\mathcal{A}_j$ and the $\mathcal{C}$ might be only 0-extensional, and so with a third quantifier, embeddings from $\mathcal{A}_j$ into various $\mathcal{A}_k$ would not necessarily correspond to extensions of $\mathcal{C}$. The same applies to other values of m, and for $m = 0$ a specific counterexample appears in Proposition 3.5. More counterexamples can be derived from Proposition 8.2 and Theorem 8.7, where the Σ_2-theory and the Σ_6-theory, respectively, can be arbitrarily complex.

Corollary 3.9. *Any two structures with the same m-extensionally locally computable cover are elementarily* $(m+1)$*-equivalent. Specifically, any two structures with the same uniformly computable cover are elementarily* 1*-equivalent, and any two structures with the same computable cover are elementarily* 0*-equivalent.*

Proof Given any sentence φ, the proof of Proposition 3.8 shows that $\models_{\mathcal{S}} \varphi$ is equivalent to a statement about the cover of $\mathcal{S}$. In fact, this would hold even if the computability-theoretic requirements were dropped from the definition of perfect cover. The other results of the corollary are proven similarly. ⊣

Proposition 3.8 is ostensibly only a result about the theory of the structure $\mathcal{S}$, not about its elementary diagram: the sentences considered there are not allowed to contain constants from the domain of $\mathcal{S}$. However, the intention of local computability is to talk about individual elements of the structure $\mathcal{S}$, not just about the theory, and it is not hard to adapt Proposition 3.8 to do so. We view the following result as the crux of our discussion of extensionality. We also extend it to include hyperarithmetical formulas.

Theorem 3.10. *For any computable ordinal* θ*, any* θ*-extensionally locally computable structure* $\mathcal{S}$*, any finite tuple* $\vec{p}$ *of parameters from* $\mathcal{S}$*, and any* $\zeta \leq \theta$*, the*

Σ_ζ-theory of $\mathcal{S}$ over $\vec{p}$,

$$\{\varphi \in Th\mathcal{S}, \vec{p} : \varphi \text{ is a } \Sigma_\zeta \text{ sentence}\},$$

is itself a Σ^0_ζ in the hyperarithmetic hierarchy. For a fixed computable presentation of a single θ, this holds uniformly for all $\zeta \leq \theta$, and uniformly in an appropriate description of the parameters (as discussed after the proof).

Proof Let $\mathcal{B}$ be generated by $\vec{p}$ in $\mathcal{S}$, and fix an θ-extensional match $\beta : \mathcal{A}_l \twoheadrightarrow \mathcal{B}$ for some $\mathcal{A}_l \in \mathfrak{A}$. As before, we give an example by evaluating the truth in $\mathcal{S}$ of an arbitrary Σ_2 sentence with the parameters $\vec{p}$, assuming now that $\theta \geq 2$. By an argument similar to that in the proof of Proposition 3.8, the Σ_2 sentence $(\exists\vec{x})(\forall\vec{y})\varphi(\vec{p}, \vec{x}, \vec{y})$ holds in $\mathcal{S}$ iff

$$(\exists i)(\exists h \in I^{\mathfrak{A}}_{li})(\exists\vec{b} \in \mathcal{A}^k_i)(\forall j)(\forall f \in I^{\mathfrak{A}}_{ij})(\forall\vec{c} \in \mathcal{A}^m_j)$$
$$\models_{\mathcal{A}_j} \varphi(f(h(\vec{a})), f(\vec{b}), \vec{c}),$$

which is a Σ^0_2 condition, uniformly in $\vec{a}$ and l. The obvious iteration works for any $\zeta \leq \theta$, but no longer applies when $\zeta = \theta + 1$, whereas Proposition 3.8 did hold when $n = m + 1$. In the example above, as long as $\mathcal{S}$ is 2-extensional, we may assume that β is 2-extensional, that the h we find lifts to an inclusion in $\mathcal{S}$ via β and some 1-extensional $\gamma : \mathcal{A}_i \hookrightarrow \mathcal{S}$, and therefore that every inclusion of $\gamma(\mathcal{A}_i)$ into any larger finitely generated substructure of $\mathcal{S}$ must lift some f in some $I^{\mathfrak{A}}_{ij}$. If β were only 1-extensional, this argument would not suffice. Adding the parameters forces us to start by fixing an $\mathcal{A}_l \in \mathfrak{A}$ and a β, whereas in Proposition 3.8 we were allowed simply to search for any $\mathcal{A}_i$ and a single embedding into an $\mathcal{A}_j$. Hence parameters require one more level of extensionality.

The extension of this argument to hyperarithmetical formulas (of complexity Σ_ζ with $\zeta \leq \theta$, of course) is intuitively reasonable; the argument is by induction on ordinals, with the paragraph above covering the case of a successor ordinal. When ζ is a limit ordinal, the natural argument applies, using the uniformity of the disjuncts in the Σ_ζ formula. Writing out the details becomes very messy, and we leave that task to the reader. Notice that it is *not* necessary to quantify over the isomorphisms β, γ, etc.; one simply fixes an appropriate θ-extensional β at the beginning of the argument, and then translates the formula into a new hyperarithmetic statement in which all quantification is over effectively given objects such as domains $\mathcal{A}_i$ or sets $I^{\mathfrak{A}}_{ij}$ of maps.

Of course, knowing an original parameter $p_i \in \mathcal{B}$ is useless to us; we need to know l and the value $a_i = \beta^{-1}(p_i)$ in $\mathcal{A}_l$. For finitely many parameters, this constitutes only finitely much information, but we also wish to consider uniformity. Of course, it does not make sense to ask for parameters from a potentially uncountable structure $\mathcal{S}$ to be given uniformly. Instead, our formal statement of uniformity is that if we are given an l and finitely many parameters $\vec{a}$ from $\mathcal{A}_l$, then for any $\zeta \leq \theta$ and any ζ-extensional match β mapping $\mathcal{A}_l$ into $\mathcal{S}$, the Σ_ζ-theory of $\mathcal{S}$

over the parameters $\beta(a_1), \dots, \beta(a_n)$ is a Σ^0_ζ set in the hyperarithmetical hierarchy, uniformly in l and $\vec{a}$. ⊣

We view Theorem 3.10 as the strongest argument yet that local computability, and in particular these ordinal levels of extensional local computability, form the correct analogue in uncountable structures to computable presentability in countable structures. The point of a computable presentation of a structure is not just that it allows us to compute the atomic theory and enumerate the Σ_1-theory and so on, but that it actually allows us to do so over specific elements of the structure: the atomic *diagram* is computable, and the Σ_ζ diagram is Σ^0_ζ, uniformly in $\zeta \leq \theta$ (for a computable presentation of θ). For an uncountable $\mathcal{S}$, of course, there is no effective way to name all individual elements, so it is hopeless to expect the entire atomic diagram to be computable. A θ-extensional cover, however, gives us a way of describing individual elements and tuples of them: using the cover, we name an $\mathcal{A}_l$ which θ-extensionally matches the substructure of $\mathcal{S}$ generated by the tuple, and specify which elements of $\mathcal{A}_l$ correspond to the tuple.

To state the same fact differently, having a θ-extensional cover tells us exactly what information we need about the tuple $\vec{p}$ from $\mathcal{S}$ in order to compute the atomic theory of $\mathcal{S}$ over $\vec{p}$, or to enumerate its Σ_1 theory over $\vec{p}$, etc. For the field of complex numbers, for instance, an $\mathcal{A}_i$ is given by its transcendence degree and the minimal polynomial of a single additional element generating the rest of $\mathcal{A}_i$ over a transcendence basis. If we can determine this information for the subfield $\mathbb{Q}(\vec{p}) \subset \mathbb{C}$, and know which elements correspond to $\vec{p}$, then without further information we can give a Σ^0_ζ description of the Σ_ζ-theory of $(\mathbb{C}, \vec{p})$. Each θ-extensional cover of any $\mathcal{S}$ says, "if you tell me this particular information about your tuple $\vec{p}$ from $\mathcal{S}$, I will give you a Σ^0_ζ-presentation of the Σ_ζ facts about $\vec{p}$ in $\mathcal{S}$, for each $\zeta \leq \theta$."

For a useful example of the foregoing (rather abstract) remarks, we urge the reader to examine the discussion in Section 8 of the lexicographic order on Cantor space 2^ω. Some further philosophical discussion takes place there as well, in the context of that example.

§4. Perfect Local Computability. Since local computability is conceived as a generalization of the notion of computable structure, it is natural to ask about the relationship between local computability and computable presentability for countable structures. All computably presentable structures are readily seen to be locally computable, and indeed θ-extensionally locally computable for all $\theta < \omega_1^{CK}$; a proof appears below. However, the converse fails. The attempt to find a version of extensionality equivalent to computable presentability for countable structures leads one to the following definition, which can be viewed as a kind of ∞-extensionality, stronger than θ-extensionality for every computable θ.

Definition 4.1. Let $\mathfrak{A}$ be a uniformly computable cover for a structure $\mathcal{S}$. A set M is a *correspondence system* for $\mathfrak{A}$ and $\mathcal{S}$ if it satisfies the following five conditions:

(1) Each element of M is an embedding of some $\mathcal{A}_i \in \mathfrak{A}$ into $\mathcal{S}$; and
(2) For every $\mathcal{A}_i \in \mathfrak{A}$, there exists a $\beta \in M$ with domain $\mathcal{A}_i$; and
(3) For every finitely generated substructure $\mathcal{B}$ of $\mathcal{S}$, there exists a $\beta \in M$ with image $\mathcal{B}$; and
(4) For every $\mathcal{A}_i \in \mathfrak{A}$, every $\beta \in M$ with domain $\mathcal{A}_i$, and every finitely generated $\mathcal{C} \subseteq \mathcal{S}$ such that $\beta(\mathcal{A}_i) \subseteq \mathcal{C}$, there exists an $\mathcal{A}_j \in \mathfrak{A}$, a $\gamma \in M$ with domain $\mathcal{A}_j$ and image $\mathcal{C}$, and an $f \in I_{ij}^{\mathfrak{A}}$ such that $\beta = \gamma \circ f$; and
(5) For every $\mathcal{A}_i \in \mathfrak{A}$, every $\beta \in M$ with domain $\mathcal{A}_i$, every $\mathcal{A}_m \in \mathfrak{A}$, and every $g \in I_{im}^{\mathfrak{A}}$, there exists an $\epsilon \in M$ with domain $\mathcal{A}_m$ such that $\beta = \epsilon \circ g$ (and hence $\beta(\mathcal{A}_i) \subseteq \epsilon(\mathcal{A}_m)$).

If $\mathcal{S}$ has a uniformly computable cover $\mathfrak{A}$ with a correspondence system M, then we say $\mathcal{S}$ is *∞-extensionally locally computable*, and refer to elements of M as *∞-extensional matches*.

A correspondence system M is *perfect* if it also satisfies:

(6) For every finitely generated $\mathcal{B} \subseteq \mathcal{S}$, if $\beta : \mathcal{A}_i \twoheadrightarrow \mathcal{B}$ and $\gamma : \mathcal{A}_j \twoheadrightarrow \mathcal{B}$ both lie in M, then $\gamma^{-1} \circ \beta \in I_{ij}^{\mathfrak{A}}$.

If a perfect correspondence system exists, then its elements are called *perfect matches* between their domains and their images. The uniformly computable cover $\mathfrak{A}$ is then called a *perfect cover* for $\mathcal{S}$, and $\mathcal{S}$ itself is said to be *perfectly locally computable*.

Once again, the diagrams for conditions (4) and (5) are exactly those from Definition 3.1; the only difference is that now the isomorphisms γ and ϵ are required to lie in M.

This concept is related to extensionality, clearly, and any correspondence system M is quickly seen (by induction on θ) to contain only θ-extensional matches, for every ordinal θ. So every ∞-extensionally locally computable structure is θ-extensionally locally computable for every θ. This justifies the terminology and, using Corollary 3.9, also yields:

Corollary 4.2. *Any two structures with the same perfect cover are elementarily equivalent, and indeed have the same hyperarithmetical theory.* ⊣

However, the definition of ∞-extensionality is stronger than that of θ-extensionality. For the map β to be a θ-extensional match, we only needed the existence of ζ-extensional matches γ (with $\zeta < \theta$) to relate the embeddings $f \in I_{ij}^{\mathfrak{A}}$ (for all j) to the finitely generated extensions of the image of β in $\mathcal{S}$, and for different values of ζ, we could use different maps γ. Here Conditions (4) and (5) require that the isomorphisms γ be in M themselves, hence that they satisfy the same conditions.

For perfect covers, Condition (6) creates a second difference, which will be important in Theorem 6.3, but is not related to Definition 3.2. In fact a converse of Condition (6) follows from the first five conditions: if $f \in I_{ij}^{\mathfrak{A}}$ is an isomorphism of $\mathcal{A}_i$ onto $\mathcal{A}_j$, then M contains maps β with domain $\mathcal{A}_i$ and γ with domain $\mathcal{A}_j$, with the same image in $\mathcal{S}$ and with $f = \gamma^{-1} \circ \beta$. Indeed, the $\beta : \mathcal{A}_i \hookrightarrow \mathcal{S}$ given by Condition (2) and the $\gamma : \mathcal{A}_j \hookrightarrow \mathcal{S}$ subsequently given by Condition (4) have the same image in $\mathcal{S}$, since if $y \in \gamma(\mathcal{A}_j)$, then $y = \gamma(f(a))$ for some $a \in \mathcal{A}_i$, but $\gamma(f(a)) = \beta(a)$ lies in the image of β. Thus Condition (6) is essentially a characterization of the isomorphisms in $I^{\mathfrak{A}}$: there is an isomorphism $f \in I^{\mathfrak{A}}_{ij}$ iff $\mathcal{A}_i$ and $\mathcal{A}_j$ describe the same substructure of $\mathcal{S}$ under the perfect correspondence system M, in which case f factors through the relevant maps in M.

Condition (6) is most easily met by requiring the map β in Condition (3) to be unique. We refer to this as the *Uniqueness Condition*:

(3′) For every finitely generated substructure $\mathcal{B}$ of $\mathcal{S}$, there exists a unique $\beta \in M$ with image $\mathcal{B}$.

Then Condition (6) only requires that each $I^{\mathfrak{A}}_{ii}$ contain the identity map; we will discuss this further in Section 5. If the Uniqueness Condition holds, then each finitely generated $\mathcal{B} \subseteq \mathcal{S}$ has a unique $\mathcal{A}_i \in \mathfrak{A}$ to describe it.

Condition (6) itself does not quite require this uniqueness, but it comes close to doing so. If we build the equivalence relation $\equiv$ on $\{\mathcal{A}_i : i \in \omega\}$ generated by $\{\langle \mathcal{A}_i, \mathcal{A}_j \rangle : (\exists f \in I^{\mathfrak{A}}_{ij}) \text{ range}(f) = \mathcal{A}_j\}$, and define the new cover $\mathfrak{A}/\equiv$, with an appropriate adjustment to $I^{\mathfrak{A}}$, then we would have uniqueness. Of course, in a uniformly computable cover $\mathfrak{A}$ of $\mathcal{S}$, it is Σ_1 but not necessarily decidable whether the image of an $f \in I^{\mathfrak{A}}_{ij}$ contains all of $\mathcal{A}_j$ or not, and therefore $\mathfrak{A}/\equiv$ might not be a uniformly computable cover.

We note that it is not reasonable to replace Condition (2) in Definition 4.1 by any uniqueness condition dual to Condition (3′) above. Since $\mathfrak{A}$ is countable, the uniqueness of the maps β in Condition (2) would force $\mathcal{S}$ also to be countable, and of course we wish our analysis to apply to uncountable structures as well as countable ones.

Corollary 4.3. *Every algebraically closed field of characteristic* 0 *is perfectly locally computable.*

Proof We refer to the proof of Proposition 3.3. The necessary correspondence system M is the set of all embeddings of structures of $\mathfrak{A}$ into $\mathcal{C}$. Lemma 3.4 showed that every such isomorphism is an extensional match, and so the first five conditions of Definition 4.1 are quickly satisfied. Finally, if β and γ are as in Condition (6), then $\gamma^{-1} \circ \beta$ is an embedding of $\mathcal{A}_i$ into $\mathcal{A}_j$, and therefore must lie in $I^{\mathfrak{A}}_{ij}$. ⊣

The situation of Corollary 4.3 generalizes to a further result:

Proposition 4.4. *Suppose that* $\mathfrak{A}$ *is a perfect cover for* $\mathcal{S}$*, with a correspondence system* M *such that for all* $\mathcal{A}_i \in \mathfrak{A}$*, every embedding of* $\mathcal{A}_i$ *into* $\mathcal{S}$ *is an element of* M*. Then* $\mathcal{S}$ *is* ω*-homogeneous.*

This result is purely model-theoretic: the same proof would hold for any cover $\mathfrak{A}$ for which such a correspondence system M exists, regardless of whether $\mathfrak{A}$ were computable. For an introduction to ω-homogeneity, see p. 212 of [8].

Proof Let $\vec{x}, \vec{y} \in \mathcal{S}^m$ be two finite sequences of elements of $\mathcal{S}$ such that $(\mathcal{S}, \vec{x}) \equiv (\mathcal{S}, \vec{y})$. (The notation $\equiv$ denotes elementary equivalence, as usual.) We must show that for every $z \in \mathcal{S}$ there exists $w \in \mathcal{S}$ such that $(\mathcal{S}, \vec{x}, z) \equiv (\mathcal{S}, \vec{y}, w)$. To see this, notice that the substructures $\mathcal{B}_x$ and $\mathcal{B}_y$ of $\mathcal{S}$ generated by $\vec{x}$ and $\vec{y}$ must be isomorphic, say via an isomorphism $g : \mathcal{B}_x \twoheadrightarrow \mathcal{B}_y$. Now there exists an $\mathcal{A}_i \in \mathfrak{A}$ and a $\beta \in M$ mapping $\mathcal{A}_i$ isomorphically onto $\mathcal{B}_x$, and also an $\mathcal{A}_j \in \mathfrak{A}$, an $f \in I^{\mathfrak{A}}_{ij}$, and a $\gamma \in M$ mapping $\mathcal{A}_j$ isomorphically onto the substructure $\mathcal{B}_z$ generated by $\mathcal{B}_x$ and z, such that $\gamma \circ f = \beta$. But the isomorphism $g \circ \beta$ from $\mathcal{A}_i$ onto $\mathcal{B}_y$ must lie in M, by the assumption of the Proposition. So there also exists some $\mathcal{C} \subseteq \mathcal{S}$ and some isomorphism $\alpha : \mathcal{A}_j \twoheadrightarrow \mathcal{C}$ such that $\alpha \circ f = g \circ \beta$. Then $\alpha \circ \gamma^{-1}$ is an isomorphism from $\mathcal{B}_z$ onto $\mathcal{C}$ extending β, and we let $w = \alpha(\gamma^{-1}(z))$. Definition 4.1 shows that every extension of $\mathcal{B}_z$ in $\mathcal{S}$ corresponds to an embedding of $\mathcal{A}_j$ into some other element of $\mathfrak{A}$, which in turn corresponds to an extension of $\mathcal{C}$, and conversely. Hence $(\mathcal{S}, \vec{x}, z) \equiv (\mathcal{S}, \vec{y}, w)$ as required.

(Alternatively, having found $\mathcal{A}_j$ and w as above, work in the language augmented by constants $\vec{x}$ and z. Apply Corollary 3.9 to the structures $(\mathcal{S}, \vec{x}, z)$ and $(\mathcal{S}, \vec{y}, w)$, for each of which $\mathfrak{A}' = \{A_m \in \mathfrak{A} : I^{\mathfrak{A}}_{jm} \neq \emptyset\}$ is a perfect cover, with all $I^{\mathfrak{A}'}_{mn} = I^{\mathfrak{A}}_{mn}$.) ⊣

§5. A Dash of Category Theory. Here we define the *natural cover* of a structure $\mathcal{B}$. If $\mathcal{B}$ is a (globally) computable structure, then its natural cover will be uniformly computable and perfect. We give the definition under the assumption that $\mathcal{B}$ is computable.

Let $\mathcal{B}_i$ be the substructure of $\mathcal{B}$ generated by the ith tuple $(\vec{b})_i$ of $\omega^{<\omega}$. Then the domain of $\mathcal{B}_i$ is a computably enumerable set. Enumerate its elements, and let A_i be the domain of the enumeration. (That is, A_i is an initial segment of ω from which we have a bijection β_i onto the domain of $\mathcal{B}_i$, all computable uniformly in i.) Then we may define computable structures $\mathcal{A}_i$, uniformly in i and each with domain A_i, isomorphic to each $\mathcal{B}_i$ via the map β_i. Next, we define $I^{\mathfrak{A}}_{ij}$ to be the set of functions

$$\{\beta_j^{-1} \circ \beta_i : \mathcal{B}_i \hookrightarrow \mathcal{B}_j\},$$

i.e. the inclusion maps in $\mathcal{B}$, lifted to the cover. Clearly these are embeddings, and the condition $(\vec{b})_i \subseteq \mathcal{B}_j$ is Σ_1, so this is a c.e. set, uniformly in i and j.

If $\mathcal{B}$ is countable but not computable, then the $\mathfrak{A}$ built here is still a cover of $\mathcal{B}$, but may fail to be uniformly computable, or even to be computable at all. If $\mathcal{B}$ is uncountable, then the indices i range over the uncountable set containing all finite tuples of elements of the domain of $\mathcal{B}$, and so the cover $\mathfrak{A}$ itself is uncountable. In this case, of course, effectiveness considerations do not apply.

Modulo the pullback to the domain ω, the natural cover of $\mathcal{B}$ really just the category **FGSub**($\mathcal{B}$) of all finitely generated substructures of $\mathcal{S}$, known long before now to model theorists and category theorists. The objects of **FGSub**($\mathcal{B}$) are precisely the finitely generated substructures of $\mathcal{B}$, and the morphisms are the inclusions among these substructures. (Fraïssé referred to the set of objects of **FGSub**($\mathcal{B}$) as the *age of* $\mathcal{B}$.) Our definition of the natural cover pulls each substructure back to the domain ω, or an initial segment thereof, since we are concerned with issues of computability, but in pure category theory one normally uses just the substructure itself.

This raises the question of whether all covers are themselves categories. As collections of objects with maps among them, covers seem ripe for consideration as categories. Definition 2.9 does come up short in two respects. The first is trivial: for a cover $\mathfrak{A}$, it does not require that $I^{\mathfrak{A}}_{ii}$ contain the identity map from $\mathcal{A}_i$ to itself. The second is less so: it does not require that the sets $I^{\mathfrak{A}}_{ij}$ of embeddings be closed under composition. This may be rectified in the case of an ∞-extensional cover.

LEMMA 5.1. *Let* $\mathfrak{A}$ *be a uniformly computable cover of a structure* $\mathcal{S}$.

(1) *Define* $\mathfrak{A}_{id}$ *to consist of the same simple cover as* $\mathfrak{A}$, *with*

$$I^{\mathfrak{A}_{id}}_{ij} = \begin{cases} I^{\mathfrak{A}}_{ij}, & \text{if } i \neq j \\ I^{\mathfrak{A}}_{ii} \cup \{id_{\mathcal{A}_i}\}, & \text{if } j = i. \end{cases}$$

Then $\mathfrak{A}_{id}$ *is also a uniformly computable cover of* $\mathcal{S}$. *Moreover, each of these covers is* θ*-extensional (resp.* ∞*-extensional) iff the other is. If* $\mathfrak{A}$ *was perfect, then* $\mathfrak{A}_{id} = \mathfrak{A}$.

(2) *Define* $\overline{\mathfrak{A}}$ *to consist of the same simple cover as* $\mathfrak{A}$, *with* $I^{\overline{\mathfrak{A}}}_{ij}$ *containing all compositions*

$$\mathcal{A}_i = \mathcal{A}_{i_0} \hookrightarrow \mathcal{A}_{i_1} \hookrightarrow \cdots \hookrightarrow \mathcal{A}_{i_n} \hookrightarrow \mathcal{A}_{i_{n+1}} = \mathcal{A}_j$$

with the intermediate maps each from the appropriate $I^{\mathfrak{A}_{id}}_{i_k i_{k+1}}$. *The* $\overline{\mathfrak{A}}$ *is another uniformly computable cover of* $\mathcal{S}$. *If* $\mathfrak{A}$ *was* ∞*-extensional (resp. perfect), then so is* $\overline{\mathfrak{A}}$, *and if* $\mathfrak{A}$ *has the Amalgamation Property (as described in Definition 6.1 below), then so does* $\overline{\mathfrak{A}}$.

Proof Item (1) is immediate from the definitions, since Condition (6) of Definition 4.1 shows that in a perfect cover, each $I^{\mathfrak{A}}_{ii}$ already contains the identity embedding. Likewise, it is readily seen that $\overline{\mathfrak{A}}$ is a uniformly computable cover of $\mathcal{S}$: one need only remark that if $f \in I^{\mathfrak{A}}_{ij}$ and $g \in I^{\mathfrak{A}}_{jk}$, then g lifts to an inclusion $\mathcal{B} \subseteq \mathcal{C}$ via some β and γ, and $(g \circ f)$ lifts to the inclusion $\beta(f(\mathcal{A}_i)) \subseteq \mathcal{C}$. Now suppose that M is a correspondence system for $\mathfrak{A}$. Only Condition (5) of Definition 4.1 warrants consideration, and it is not difficult: if $\beta \in M$ with domain $\mathcal{A}_i$ and $g \in I^{\overline{\mathfrak{A}}}_{im}$, then g is the composition of a finite chain of maps from sets $I^{\mathfrak{A}}_{i_k i_{k+1}}$, and the result follows by induction on the length of this chain. Condition (6) is also immediate, giving the result for perfect covers. A similar induction shows the the Amalgamation Property holds for $\overline{\mathfrak{A}}$ as well, assuming it held for $\mathfrak{A}$. ⊣

For a θ-extensional cover $\mathfrak{A}$ with $\theta > 0$, our $\overline{\mathfrak{A}}$ can fail to be θ-extensional. With $\theta = 1$, for example, suppose that β is a 1-extensional match with respect to $\mathfrak{A}$, and fix $f \in I^{\mathfrak{A}}_{ij}$ and $g \in I^{\mathfrak{A}}_{jk}$. 1-extensionality guarantees the existence of the $\mathcal{C}$ and γ in the diagram below, and $\mathcal{A}_j$ in turn will have a 1-extensional match onto some substructure of $\mathcal{S}$, but not necessarily onto $\mathcal{C}$, let alone via γ. There may not exist any $\mathcal{D} \supseteq \mathcal{B}$ and 0-extensional $\delta : \mathcal{A}_k \to \mathcal{D}$ corresponding to $(g \circ f)$ and β, in which case $(g \circ f)$ shows β not to be 1-extensional with respect to $\overline{\mathfrak{A}}$.

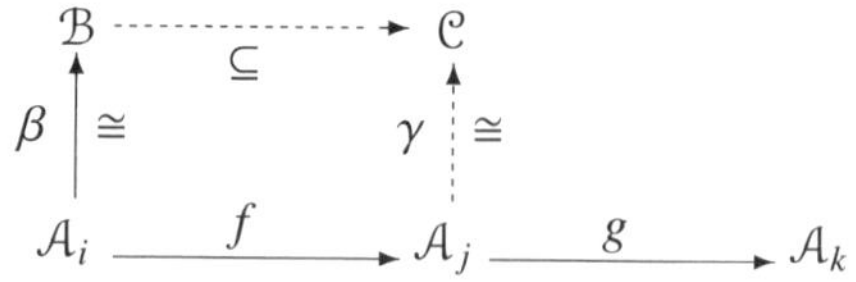

It remains open whether, for $0 < \theta < \infty$, a structure $\mathcal{S}$ can have a θ-extensional cover without having a θ-extensional cover closed under composition and identity embeddings.

The following appears as Proposition 4.1 in [12].

Proposition 5.2 (Miller-Mulcahey). *If $\mathcal{S}$ is perfectly locally computable, then there exists a faithful functor R mapping* **FGSub**$(\mathcal{S})$ *into a perfect cover $\mathfrak{A}$ of $\mathcal{S}$ which is closed under composition and identity embeddings. Moreover, there exists a natural isomorphism $\beta : (I_{\mathfrak{A}} \circ R) \to I_{\mathbf{FGSub}(\mathcal{S})}$ where the I_{-} denote the appropriate inclusions into the category of all L-structures under embeddings.* ⊣

As shown there, this functor R : **FGSub**$(\mathcal{S}) \to \mathfrak{A}$ is essentially surjective, although it need not be onto. It also follows that $colim(I_{\mathfrak{A}} \circ R) \simeq \mathcal{S}$.

§6. Countable Structures. The general intention of local computability is to apply computability theory to uncountable structures. Nevertheless, we can learn a good deal about our definitions by asking which countable structures satisfy them. In particular, our first result makes clear (especially when seen in concert with Theorem 3.10) that among all the concepts we have defined for uncountable structures, perfect local computability is the most apt generalization of the notion of computable presentability for countable structures. First we require a notion from model theory, adapted to our concept of a cover.

Definition 6.1. A cover $\mathfrak{A}$ has the *Amalgamation Property*, abbreviated AP, if for all $i, j, k \in \omega$ and all maps $e \in I^{\mathfrak{A}}_{ij}$ and $f \in I^{\mathfrak{A}}_{ik}$, there exists an m and maps $g \in I^{\mathfrak{A}}_{jm}$ and $h \in I^{\mathfrak{A}}_{km}$ for which $h \circ f = g \circ e$:

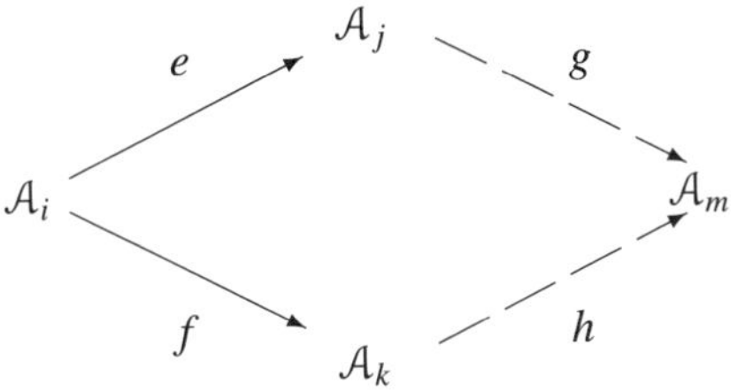

We say that we can *amalgamate* $\mathcal{A}_j$ and $\mathcal{A}_k$ over $\mathcal{A}_i$, relative to the maps e and f.

LEMMA 6.2. *Every perfect cover $\mathfrak{A}$ of any structure $\mathcal{S}$ has the Amalgamation Property.*

Proof Let M be a perfect correspondence system for $\mathcal{S}$ over $\mathfrak{A}$, and fix maps $e \in I_{ij}^{\mathfrak{A}}$ and $f \in I_{ik}^{\mathfrak{A}}$. We use the six conditions from Definition 4.1, and suggest that the reader follow the diagram below (which shows only the maps among structures in the cover, omitting their images in $\mathcal{S}$).

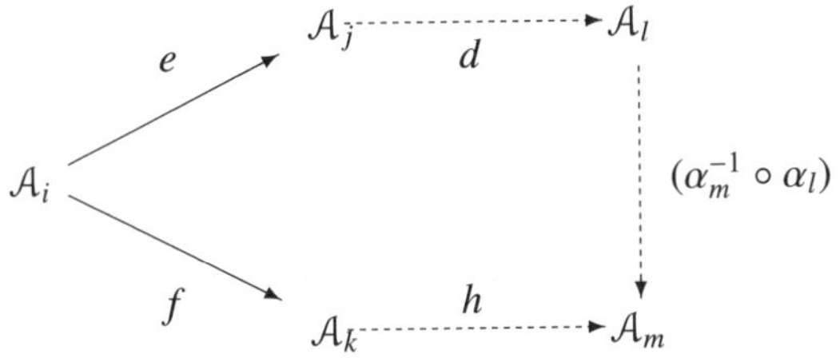

Now there is an $\alpha_i : \mathcal{A}_i \hookrightarrow \mathcal{S}$ in M, by Condition (2), and Condition (5) yields maps $\alpha_j : \mathcal{A}_j \hookrightarrow \mathcal{S}$ and $\alpha_k : \mathcal{A}_k \hookrightarrow \mathcal{S}$ in M with $\alpha_k \circ f = \alpha_j \circ e = \alpha_i$. Let $\mathcal{C} \subseteq \mathcal{S}$ be generated by the images of α_j and α_k together. Then $\mathcal{C}$ is finitely generated, so by Condition (4) we have elements $l, m \in \omega$, embeddings $d \in I_{jl}^{\mathfrak{A}}$ and $h \in I_{km}^{\mathfrak{A}}$, and maps $\alpha_l, \alpha_m \in M$ with domains $\mathcal{A}_l$ and $\mathcal{A}_m$, respectively, and both with image $\mathcal{C}$, such that $\alpha_l \circ d = \alpha_j$ and $\alpha_m \circ h = \alpha_k$. But Condition (6) also holds, since M is perfect, and so $\alpha_m^{-1} \circ \alpha_l \in I_{lm}^{\mathfrak{A}}$. Lemma 5.1 allows us to assume closure of the sets of embeddings under composition, In particular, we set

$$g = \alpha_m^{-1} \circ \alpha_j = (\alpha_m^{-1} \circ \alpha_l) \circ d \in I_{jm}^{\mathfrak{A}}.$$

But then

$$g \circ e = \alpha_m^{-1} \circ (\alpha_j \circ e) = \alpha_m^{-1} \circ \alpha_i = \alpha_m^{-1} \circ (\alpha_k \circ f) = \alpha_m^{-1} \circ (\alpha_m \circ h) \circ f = h \circ f,$$

as required by the Amalgamation Property. ⊣

THEOREM 6.3. *Let $\mathcal{S}$ be any countable structure. Then the following are equivalent.*

1. $\mathcal{S}$ *is computably presentable.*
2. $\mathcal{S}$ *is perfectly locally computable.*
3. $\mathcal{S}$ *is* ∞*-extensional over a cover* $\mathfrak{A}$ *with the Amalgamation Property.*

Proof ($1 \Longrightarrow 2$) is not difficult. Let $\mathfrak{A}$ be the natural cover of a computable presentation $\mathcal{B}$ of $\mathcal{S}$, as defined in Section 5, and fix the maps β_i defined there as well. It is quickly seen that this $\mathfrak{A}$ is a perfect cover of $\mathcal{B}$, under the correspondence system $M = \{\beta_i : i \in \omega\}$. Conditions (1)–(3) of Definition 4.1 are immediate, as is Condition (5), given our definition of $I^{\mathfrak{A}}_{ij}$. For Condition (4), when $\beta_i(\mathcal{A}_i) \subseteq \mathcal{C}$, we know that $\mathcal{C} = \mathcal{B}_j$ for some j, that the map β_j lies in M, and that $(\beta_j^{-1} \circ \beta_i) \in I^{\mathfrak{A}}_{ij}$ has the necessary properties. Finally, if $\beta_i, \beta_j \in M$ have the same image, then $\mathcal{B}_i = \mathcal{B}_j$, and so $(\beta_j^{-1} \circ \beta_i) \in I^{\mathfrak{A}}_{ij}$ as required by Condition (6). Thus $\mathcal{B}$ is perfectly locally computable. To see that $\mathcal{S}$ is perfectly locally computable, take the same cover $\mathfrak{A}$ and use the correspondence system $\{\gamma \circ \beta_i : \beta_i \in M\}$, where γ is an isomorphism from $\mathcal{B}$ onto $\mathcal{S}$.

($2 \Longrightarrow 3$) follows from Lemma 6.2, so it remains to prove ($3 \Longrightarrow 1$). Suppose that $\mathfrak{A}$ is a uniformly computable cover for $\mathcal{S}$, with a correspondence system M satisfying AP. Using Lemma 5.1, we may assume that $\mathfrak{A}$ is closed under composition. In category-theoretic terms, $\mathcal{S}$ is just the inverse limit of the category $\mathfrak{A}$, but we need to build a computable presentation $\mathcal{B}$ of this inverse limit. We will define i_s recursively so that $\mathcal{B}_s = \mathcal{A}_{i_s}$, with a map $g_s : \mathcal{B}_s \hookrightarrow \mathcal{B}_{s+1}$ from $I^{\mathfrak{A}}_{i_s, i_{s+1}}$. To start, we let $\mathcal{B}_0 = \mathcal{A}_0$ and $i_0 = 0$.

Given $\mathcal{B}_s$, let $s = \langle t, u, v\rangle$, and fix $i = i_t$ and $k = i_s$, so that $\mathcal{B}_t = \mathcal{A}_i$ and $\mathcal{B}_s = \mathcal{A}_k$. We begin stage $s+1$ by listing out those maps in $I^{\mathfrak{A}}$ with domain $\mathcal{B}_t$ until we find the uth element on this list. Let $f \in I^{\mathfrak{A}}_{ij}$ (for some j) be this element. (Of course $t \leq s$, so $\mathcal{B}_t$ is defined. Also, if there were only finitely many maps in $I^{\mathfrak{A}}$ with domain $\mathcal{A}_i$, then Definition 4.1 would imply that $\mathcal{S}$ is itself finitely generated, hence isomorphic to some $\mathcal{A}_i \in \mathfrak{A}$, hence computably presentable. Therefore we may assume that we do find such an f.)

We will incorporate this f into $\mathcal{B}_{s+1}$ as follows. By the Amalgamation Property, there exists some m and embeddings $g \in I^{\mathfrak{A}}_{km}$ and $p \in I^{\mathfrak{A}}_{jm}$ such that

$$p \circ f = g \circ (g_{s-1} \circ \cdots \circ g_t).$$

(Recall that we closed the cover under composition of maps, so that $g_{s-1} \circ \cdots \circ g_t \in I^{\mathfrak{A}}_{ik}$.) We find the least such triple $\langle m, g, p\rangle$, and define $i_{s+1} = m$ and $\mathcal{B}_{s+1} = \mathcal{A}_m$, with $f_s = p \circ f : \mathcal{B}_{t_{s+1}} \hookrightarrow \mathcal{B}_{s+1}$ and $g_s = g : \mathcal{B}_s \hookrightarrow \mathcal{B}_{s+1}$. This completes stage $s+1$, and we say that f has been incorporated into $\mathcal{B}$ at this stage. Notice that every $f \in I^{\mathfrak{A}}$ whose domain is $\mathcal{A}_{i_t}$ ($= \mathcal{B}_t$) for any t will be incorporated into $\mathcal{B}$ at infinitely many stages.

The structure $\mathcal{B}$ itself is the union of the chain of structures $\mathcal{B}_s$ under the embeddings $g_{s+1} : \mathcal{B}_s \hookrightarrow \mathcal{B}_{s+1}$. To build this $\mathcal{B}$ computably, at stage $\langle u, v\rangle$ we consider the element t (if any) which is enumerated into the domain of $\mathcal{B}_u$ at stage v. (Note that $\mathcal{B}_u$ might be finite, so we cannot just wait for its next element to appear.) For each of the finitely many elements x which has already entered $\mathcal{B}$ at some stage $\langle u', v' \rangle < \langle u, v \rangle$, we check whether either $x = g'_u \circ g_{u'-1} \circ \cdots \circ g_{u+1}(t)$ (if $u' > u$) or $t = g_u \circ g_{u-1} \circ \cdots \circ g_{u'+1}(x)$ (if $u' < u$). If either of these holds, we do nothing at this stage; if neither holds, then we add a fresh element to $\mathcal{B}$ and identify it with $t \in \mathcal{B}_u$. Iterating this process over all stages $\langle u, v \rangle$ builds the domain of $\mathcal{B}$, and we define the functions and relations on it in the obvious way. Notice that for every s the embeddings of $\mathcal{B}_s$ and $\mathcal{B}_{s+1}$ into $\mathcal{B}$ are compatible with the map $g_s : \mathcal{B}_s \hookrightarrow \mathcal{B}_{s+1}$, and that we can compute these embeddings uniformly in s, so from here on we will view each $\mathcal{B}_s$ as a substructure of $\mathcal{B}$, with $g_s : \mathcal{B}_s \hookrightarrow \mathcal{B}_{s+1}$ as an inclusion.

Next we build the isomorphism $\alpha : \mathcal{B} \to \mathcal{S}$. Of course, α need not be computable. At every stage $s + 1$, the image $\mathcal{S}_s$ of the current approximation α_s will be a finitely generated substructure of $\mathcal{S}$, its domain will be some $\mathcal{B}_t$, and α_s will lie in M. We will ensure that the extension of α_s is compatible with the inclusion maps $g_t : \mathcal{B}_t \hookrightarrow \mathcal{B}_{t+1}$, so that at the end of the construction, we may define $\alpha(x)$ for $x \in \mathcal{B}$ simply by finding some s and t with $x \in \mathcal{B}_t = \operatorname{dom}(\alpha_s)$ and letting $\alpha(x) = \alpha_s(x)$. We start by taking $\alpha_0 \in M$ to be any embedding of $\mathcal{B}_0 = \mathcal{A}_{i_0}$ into $\mathcal{S}$.

Suppose that $\operatorname{dom}(\alpha_s) = \mathcal{B}_t = \mathcal{A}_i$ and that we wish to extend α_s to α_{s+1} by adding a new element $y \in \mathcal{S}$ to $\mathcal{S}_s = \operatorname{range}(\alpha_s)$. Since M is a correspondence system containing α_s, there exists $A_j \in \mathfrak{A}$, $f \in I_{ij}^{\mathfrak{A}}$, and a map $\beta \in M$ such that β maps $\mathcal{A}_j$ onto the substructure of $\mathcal{S}$ generated by y and $\mathcal{S}_s$, with $\beta \circ f = \alpha_s$. Now since $f \in I^{\mathfrak{A}}$, there is some stage $s' > t$ at which f is incorporated into $\mathcal{B}$. When this happened, we defined an embedding $f_{s'} = p \circ f : \mathcal{B}_t \hookrightarrow \mathcal{B}_{s'+1}$, with $p \in I_{jm}^{\mathfrak{A}}$ for some m, and with $f_{s'} \in I_{im}^{\mathfrak{A}}$ because $I^{\mathfrak{A}}$ is closed under composition. But since β lies in the correspondence system M, there exists $\gamma \in M$ with domain $\mathcal{B}_{s'+1}$ such that $\beta = \gamma \circ p$. We define $\alpha_{s+1} = \gamma$, with domain $\mathcal{B}_{s'+1}$ and let $\mathcal{S}_{s+1} = \operatorname{range}(\alpha_{s+1})$, noting that for $x \in \operatorname{dom}(\alpha_s)$,

$$\alpha_{s+1}(g_{s'} \circ \cdots \circ g_t(x)) = \gamma(p \circ f(x)) = \beta(f(x)) = \alpha_s(x).$$

Finally, $y \in \operatorname{range}(\beta) \subseteq \operatorname{range}(\gamma) = \operatorname{range}(\alpha_{s+1})$. This completes the construction. Notice that for all s, $\operatorname{dom}(\alpha_s) = \mathcal{B}_t \supseteq \mathcal{B}_s$, so that $\operatorname{dom}(\alpha)$ is all of $\mathcal{B}$, and $\operatorname{range}(\alpha) = \mathcal{S}$ by construction. Thus $\mathcal{S}$ and $\mathcal{B}$ really are isomorphic, and $\mathcal{S}$ is computably presentable. ⊣

Recently Franklin, Kach, Miller, and Solomon, in [5], have considered the local computability properties of ordinals (viewed as linear orders). They argue there that the Amalgamation Property (Definition 6.1) is so natural and useful that it should be included in the definition of a cover. In this case, the key result about extensionality, Theorem 3.10, would still hold, since a uniformly computable cover with the AP is still a uniformly computable cover. However, every ∞-extensional cover would automatically be perfect, and so, for countable structures, having an ∞-extensional cover would be equivalent to computable presentability. They also make a philosophical point about Definition 2.9: if the point of the embeddings is to explain how any two finitely generated substructures $\mathcal{B}$ and $\mathcal{C}$ of $\mathcal{S}$ "fit together," then both the AP and a similar *Joint Embedding Property* (as described by Fraïssé; see e.g. [8, Section 6.1]) should be required. When we know an $\mathcal{A}_j$ and an $\mathcal{A}_k$ in $\mathfrak{A}$ isomorphic to $\mathcal{B}$ and $\mathcal{C}$ respectively, Definition 6.1 does not ensure that we can find any amalgamation of $\mathcal{A}_j$ and $\mathcal{A}_k$ within our cover: the substructure $\mathcal{D} \subseteq \mathcal{S}$ generated by $\mathcal{B}$ and $\mathcal{C}$ together certainly yields $\mathcal{A}_m$ and $\mathcal{A}_n$ in $\mathfrak{A}$, both isomorphic to $\mathcal{D}$, and embeddings $f \in I^{\mathfrak{A}}_{jm}$ and $g \in I^{\mathfrak{A}}_{kn}$ which make the corresponding diagrams commute, but unless $n = m$, this tells us effectively nothing about how $\mathcal{B}$ and $\mathcal{C}$ actually fit together within $\mathcal{S}$. With the JEP, we would be certain of being able to find a single m and embeddings of both $\mathcal{A}_j$ and $\mathcal{A}_k$ into $\mathcal{A}_m$. Likewise, if we had embeddings of one $\mathcal{A}_i$ into both an $\mathcal{A}_j$ and an $\mathcal{A}_k$, the AP would ensure that we could figure out how these two extensions of $\mathcal{A}_i$ fit together over $\mathcal{A}_i$ – which is to say, how the corresponding substructures of $\mathcal{S}$ fit together over the orgiinal substructure isomorphic to $\mathcal{A}_i$. It is quite possible that the notion of a uniformly computable cover with AP (and perhaps JEP as well) will come in time to supplant Definition 2.9 as the basis for local computability.

§7. Computable Simulations. We now show that perfectly locally computable structures can be "simulated" in a strong way by globally computable (and hence countable) structures. This can be viewed either as an indictment of perfect local computability, saying that it is such a strong condition that the only uncountable structures satisfying it are those which are very closely related to computable structures; or as further evidence that perfect local computability is the correct analogue, in the uncountable setting, to computable presentability in the countable setting. We leave this judgment to the reader. Work in this section is joint between Dustin Mulcahey and the author.

Definition 7.1. Let $\mathcal{S}$ be any structure. A *simulation of* $\mathcal{S}$ is an elementary substructure $\mathcal{B} \preceq \mathcal{S}$ such that $\mathcal{B}$ and $\mathcal{S}$ realize exactly the same finitary types. We often refer to any $\mathcal{A}$ isomorphic to such a $\mathcal{B}$ as a simulation of $\mathcal{S}$, even if $\mathcal{A}$ is not itself a substructure of $\mathcal{S}$. Hence a *computable simulation of* $\mathcal{S}$ is a computable structure isomorphic to a simulation of $\mathcal{S}$.

Lemma 7.2. *Let $\mathcal{S}$ be locally computable, with a correspondence system N over a uniformly computable cover $\mathfrak{A}$. Then $\mathcal{S}$ has a countable substructure $\mathcal{B}$ for which $\mathfrak{A}$ is also a cover, and which has its own correspondence system $M \subseteq N$ over $\mathfrak{A}$. If N was a perfect correspondence system for $\mathcal{S}$, then M is perfect for $\mathcal{B}$ as well.*

Proof $\mathcal{B}$ will be a countable union of countable substructures $\mathcal{B}_s$ of $\mathcal{S}$. To start, we fix for each $i \in \omega$ one map $\alpha_i \in N$ with domain $\mathcal{A}_i$, Let $M_0 = \{\alpha_i : i \in \omega\}$, and let $\mathcal{B}_0$ be the substructure of $\mathcal{S}$ generated by the union of all the images of these α_i. The conditions for a perfect cover are $\forall\exists$ conditions, so now we will be able to keep $\mathcal{B}$ countable as we close M under those conditions, using the analogous conditions in the correspondence system N.

Assume we have defined a countable $\mathcal{B}_s$ and M_s. First, for every i, every $\alpha \in M_s$ with domain $\mathcal{A}_i$, and every $f \in I_{ij}^{\mathfrak{A}}$ (for any j), there exists some $\gamma \in N$ with domain $\mathcal{A}_j$ such that f lifts via α and γ to the inclusion $\alpha(\mathcal{A}_i) \subseteq \gamma(\mathcal{A}_j)$. Form $M'_s \supseteq M_s$ by adjoining one such γ to M_s for each such i, α and f. Also, let $\mathcal{B}'_s$ be generated by the union of the images of the maps in M'_s. Clearly both $\mathcal{B}'_s$ and M'_s remain countable.

Next, for every i, every $\alpha \in M_s$ with domain $\mathcal{A}_i$, and every finitely generated $\mathcal{C} \subseteq \mathcal{B}_s$ with $\alpha(\mathcal{A}_i) \subseteq \mathcal{C}$, there exists some j, some $f \in I_{ij}^{\mathfrak{A}}$, and some $\gamma \in N$ with domain $\mathcal{A}_j$ such that f lifts to the inclusion $\alpha(\mathcal{A}_i) \subseteq \mathcal{C}$ via α and γ. Adjoin to M'_s one such γ for each such i, α, and $\mathcal{C}$, to form M''_s.

Finally, for every finitely generated substructure $\mathcal{C} \subseteq \mathcal{B}_s$, there exists a $\gamma \in N$ with image $\mathcal{C}$ (since $\mathcal{C} \subseteq \mathcal{S}$). Form M_{s+1} by adjoining to M''_s one such γ for each such $\mathcal{C}$. Since $\mathcal{B}_s$ was countable, it has only countably many finitely generated substructures, and so M_{s+1} is still countable.

It is clear that the union $\mathcal{B} = \bigcup_s \mathcal{B}_s$ is a countable substructure of $\mathcal{S}$, with cover $\mathfrak{A}$, and that $M = \bigcup_s M_s$ is a correspondence system for this $\mathcal{B}$ over $\mathfrak{A}$. Our $\mathcal{B}_0$ already satisfied item (2) of Definition 4.1, and our ensuing adjoinments satisfied (4), (5), and (3), in that order, without ever violating (1). (Of course $\mathfrak{A}$ is still uniformly computable as well; that definition has nothing to do with the structure covered by $\mathfrak{A}$.)

It remains to see that this M is perfect for $\mathcal{B}$ whenever N is perfect for $\mathcal{S}$. But this is easy: if α and γ lie in M and have the same image in $\mathcal{B}$, then they lie in N and have the same image in $\mathcal{S}$. Since N is perfect, $\gamma^{-1} \circ \alpha$ must then lie in the appropriate $I_{ij}^{\mathfrak{A}}$. ⊣

In this situation, $\mathcal{B}$ will be an elementary substructure of $\mathcal{S}$. The next lemma extends this observation. (If $\mathcal{B}$ is as in Lemma 7.2, and P is empty, then in the proof of Lemma 7.3 we may show that at every step ψ_s is just inclusion.)

Lemma 7.3. *Let $\mathcal{B}$ and $\mathcal{S}$ be two structures, each with a correspondence system over the same uniformly computable cover. Assume that $\mathcal{B}$ is countable. Then $\mathcal{B}$ is a simulation of $\mathcal{S}$. Indeed, for any countable set $P \subseteq \mathcal{S}$ of parameters, we can elementarily embed $\mathcal{B}$ into $\mathcal{S}$ so that its image contains P and realizes the same finitary types as $\mathcal{S}$ over every finite $P_0 \subseteq P$.*

Proof Let $\mathfrak{A}$ be a common uniformly computable cover of $\mathcal{S}$ and $\mathcal{B}$, with correspondence systems M for $\mathcal{B}$ and N for $\mathcal{S}$. Our embedding is built step by step, so we start by enumerating the domain of $\mathcal{B}$ as $\{b_0, b_1, \dots\}$, and P as $\{p_0, p_1, \dots\}$. Fix an $\alpha \in M$ whose image is the substructure $\mathcal{B}_0 \subseteq \mathcal{B}$ generated by b_0, and a $\gamma \in N$ with the same domain as α, and define ψ_0 to be $\gamma \circ \alpha^{-1}$, with $\mathcal{B}_0 = \mathrm{dom}(\psi_0) \subseteq \mathcal{B}$ and $\mathcal{C}_0 = \mathrm{range}(\psi_0) \subseteq \mathcal{S}$.

At stage $t+1 = 2s+1$, we extend ψ_t so that its range contains p_s. By induction $\psi_t = \gamma \circ \alpha^{-1}$ for some $\gamma \in N$ and $\alpha \in M$ with common domain $\mathcal{A}_i$ in $\mathfrak{A}$. Let $\mathcal{C}_{t+1}$ be the substructure of $\mathcal{S}$ generated by $\mathcal{C}_t$ and p_s. By induction $\mathcal{C}_t$ is finitely generated, so there is a $\delta \in N$ with some domain $\mathcal{A}_j \in \mathfrak{A}$, and an $f \in I^{\mathfrak{A}}_{ij}$, such that f lifts via γ and δ to the inclusion $\mathcal{C}_t \subseteq \mathcal{C}_{t+1}$. In turn there is a $\beta \in M$ with domain $\mathcal{A}_j$ such that f lifts via α and β to the inclusion $\mathcal{B}_t \subseteq \beta(\mathcal{A}_j)$. Set $\mathcal{B}_{t+1} = \text{range}(\beta)$ and $\psi_{t+1} = \delta \circ \beta^{-1}$.

At stage $t+1 = 2s+2$, we extend the embedding ψ_t from its current domain $\mathcal{B}_t$ to the structure $\mathcal{B}_{t+1}$ generated by $\mathcal{B}_t$ and b_s. By induction $\mathcal{B}_t$ is finitely generated, and $\psi_t = \gamma \circ \alpha^{-1}$ for some $\gamma \in N$ and $\alpha \in M$ with common domain $\mathcal{A}_i$ in $\mathfrak{A}$. So there is a $\beta \in M$ with some domain $\mathcal{A}_j \in \mathfrak{A}$, and an $f \in I^{\mathfrak{A}}_{ij}$, such that f lifts via α and β to the inclusion $\mathcal{B}_t \subseteq \mathcal{B}_{t+1}$. In turn there is a $\delta \in M$ with domain $\mathcal{A}_j$ such that f lifts via γ and δ to the inclusion $\mathcal{C}_t \subseteq \delta(\mathcal{A}_j)$. Set $\mathcal{C}_{t+1} = \text{range}(\delta)$ and $\psi_{t+1} = \delta \circ \beta^{-1}$.

Now we define $\psi = \bigcup_t \psi_t$. Clearly ψ has domain $\mathcal{B}$ and range $\subseteq \mathcal{S}$ containing P, and ψ must be an embedding. To see that it is elementary, suppose that $\exists x \theta(\psi(b_0), \dots, \psi(b_s), x)$ is an existential formula true in $\mathcal{S}$. Now $\psi_s = \gamma \circ \alpha^{-1}$ for some $\alpha \in M$ and $\gamma \in N$ with common domain $\mathcal{A}_i \in \mathfrak{A}$. Since N is perfect, there is a $\delta \in N$ (with some domain $\mathcal{A}_j$) and an $a \in \mathcal{A}_j$ and an $f \in I^{\mathfrak{A}}_{ij}$, such that $\theta(f(\alpha^{-1}(b_0)), \dots, f(\alpha^{-1}(b_s)), a)$ holds in $\mathcal{A}_j$. But since M is also a perfect cover, there is a $\beta \in M$ with the same domain $\mathcal{A}_j$ such that f lifts to the inclusion $\mathcal{B}_s \subseteq \beta(\mathcal{A}_j)$ via α and β. Therefore $\theta(b_0, \dots, b_s, \beta(a))$ holds in $\mathcal{B}$. Thus ψ is an elementary embedding.

Finally, given any n-type Γ over any finite parameter set $P_0 \subseteq P$, such that Γ is realized in $\mathcal{S}$ by a tuple $(d_1, \dots, d_n)$, we start with the substructure $\mathcal{P}_0 \subseteq \mathcal{S}$ generated by P_0. Since $P_0 \subseteq \text{range}(\psi)$, we have a t for which $P_0 \subseteq \text{range}(\psi_t)$. Let $\psi_t = \gamma \circ \alpha^{-1}$ with $\alpha \in M$ and $\gamma \in N$. There must be a $\delta \in N$ with some domain $\mathcal{A}_j \in \mathfrak{A}$ and an $f \in I^{\mathfrak{A}}_{ij}$ such that f lifts via γ and δ to the inclusion of $\mathcal{P}_0$ into the substructure generated by $\mathcal{P}_0$ and $d_0, \dots, d_n$. But now there is also some $\beta \in M$ with domain $\mathcal{A}_j$ such that f lifts via α and β to the inclusion $\mathcal{B}_t \subseteq \beta(\mathcal{A}_j)$, and we set $b_i = \beta(\delta^{-1}(d_i))$ and $c_i = \psi(b_i)$ for each i. Then $(c_1, \dots, c_n)$ is an n-tuple within the image of ψ which realizes the type Γ over P_0, by standard arguments using M and N. ⊣

When we have a parameter set P as in Lemma 7.3, we refer to the image of $\mathcal{B}$ as a *simulation of* $\mathcal{S}$ *over* P. We might also refer to $\mathcal{B}$ itself the same way, but only when the embedding $\psi : \mathcal{B} \hookrightarrow \mathcal{S}$ is clear, because we need to know which elements $\psi^{-1}(p) \in \mathcal{B}$ correspond to the elements of P in this simulation. Later we will discuss the extent to which $\mathcal{B}_P$ can be said to be uniform in P.

Corollary 7.4. *Two countable structures with correspondence systems over the same uniformly computable cover are isomorphic.*

Proof Since $\mathcal{S}$ is countable, we simply set $P = \mathcal{S}$ and apply Lemma 7.3, whose proof may now be regarded as a back-and-forth construction of an isomorphism from $\mathcal{B}$ onto $\mathcal{S}$. ⊣

We are now ready for the main result of this section.

THEOREM 7.5. *Every perfectly locally computable structure $\mathcal{S}$ has a computable simulation $\mathcal{A}$, which can be embedded into $\mathcal{S}$ so as to simulate $\mathcal{S}$ over arbitrary countable parameter sets. Specifically, there is a set of elementary embeddings $\psi_p : \mathcal{A} \hookrightarrow \mathcal{S}$, one for each function $p : \omega \to \mathcal{S}$ which enumerates a countable parameter set $Q_p = range(p) \subseteq \mathcal{S}$, such that:*

- $Q_p \subseteq \psi_p(\mathcal{A})$*; and*
- $\psi_p(\mathcal{A})$ *is a simulation of $\mathcal{S}$ over Q_p; and*
- *if p and p' are two such functions and $p\restriction n = p'\restriction n$, then for all $k < n$,*

$$\psi_p^{-1}(p(k)) = \psi_{p'}^{-1}(p'(k)).$$

As a partial converse, every structure which has a computable simulation $\mathcal{A}$ with embeddings ψ_p satisfying these conditions has a uniformly computable cover with a correspondence system.

To make this last claim an actual converse, we would need to show that the correspondence system for $\mathcal{S}$ is perfect. Whether this is true remains open. We also note that it would be equivalent to give the same statement only for finite parameter enumerations p, since the last condition would allow a simulation over a countable parameter set P to be built by taking successive nested finite enumerations $p_m \subseteq p_{m+1}$ with $P = \bigcup_m \text{range}(p_m)$, and setting $\psi = \lim_m \psi_{p_m}$

Proof When we assume that $\mathcal{S}$ is perfectly locally computable, the existence of a computable simulation of $\mathcal{S}$ follows from Lemma 7.2, which also ensures that $\mathcal{S}$ and its computable simulation $\mathcal{A}$ both have perfect correspondence systems over the same uniformly computable cover. Therefore Lemma 7.3 shows that $\mathcal{A}$ can be elementarily embedded into $\mathcal{S}$ so as to simulate $\mathcal{S}$ over any parameter set Q_p enumerated by a function $p : \omega \to \mathcal{S}$. Moreover, an examination of the proof of Lemma 7.3 shows that the embedding chooses the $a \in \mathcal{A}$ with $\psi_p(a) = p(k)$ using only the common cover $\mathfrak{A}$, its correspondence systems for $\mathcal{A}$ and $\mathcal{S}$, and the elements $p(0), p(1), \ldots, p(k)$ in $\mathcal{S}$. This proves the claim about parameter enumerations p and p' which agree up to n.

Next we show our partial converse: that the existence of such an A implies that $\mathcal{S}$ has a uniformly computable cover with a correspondence system. $\mathcal{A}$ has a perfect cover $\mathfrak{A}$, by Theorem 6.3. Let M be a perfect correspondence system for $\mathcal{A}$ and $\mathfrak{A}$. The correspondence system N will consist of all maps of the form $\psi_p \circ \alpha$, for all finite $p : n \to \mathcal{S}$ and all $\alpha \in M$ such that range(α) is generated by $\{\psi_p^{-1}(p(i)) : i < n\}$. (Here we think of a finite function $p : n \to \mathcal{S}$ as a function from ω into $\mathcal{S}$ by repeating its image over and over: $p(k+nm) = p(k)$ for all k and m.)

Now each finitely generated $\mathcal{C} \subseteq \mathcal{S}$ with generators enumerated by p lies within the image of ψ_p, and the finitely generated substructure $\psi_p^{-1}(\mathcal{C}) \subseteq \mathcal{A}$ must be the image of some $\mathcal{A}_i \in \mathfrak{A}$ under some $\alpha \in M$, since M is a perfect cover of $\mathcal{A}$. Hence $\mathcal{C} = (\psi_p \circ \alpha)(\mathcal{A}_i)$ is the image of some map in N. Likewise, each $\mathcal{A}_i \in \mathfrak{A}$ is the domain of some $\alpha \in M$, hence also of some map in N. Moreover, each $f \in I_{ij}^{\mathfrak{A}}$, for any i and j, lifts to an inclusion map within $\mathcal{A}$, and then lifts further to an inclusion map within $\mathcal{S}$, via any ψ_p we like. Conversely, any inclusion $\mathcal{C}' \subseteq \mathcal{C}$ of finitely-generated substructures of $\mathcal{S}$ is the lift (via ψ_p, where p enumerates first the generators of $\mathcal{C}'$, and then the generators of $\mathcal{C}$) of an inclusion in $\mathcal{A}$, which in turn is the lift of some f in some $I_{ij}^{\mathfrak{A}}$ via some $\alpha, \beta \in M$. If p' is the restriction of p to the generators of $\mathcal{C}'$, then the inclusion $\mathcal{C}' \subseteq \mathcal{C}$ is the lift of f via $(\psi_{p'} \circ \alpha)$ and $(\psi_p \circ \beta)$, which both lie in N. Thus $\mathfrak{A}$ is a uniformly computable cover of $\mathcal{S}$.

The preceding remarks also proved the first three conditions in Definition 4.1. For Condition (5), fix any $f \in I_{ij}^{\mathfrak{A}}$ for any i and j, along with any $\beta \in N$ with domain $\mathcal{A}_i$. Then $\beta = \psi_p \circ \alpha$ for some $\alpha \in M$ and some $p : n \to \mathcal{S}$ for which $\{\psi_p^{-1}(p(k)) : k < n\}$ generates range(α). Since M is a correspondence system, there is a $\gamma \in M$ with domain $\mathcal{A}_j$ such that $\alpha = \gamma \circ f$. But now there is a finite q such that $q \restriction n = p$ and $q(n + k) = \psi_p(a_k)$, where $a_0, \ldots, a_m$ generate $\gamma(\mathcal{A}_j)$ within $\mathcal{A}$. So $(\psi_q \circ \gamma) \in N$, and

$$(\psi_q \circ \gamma \circ f) = (\psi_q \circ \alpha) = (\psi_p \circ \alpha),$$

with the last equality following because $p \restriction n = q \restriction n$ and range(α) is generated by the elements $\psi_p^{-1}(p(k)) = \psi_q^{-1}(q(k))$ for $k < n$. This proves Condition (5).

For Condition (4) of Definition 4.1, fix any $\beta \in N$ with domain $\mathcal{A}_i$ and any finitely generated $\mathcal{C} \subseteq \mathcal{S}$ with $\beta(\mathcal{A}_i) \subseteq \mathcal{C}$. Now $\beta = \psi_p \circ \alpha$ for some $\alpha \in M$ and some finite $p : n \to \mathcal{S}$, with the elements $\psi_p^{-1}(p(k))$ generating range(α). Let $q(k) = p(k)$ for $k < n$, and let $q(n), \ldots, q(n + m - 1)$ enumerate the generators of $\mathcal{C}$ in $\mathcal{S}$. By assumption, ψ_q is an elementary embedding of $\mathcal{A}$ into $\mathcal{S}$ whose image contains range(q). Let $\mathcal{D} = \langle \psi_q^{-1}(q(k)) : k < m \rangle \subseteq \mathcal{A}$. Since $\psi_q^{-1}(q(k)) = \psi_p^{-1}(p(k))$ for all $k < n$, we know that $\alpha(\mathcal{A}_i) \subseteq \mathcal{D}$, and since M is a correspondence system, there is some $\beta \in M$ and some j and $f \in I_{ij}^{\mathfrak{A}}$ with $\mathcal{D} = \text{range}(\beta)$ and $\beta \circ f = \alpha$. But then

$$(\psi_q \circ \beta \circ f) = (\psi_q \circ \alpha) = (\psi_p \circ \alpha),$$

proving Condition (4), since $(\psi_q \circ \beta) \in N$. Thus $\mathcal{A}$ is a uniformly computable cover of $\mathcal{S}$ with correspondence system N. ⊣

We can think of $\mathcal{B}_P$ as being built uniformly in the parameter set P if the elements of P are named as elements in different $\mathcal{A}_i$ in the cover $\mathfrak{A}$ of $\mathcal{S}$. That is, suppose that we are given a computable enumeration $\langle (i_k, a_k, f_k) \rangle_{k \in \omega}$ for which there exist maps $\beta_k \in N$ with $a_k \in \mathcal{A}_{i_k} = \text{dom}(\beta_k)$ such that

- each $f_k \in I_{i_k, i_{k+1}}^{\mathfrak{A}}$; and
- $\beta_{k+1} \circ f_k = \beta_k$; and

- $\{\beta_k(a_k) : k \in \omega\} = P$.

Then we could build a computable copy of the simulation $\mathcal{B}_P$ of $\mathcal{S}$ over P, uniformly in the perfect cover of $\mathcal{S}$ and the enumeration $\langle(i_k, a_k, f_k)\rangle_{k\in\omega}$, and enumerate the image of P in $\mathcal{B}_P$. More generally, if the enumeration $\langle(i_k, a_k, f_k)\rangle_{k\in\omega}$ has Turing degree $\boldsymbol{d}$, then with a $\boldsymbol{d}$-oracle we can build a copy of $\mathcal{B}_P$ in which the image of P will be computably enumerable in $\boldsymbol{d}$. It is awkward to think of the set P itself as having Turing degree $\boldsymbol{d}$, because an infinite set P will have distinct enumerations with distinct Turing degrees, but within the cover $\mathfrak{A}$ of $\mathcal{S}$, we can view P as being computably enumerable in $\boldsymbol{d}$, as well as in the degrees of other enumerations. Of course, P itself, viewed as a subset of $\mathcal{S}$, does not admit effective enumeration in any obvious way.

It is immediate from Definition 4.1 that if M is a correspondence system for a cover $\mathfrak{A}$ of $\mathcal{S}$, then likewise M is a correspondence system for the cover $\overline{\mathfrak{A}}$ defined in Section 5. in which we enumerate the identity map on each $\mathcal{A}_i$ into the appropriate set $I_{ii}^{\mathfrak{A}}$ and then close the set $I^{\mathfrak{A}}$ under composition. That is, for every $f \in I_{ij}^{\mathfrak{A}}$ and $g \in \mathcal{I}_{jk}^{\mathfrak{A}}$, we enumerate $(g \circ f)$ into $\mathcal{I}_{ik}^{\overline{\mathfrak{A}}}$. If $\mathfrak{A}$ is uniformly computable, then so is $\overline{\mathfrak{A}}$, since we can build $\overline{\mathfrak{A}}$ from $\mathfrak{A}$ just by applying this rule as we enumerate the sets $I_{ij}^{\overline{\mathfrak{A}}}$. ($\overline{\mathfrak{A}}$ has the same simple cover of $\mathcal{S}$ that $\mathfrak{A}$ has, of course.) It is clear that M is now a correspondence system for the derived cover as well. Moreover, this derived cover $\mathfrak{A}'$ may be viewed as a category, with the elements $\mathcal{A}_i$ of the simple cover as the objects, and with each $I_{ij}^{\mathfrak{A}'}$ as the set of morphisms from $\mathcal{A}_i$ into $\mathcal{A}_j$.

§8. Examples. Several examples, mainly involving fields, have been given in the preceding sections. In particular, see Proposition 2.3, Proposition 2.8, Proposition 3.3, Proposition 3.5, and Corollary 4.3. Now we provide an assortment of further examples.

As an example of perfect local computability, we propose the linear order $\mathcal{L}$ on Cantor space. The domain of $\mathcal{L}$ is the uncountable set 2^ω, with the relation $<$ being simply the lexicographic order. This structure is well known in mathematics, and fairly straightforward to describe via the "middle thirds" construction, but we know (from Proposition 3.5, for instance) that such characteristics do not always ensure even 1-extensional local computability.

Of course, a subset $S \subseteq 2^\omega$ generates only the substructure $(S, <)$, and so the finitely generated substructures of $\mathcal{L}$ are just the finite linear orders. (This would hold with any infinite linear order in place of $\mathcal{L}$.) So it is trivial to show that $\mathcal{L}$ is locally computable. Building a perfect cover $\mathfrak{A}$, on the other hand, will require some description.

Consider the larger signature containing the relation $<$, one other binary relation G, and four unary relations L, R, L_G, and R_G. The intuition is that we use L and R to designate the left and right end points of the order, and L_G and R_G to name left and right end points of the "gaps" in $\mathcal{L}$, i.e. the end points of the missing middle thirds. Gxy holds iff x and y are the left and right end points of the same gap. One can extend $\mathcal{L}$ itself to be a structure in this signature, indeed with the new relations all definable from $<$: $G^{\mathcal{L}}$ is just the adjacency relation in $\mathcal{L}$, while $L_G^{\mathcal{L}}x$ holds iff $(\exists y)G^{\mathcal{L}}xy$ and $R_G^{\mathcal{L}}x$ iff $(\exists y)G^{\mathcal{L}}yx$. $L^{\mathcal{L}}x$ and $R^{\mathcal{L}}y$ hold only of the left and right end points x and y of the entire order, respectively. We use this new signature to build our cover.

First, for any finite structure $\mathcal{B}$ in the new signature, it is decidable whether $\mathcal{B}$ satisfies the following axioms.

1. $(\mathcal{B}, <)$ is a linear order.
2. For each $x \in \mathcal{B}$, at most one of Lx, Rx, L_Gx, or R_Gx holds.
3. $(\forall x)[(Lx \to \forall y\, x \leq y)\ \&\ (Rx \to \forall y\, y \leq x)]$.
4. $(\forall x \forall y)[Gxy \to (x < y\ \&\ L_Gx\ \&\ R_Gy\ \&\ \forall z \neg(x < z < y))]$.

The idea is that all finite substructures of $\mathcal{L}$ should satisfy these axioms. We do not require $\mathcal{B}$ to have any x with $L^{\mathcal{B}}x$, nor any y with $R^{\mathcal{B}}y$, because an arbitrary finite subset of $\mathcal{L}$ will not necessarily contain the end points of $\mathcal{L}$. Similarly, it is allowed for $L_G^{\mathcal{B}}x$ to hold even if there is no $y \in \mathcal{B}$ with $G^{\mathcal{B}}xy$, and likewise for $R_G^{\mathcal{B}}$.

Now we may compute, uniformly, a list $\mathcal{B}_0, \mathcal{B}_1, \ldots$ (with no repetitions) of all models $\mathcal{B}$ of these axioms such that the domain of $\mathcal{B}$ is an initial segment of ω and $<^{\mathcal{B}}$ is just the standard relation $<$ on that domain. The simple cover $\mathfrak{A}$ consists of all $\mathcal{A}_i$ with $i \in \omega$, where $\mathcal{A}_i$ is the reduct of $\mathcal{B}_i$ to the signature with just $<$. (So in fact $\mathcal{A}_i$ is just the linear order $0 < 1 < \cdots < |\mathcal{A}_i| - 1$.) Of course, the generating set of $\mathcal{A}_i$ is its entire domain, and clearly this does form a uniformly computable simple cover of $\mathcal{L}$.

The maps in $I_{ij}^{\mathfrak{A}}$ will be defined as embeddings of $\mathcal{B}_i$ into $\mathcal{B}_j$, since such an embedding is clearly also an embedding $\mathcal{A}_i \hookrightarrow \mathcal{A}_j$. We enumerate into $I_{ij}^{\mathfrak{A}}$ all strong homomorphisms $f : \mathcal{B}_i \to \mathcal{B}_j$ in the (larger) signature of these structures. Such an f must be injective, since it is a homomorphism of strict linear orders, and must also satisfy $\forall x(L^{\mathcal{B}_i}x \leftrightarrow L^{\mathcal{B}_j}f(x))$ and $(\forall x \forall y)[G^{\mathcal{B}_i}xy \to G^{\mathcal{B}_j}f(x)f(y)]$, and likewise for the other symbols in the larger signature.. (However, it is allowed for $G^{\mathcal{B}_j}$ to hold of the pair $\langle f(x), z\rangle$, even if there was no $y \in \mathcal{B}_i$ such that $G^{\mathcal{B}_i}$ held of $\langle x, y\rangle$. On the other hand, if such a y did exist, then $G^{\mathcal{B}_j}$ holds of $\langle f(x), f(y)\rangle$, by our rules above, and the axioms for G and $<$ then ensure that $z = f(y)$.)

The perfect correspondence system M will contain all maps $\alpha : \mathcal{A}_i \to \mathcal{L}$ which are homomorphisms when viewed as maps from $\mathcal{B}_i$ into $\mathcal{L}$ in the larger signature (to which $\mathcal{L}$ was extended above). That is, they must satisfy:

- $\alpha(x) = 0000\cdots$ iff $L^{\mathcal{B}_i}(x)$.
- $\alpha(x) = 1111\cdots$ iff $R^{\mathcal{B}_i}(x)$.
- $\alpha(x)$ has tail $0000\cdots$ (that is, $\alpha(x)$ contains only finitely many 1's) iff $R_G^{\mathcal{B}_i}(x)$ or $L^{\mathcal{B}_i}(x)$.
- $\alpha(x)$ has tail $1111\cdots$ iff $L_G^{\mathcal{B}_i}(x)$ or $R^{\mathcal{B}_i}(x)$.

- $G^{\mathcal{B}_i}(x, y)$ iff there is some $\sigma \in 2^{<\omega}$ with $\alpha(x) = \sigma\hat{\ }01111\cdots$ and $\alpha(y) = \sigma\hat{\ }10000\cdots$.

Now for any finite substructure $\mathcal{C}$ of $\mathcal{L}$, there is a unique i such that an $\alpha \in M$ maps $\mathcal{A}_i$ onto $\mathcal{C}$. The elements of $\mathcal{B}_i$ must be $0, 1, \ldots, |\mathcal{C}| - 1$, under the usual $<$ relation, and the other relations on $\mathcal{B}_i$ are determined by these conditions. Our enumeration $\mathcal{B}_0, \mathcal{B}_1, \ldots$ clearly must include some such $\mathcal{B}_i$, and conversely, it is uniquely determined by the choice of $\mathcal{C}$. On the other hand, given any i, the conditions above show how to pick out a substructure $\mathcal{C} \subseteq \mathcal{L}$ and a bijection $\alpha \in M$ from $\mathcal{B}_i$ onto $\mathcal{C}$. There will almost always be more than one such $\mathcal{C}$, of course: $\mathcal{C}$ is uniquely determined only if every $x \in \mathcal{B}_i$ satisfies either $L^{\mathcal{B}_i}$ or $R^{\mathcal{B}_i}$. This shows that our M satsifies the first three conditions of Definition 4.1, and also satisfies the Uniqueness Condition described subsequently.

Now suppose that $\mathcal{C} \subset \mathcal{D}$ are finite substructures of $\mathcal{L}$, and that $\alpha \in M$ with image $\mathcal{C}$. As argued above, there is a unique $\beta \in M$ with image $\mathcal{D}$, and $(\beta^{-1} \circ \alpha)$ defines a homomorphism (in the larger signature) from $\mathcal{B}_i = \mathrm{dom}(\alpha)$ onto $\mathcal{B}_j = \mathrm{dom}(\beta)$. This homomorphism must lie in $I_{ij}^{\mathfrak{A}}$, so Condition 4 is fulfilled. Conversely, if we have $\alpha \in M$, with domain $\mathcal{A}_i$ and image $\mathcal{C}$, and are given j and $f \in I_{ij}^{\mathfrak{A}}$, then we define $\beta : \mathcal{A}_j \to \mathcal{L}$ as follows, starting with $\beta \upharpoonright f(\mathcal{A}_i) = \alpha \circ f^{-1}$. For each element $x \in \mathcal{A}_j - f(\mathcal{A}_i)$, starting with the smallest, we ask first whether there is a $y \in \mathcal{A}_j$ such that $G^{\mathcal{B}_j}$ holds of $\langle x, y\rangle$ or of $\langle y, x\rangle$. If so, then either $\beta(x)$ is determined by $\alpha(f^{-1}(y))$ (if $y \in f(\mathcal{A}_i)$), or else we choose $\beta(x)$ and $\beta(y)$ to be the end points of an appropriate gap in $\mathcal{L}$. If there is no such y, then we simply choose $\beta(x)$ in the appropriate interval in $\mathcal{L}$ satisfying either L, R, L_G, R_G, or none of the above there, according to the relation satisfied by x in $\mathcal{B}_j$, but with $\neg G^{\mathcal{L}}xz$ and $\neg G^{\mathcal{L}}zx$ for each z already in the image of β. This β, and its image $\beta(\mathcal{A}_j) \subseteq \mathcal{L}$, satisfy Condition 5 in Definition 4.1, so M is indeed a correspondence system. Moreover, we have already noted that M satisfies the Uniqueness Condition, hence must be perfect.

The main point of this example is that building a cover of a structure $\mathcal{S}$, especially a perfect cover (or at least a highly extensional cover), usually requires us to sort out exactly what the important attributes of the various elements of $\mathcal{S}$ may be. In this case, those attributes mainly involved the gaps in $\mathcal{L}$: being the left or right end point of a gap, first of all, and recognizing the corresponding right or left end point, if this was the case. If we had used the same unary relations but omitted G from the signature, then we could have built a 1-extensional cover, but not a 2-extensional cover, since recognizing the corresponding right or left end point requires two quantifiers. Cantor space is sufficiently homogeneous that we can do all of this effectively. Of course, it could be far more difficult to build a 1- or 2-extensional cover, let alone a perfect cover, for a suborder of $\mathcal{L}$ in which certain end points of gaps were removed, while others remained.

Likewise, in the case of algebraically closed fields of characteristic 0, building a perfect cover requires recognizing the essential attributes of elements of the ACF:

their algebraic relationships to each other and to the ground field $\mathbb{Q}$. We invite the reader to build extensional covers of ordinals: by doing so, he or she will find that the salient attribute of elements there is being a limit ordinal, and in particular the specific limit level, as described by the Cantor-Bendixson rank (e.g. ω^2 is a limit of limits, ω^3 a limit of limits of limits, etc.). The extent to which one must worry about the Cantor-Bendixson rank depends on the level of extensionality one demands of the cover. Again, the process of finding a highly extensional cover leads one to an understanding of the most relevant characteristics of elements of the structure.

The odd fact is that for countable structures, this is not so much the case. As shown in Theorem 6.3, every (globally) computable structure has a perfect cover, yet there this cover is built not by recognizing specific aspects or attributes to be described. The construction of the cover there comes directly from the computable presentability of the original stucture, wherein all those attributes are wrapped up.

By Theorem 7.5, there is a computable simulation of the linear order $\mathcal{L}$ of Cantor space. We omit the details here, but this simulation may be envisioned by taking the Cantor middle-thirds set C within the real unit interval $[0, 1]$ and intersecting C with $\mathbb{Q}$. This linear order is soon seen to be computably presentable: one way is to present $(\mathbb{Q}, <)$ computably and then take the suborder of those rationals with ternary expressions using only the digits 0 and 2; while another presentation can be given by a computable set of pairs $\langle\sigma, \tau\rangle \in (2^{<\omega} \times 2^{<\omega})$, with $\langle\sigma, \tau\rangle$ representing the string $\sigma\hat{}\tau\hat{}\tau\hat{}\tau\hat{}\cdots$ in 2^ω. We leave the reader to show that both of these methods build computable simulations of $\mathcal{L}$.

We conclude this section with some examples of countable structures that distinguish various of our notions. For example, the following is our first proof that having a computable cover does not imply local computability.

Proposition 8.1. *Let $\mathcal{R}_c$ be the ordered field containing all computable real numbers. Then $\mathcal{R}_c$ has a computable cover, but no uniformly computable cover.*

A real number is computable if the lower cut which it defines in $\mathbb{Q}$ is computable; equivalently, if its binary expansion is $\sum_{n\in\omega} f(n) \cdot 2^{-n}$ for some computable function f such that $f(n) \leq 1$ for all $n > 0$. That is, f computes all bits in the binary expansion of the real. The computable reals form a countable real closed subfield of the reals, of infinite transcendence degree over $\mathbb{Q}$. (The field $\mathcal{R}_c[i]$ of computable complex numbers, on the other hand, is algebraically closed, hence cannot be extended to an ordered field, and by Corollary 4.3, it is perfectly locally computable.)

Proof For any finite set of generators of a subfield $\mathcal{B} \subseteq \mathcal{R}_c$, we already have a computable presentation of $\mathcal{B}$ as a field, given in Proposition 2.3. To compute the order $<$ on $\mathcal{B}$, we need only know the upper and lower cuts of each generator. Since these generators are all computable real numbers, there are algorithms for computing these cuts, and ordinary algebra then allows us to extend our computation of $<$ to all rational functions of these generators, hence to all of $\mathcal{B}$. Thus the (countable) set of all finitely generated substructures of $\mathcal{R}_c$ forms a computable cover of $\mathcal{R}_c$, with the inclusion maps as the embeddings for this cover.

However, suppose $\mathfrak{A}$ were a uniformly computable cover for $\mathcal{R}_c$. Then for every element x of every $\mathcal{A}_i \in \mathfrak{A}$, we could compute the binary expansion of x, since the upper and lower cuts of rationals above and below x can easily be determined by the computable relation $<$ in $\mathcal{A}_i$. Since this could be done uniformly in i and x, we could give a single algorithm which would list out all these binary expansions. However, for every computable real $r \in \mathcal{R}_c$, the ordered subfield $\mathbb{Q}(r)$ would be isomorphic to some $\mathcal{A}_i$, and since the isomorphism would preserve the order, the binary expansion of r would appear (infinitely often, in fact) on the list above. This is impossible: it is well known (via an easy diagonalization) that there is no universal computable set. ⊣

Notice, however, that $\mathcal{R}_c$ is not too far from being locally computable. In particular, we could say that $\mathcal{R}_c$ is *locally $\emptyset''$-computable*, in the sense that there is a $\emptyset''$-computable enumeration of a computable (but not uniformly computable) cover of $\mathcal{R}_c$. In fact, by relativizing Theorem 6.3 using a $\emptyset''$-oracle, we could show that the ordered field $\mathcal{R}_c$ is perfectly locally $\emptyset''$-computable, since it has a presentation computable in $\emptyset''$.

We also give a simple example to show that even for countable structures, local computability is not equivalent to extensional local computability (and hence, by Theorem 6.3, not equivalent to computable presentability either). For any nonempty set $S \subseteq \omega$, let T_S be the countable tree, in the language of strict partial orders $\prec$ with a constant r for the root, built as follows. The root of T is $r = 0$, and we put all odd numbers at level 1 in T. Then, writing $S = \{n_0 < n_1 < \cdots\}$, for each $k \in \omega$, we add a chain of n_k nodes above each of the nodes $2\langle k, i\rangle + 1$ at level 1. (If $|S| = j < \omega$, then we partition the nodes at level 1 into j countable classes instead, so that every node at level 1 has a chain above it. The chain could have length 0 if $0 \in S$.) Thus T_S branches only at the root, with countably many branches starting at level 1, and for each $n \in \omega$, n lies in S iff T_S has some branch (hence infinitely many) containing exactly $n + 1$ nodes. (Here we do not regard the root as a node on any branch.) Clearly this structure T_S can be built to have the same Turing degree as S.

Proposition 8.2. *The following are equivalent.*

1. *T_S is perfectly locally computable.*
2. *T_S is extensionally locally computable.*
3. *S is the range of a limitwise monotonic function.*

Proof Recall that a (total) function f is *limitwise monotonic* (also called *approximable from below*) if there exists a total computable binary function h which is monotonic, i.e. $h(z, s) \leq h(z, s+1)$ for all z and s, and such that $f(z) = \lim_s h(z, s)$.

If such a function exists, then it is a simple matter to use it to build a computable tree isomorphic to T_S, and then Theorem 6.3 shows that T_S is perfectly locally computable. Thus $(3 \Longrightarrow 1)$.

$(1 \Longrightarrow 2)$ is immediate. To see that $(2 \Longrightarrow 3)$, suppose that $\mathfrak{A}$ is an extensional cover of T_S. We define a total computable function $h(\langle x, i\rangle, s)$. First fix a single element $n_0 \in S$, and let $h(\langle x, i\rangle, s) = n_0$ whenever x is not a node in the structure $\mathcal{A}_i$ of $\mathfrak{A}$. (Since the language has no function symbols, $\mathcal{A}_i$ contains only its generators and r, so the domain of $\mathcal{A}_i$ is computable.) Also, $h(\langle r, i\rangle, s) = n_0$ for all i and s. If x does lie in $\mathcal{A}_i$ and $x \neq r$, then for each of the finitely many embeddings f which have appeared in any $I_{ij}^{\mathfrak{A}}$ by stage s, we ask how many other nodes lie on the branch containing $f(x)$:

$$h(\langle x, i\rangle, s) = \max\ |\{y \in \mathcal{A}_j : r \prec y \prec f(x) \text{ or } f(x) \prec y\}|,$$

taking the maximum over all $j \in \omega$ and all $f \in I_{ij,s}^{\mathfrak{A}}$. Then h is computable, increasing in s, and total (since we put $h(\langle x, i\rangle, s) = 0$ if no embeddings f have appeared yet).

Now for any $\mathcal{A}_i$ and any $x \neq r$ in $\mathcal{A}_i$, we have an extensional match β mapping $\mathcal{A}_i$ onto some finite substructure $\mathcal{B} \subseteq T_S$. Consequently, any $f \in I_{ij}^{\mathfrak{A}}$ corresponds to an extension of $\mathcal{B}$ in T_S, and so $\lim_s h(\langle x, i\rangle, s)$ is the number of other nodes on the branch in T_S containing $\beta(x)$. Thus $\lim_s h(\langle x, i\rangle, s)$ exists and lies in S. Conversely, every $n \in S$ corresponds to a branch of length $n + 1$ in T_S, and that branch has an extensional match with some $\mathcal{A}_i \in \mathfrak{A}$, so every $n \in S$ lies in the range of $\lim_s h(\,\cdot\,, s)$. ⊣

Corollary 8.3. *There exists a countable, locally computable tree T (in the language of partial orders) which is not extensionally locally computable.*

Proof Fix a set S which is not the range of any limitwise monotonic function. The finitely generated substructures of the corresponding tree T_S are just the finite substructures, and we can list these out easily, since they contain precisely those finite trees which do not branch above the root. Since these are finite objects, it is easy to enumerate all possible embeddings of one into another, and every such embedding corresponds to an extension of one finite substructure of T_S to another one. Thus we have a uniformly computable cover of T_S, but no extensional cover, by Proposition 8.2. ⊣

Now consider T_ω, the tree we build by taking $S = \omega$. That is, T_ω branches infinitely often at its root, branches nowhere else, and for each $n \in \omega$ has countably many branches of length $n + 1$. Then T_ω has exactly the same uniformly computable cover as any other tree T_S (for any infinite S). However, T_ω is computable, and moreover T_ω is not elementarily equivalent to any T_S with $S \neq \omega$. (In particular, for any $n \notin S$, the Σ_2-sentence saying that there exists a branch containing exactly $n + 1$ nodes holds in T_ω but not in T_S.) This establishes the following.

Corollary 8.4. *There exist countable structures with the same uniformly computable cover, such that one structure is computable (and hence perfectly locally computable), but the other is not computably presentable, indeed not even extensionally locally computable.* ⊣

Our next corollary, in concert with Corollary 3.9, helps distinguish local computability from perfect local computability.

Corollary 8.5. *There exist 2^ω-many countable, pairwise elementarily non-equivalent structures with the same uniformly computable cover. Indeed, these structures all have distinct Σ_2-theories.*

Proof Just consider T_S for every nonempty $S \subseteq \omega$. ⊣

Finally, we use these results to show that the converse of each statement in Proposition 3.8 and Theorem 3.10 is false.

Theorem 8.6. *There exists a tree T which is not extensionally locally computable, yet such that for every $\theta < \omega_1^{CK}$, the Σ_θ-theory of T is itself Σ^0_θ.*

Proof By a result of Hirschfeldt, Miller, and Podzorov in [7, Lemma 3.1], there exists a set S which is not the range of any limitwise monotonic function, yet which is low, i.e. its jump S' is Turing equivalent to $\emptyset'$. The corresponding tree T_S is not extensionally locally computable, by Proposition 8.2. However, the atomic diagram of T_S has the same Turing degree as S, and more generally, the Σ_θ theory of T_S has the same Turing degree as the θ jump $S^{(\theta)}$, which for $\theta > 0$ is just the degree of $\emptyset^{(\theta)}$.

Now the Σ_1-theory of T_S simply describes all finite subtrees of T_S. But these subtrees are precisely those finite trees which branch only at the root, so the Σ_1-theory is computable. Now let $m > 0$ and let $\varphi(\vec{x})$ be any Π_m formula. The Σ_{m+1} sentence $\exists\vec{x}\varphi(\vec{x})$ holds iff there exist elements $\vec{a}$ in T_S such that $\varphi(\vec{a})$ holds. But a $\emptyset^{(m)}$ oracle will decide whether $\varphi(\vec{a})$ holds in T_S, uniformly for any fixed $\vec{a}$, so the Σ_{m+1}-theory of T_S is enumerable using a $\emptyset^{(m)}$ oracle, hence is Σ_{m+1}, as required. A similar argument covers hyperarithmetical formulas. ⊣

The following analogous result shows that the hierarchy of extensionalities does not completely collapse.

Theorem 8.7. *There exist structures which are 1-extensionally locally computable but not 5-extensionally locally computable. Moreover, the Σ^0_6-theory of such a structure can be of arbitrary non-Σ^0_6 Turing degree.*

Proof Fix any set $U \subset \omega$ which is not Π^0_6. The structure $\mathcal{L}_U$ will be the computable well-order of order type ω^4, given by the lexicographic order $\prec$ on the domain ω^4, with a constant symbol 0 for the least element and an additional unary function symbol S such that for $k \in S$, the $(k+1)$-st iterate $S^{k+1}(0)$ is the least element $\succ S^k(0)$ of the form $\langle j, 0, 0, 0\rangle$; and for $k \notin U$, $S^{k+1}(0)$ is the least element $\succ S^k(0)$ of the form $\langle i, j, 0, 0\rangle$. For all $x \notin \{f^k(0) : k \in \omega\}$, we define $S(x) = x$.

Notice that $\mathcal{L}_U$ has no computable presentation, for the complement of U is definable by a Σ_6-formula in the language of $\mathcal{L}_U$, yet is not Σ^0_6. By similar reasoning, $\mathcal{L}_U$ cannot be 5-extensionally locally computable: Proposition 3.8 shows that if it were, then the Σ_6-theory of $\mathcal{L}_U$ would be Σ^0_6.

We now construct a uniformly computable cover of $\mathcal{L}_U$ and show that our cover is 1-extensional. Finitely generated substructures of $\mathcal{L}_U$ consist of all $f^k(0)$ and finitely many other points. In our cover $\mathfrak{A}$, however, we include more information. A structure $\mathcal{A}_i$ in $\mathfrak{A}$ consists of such a set of points, along with a function q_i such that

- $q_i(a,b) \in \omega \cup \{\infty\}$ for all $a, b \in \mathcal{A}_i$ with $a \prec b$; and
- $q_i(a,b) + q_i(b,c) = 1 + q_i(a,c)$ for all $a, b, c \in \mathcal{A}_i$ with $a \prec b \prec c$; and
- $(\forall k > 0)(\forall a) q_i(a, f^k(0)) = \infty$.

The last condition makes it clear that q_i constitutes finitely much information. The second condition shows that q_i is determined by its values on pairs of consecutive points in $\mathcal{A}_i$. The intuition is that $q_i(a,b)$ tells how many points are allowed to go in between a and b in extensions of $\mathcal{A}_i$. The last condition ensures that every $f^k(0)$ is a limit point. We use the symbol ∞ rather than ω to emphasize that we are not naming order types of intervals, but only cardinalities.

The set $I^{\mathfrak{A}}_{ij}$ consists of those maps $f : \mathcal{A}_i \hookrightarrow \mathcal{A}_j$ which respect the symbols $0, S$, and $\prec$ and satisfy, for every $a \prec b$ in $\mathcal{A}_i$, that $\mathcal{A}_j$ contains at most $q_i(a,b)$ elements between $f(a)$ and $f(b)$. Due to the final condition, there are only finitely many intervals to be checked, so in fact each $I^{\mathfrak{A}}_{ij}$ is computable, uniformly in i and j.

The 1-extensional match for an $\mathcal{A}_i$ will be any substructure $\mathcal{B} \subset \mathcal{L}_U$, with a bijection $\beta : \mathcal{A}_i \hookrightarrow \mathcal{B}$, satisfying the condition that for each $a \prec b$ in $\mathcal{A}_i$, the open interval $(\beta(a), \beta(b))$ in $\mathcal{L}_U$ contains exactly $q_i(a,b)$ points. This β will be the 1-extensional match. Since the interval between $S^k(0)$ and $S^{k+1}(0)$ contains an interval of type ω^2, this is possible no matter how many points $f^k(0) \prec a_1 \prec \cdots \prec a_m \prec f^{k+1}(0)$ lie in $\mathcal{A}_i$, even if all $q_i(a_j, a_{j+1}) = \infty$.

Conversely, for any finitely generated $\mathcal{B} \subset \mathcal{L}_U$, there is a function $q_{\mathcal{B}}$ telling the number of points of $\mathcal{L}_U$ in each open interval between points from $\mathcal{B}$, and the 1-extensional match for $\mathcal{B}$ will be that $\mathcal{A}_i$ with exactly the same function q_i attached to it, with the obvious β as the 1-extensional match.

It is quickly seen that in each of these cases, the map β we have described really is 1-extensional. Our use of the functions q_i and our definition of the sets $I^{\mathfrak{A}}_{ij}$ were designed to ensure this. Each finitely generated $\mathcal{C} \subset \mathcal{S}$ gives an isomorphic $\mathcal{A}_j \in \mathfrak{A}$ with an embedding from $I^{\mathfrak{A}}_{ij}$ to match the inclusion of $\mathcal{B}$ in $\mathcal{C}$, and vice versa. $\dashv$

We conjecture that similar results using ω^{m+4} can be proven which show for all m that $(2m+1)$-extensional local computability need not imply $(2m+5)$-extensional local computability. Also, it is likely that by allowing the functions q to be subcomputable, we could show that $(2m+1)$-extensional local computability need not imply $(2m+3)$-extensional local computability. We leave these ideas for a different paper.

§9. Questions. Many questions arise from the notions we have introduced in this paper, and here we list some of the most compelling ones.

1. We can consider the field $\mathcal{R}$ (and $\mathcal{C}$, the complex numbers) with additional function symbols. To what extent are these structures locally computable? For example, do $\mathcal{R}$ and $\mathcal{C}$ stay locally computable when the exponential function e^x is added to the language? Similar questions apply to the trigonometric and other standard functions on $\mathcal{R}$ and $\mathcal{C}$. We have formulas for $e^{(a+b)}$ and $e^{(a\cdot b)}$, of course, but it is possible for e^x to be algebraic for transcendental x, and similarly for the trigonometric functions, so a good deal of work needs to be done to build a uniformly computable cover $\mathfrak{A}$, including the enumeration of the embeddings in $I^{\mathfrak{A}}$. Indeed, these structures might turn out not to be locally computable. In fact, if a function $f(x)$ such as e^x or $\cos x$ were included in the language, then the existence of even a nonuniform computable cover would imply that for every $x \in \mathbb{R}$, the set
$$\{n \in \omega : f^n(x) \text{ is algebraic}\}$$
is c.e. The author is not aware of any known results along these lines.
2. Lemma 7.2 can be viewed as a sort of downwards Löwenheim-Skølem Theorem for perfectly locally computable structures. Is there an upwards version? The answer is not always positive. For example, the natural cover of the computable structure $\mathcal{S} = (\omega, 0, S)$ cannot be a perfect cover for any uncountable structure: such a structure would have to be elementarily equivalent to $\mathcal{S}$, but an uncountable model of this theory must have a nonstandard element, which would generate a substructure not isomorphic to any in the natural cover of $\mathcal{S}$. In order for an upwards version to hold, it appears necessary that the structure contain infinitely many realizations of at least one 1-type, and this should remain true even if we construe types as sets of computable infinitary formulas. Is this sufficient?
3. We have inclusions of the following classes of structures, illustrated in Figure 1.

   Certain of these inclusions are known to be strict. For instance, the ordered field $\mathcal{R}_c$ of computable real numbers and the field $\mathcal{R}$ of real numbers show that the first two inclusions do not reverse. Likewise, in Section 8 we saw countable structures which had some extensionality but were not computably presentable, hence not perfectly locally computable (by Theorem 6.3). Theorem 8.7 built a structure which showed that not all inclusions between 1-extensional and 5-extensional can reverse, and suggested similar results at other levels. We believe that the related structure $(\omega^{(\theta+1)}, <, P)$ can be useful here: the linear order on the ordinal $\omega^{(\theta+1)}$, for $\theta < \omega_1^{CK}$, with a

$$\begin{aligned}
&\{\text{structures with computable covers}\}\\
&\quad\supseteq \{\text{locally computable structures}\}\\
&\quad\supseteq \{\text{extensionally locally computable structures}\}\\
&\quad\supseteq \{\text{2-extensionally locally computable structures}\}\\
&\vdots\\
&\quad\supseteq \{\theta\text{-extensionally locally computable structures}\}\\
&\quad\supseteq \{(\theta+1)\text{-extensionally locally computable structures}\}\\
&\vdots\\
&\quad\supseteq \{\infty\text{-extensionally locally computable structures}\}\\
&\quad\supseteq \{\infty\text{-extensionally locally comp. structures with AP}\}\\
&\quad\supseteq \{\text{perfectly locally computable structures}\}.
\end{aligned}$$

FIGURE 1

unary relation P which holds exactly of those elements of the form $\omega^\theta \cdot n$ with $n \in \emptyset^{(\theta)}$. The ordinal ω_1^{CK} itself, viewed as a linear order with no further relations on it, should also distinguish some two levels of this hierarchy. In general we conjecture that there is no collapse within the diagram shown above, but this conjecture remains open. The case of θ-extensional and $(\theta+1)$-extensional structures with $\theta \geq \omega_1^{CK}$ seems especially mysterious.

4. In Section 3, we saw that for a θ-extensionally locally computable structure $\mathcal{S}$, the Σ_ζ-theory of $\mathcal{S}$ is itself Σ_ζ^0 whenever $\zeta \leq \theta + 1$. However, Theorem 8.6 showed that the converse of Proposition 3.8 can fail: even if the Σ_ζ-theory of $\mathcal{S}$ is Σ_ζ^0 for every $\zeta < \omega_1^{CK}$, $\mathcal{S}$ can still fail to be even 1-extensionally locally computable. What additional conditions on the structure $\mathcal{S}$ might yield converses to Proposition 3.8? That is, we wish to say that, if the Σ_ζ theory of $\mathcal{S}$ is Σ_ζ^0 for all $\zeta \leq \theta + 1$ and $\mathcal{S}$ satisfies some further condition, then $\mathcal{S}$ must be θ-extensionally locally computable; and similarly that if the Σ_ζ theory of $\mathcal{S}$ is Σ_ζ^0 for all ζ and $\mathcal{S}$ satisfies some (different?) further condition, then $\mathcal{S}$ must be ∞-extensionally locally computable. Related versions of these questions can be posed about the Σ_ζ-theory of $(\mathcal{S}, \vec{p})$ over all finite tuples $\vec{p}$ of parameters from $\mathcal{S}$, of course.
5. For locally (i.e. 0-extensionally) computable structures, it was proven that for any $n > 1$, the Σ_n-theory can fail to be Σ_n, since the tree T constructed in Proposition 8.2 has a Σ_2 theory which computes the set S from which T was built, and we can take S to have arbitrarily high degree. Are there analogous examples of 1-extensionally locally computable structures with

arbitrarily complex Σ_3-theory? And can this be extended to θ-extensionally locally computable structures and the $\Sigma_{\theta+2}$-theory?

6. The notion of θ-extensional local computability could be viewed as a measure of how far a countable structure is from being computably presentable. Traditionally, the spectrum of a structure has provided another measure of this distance, so one might ask how the spectrum is related to the strength of the local computability. (The spectrum of $\mathcal{S}$ is the set of Turing degrees of all structures isomorphic to $\mathcal{S}$ with domain ω.) Proposition 8.2 is a small step in this direction, showing that countable locally computable structures can have arbitrarily high Turing degree. (More exactly, the spectrum of a countable locally computable structure can have an arbitrarily high lower bound.) On the other hand, we can relativize the notion of a uniformly computable cover to any Turing degree **d**. For example, the structure $\mathcal{R}_c$ of Proposition 8.1, while not locally computable, was shown to be $\emptyset''$-computable in this sense. What relations (if any) exist between the spectrum of a countable structure and the degrees in which it is locally computable, or extensionally locally computable, or perfectly locally computable?
7. What can be said about homomorphisms or isomorphisms between locally computable structures? Do they induce embeddings or other actions on the uniformly computable covers? Or vice versa? Is there any case in which one can recover the automorphism group of a structure from a perfect cover of the structure (if one exists)? Or from a perfect cover and a correspondence system?

   The analogy to category theory suggests natural transformations as the most reasonable definition of interest when one considers "nice" maps from one cover to another. One would like these maps between covers to correspond as closely as possible to homomorphisms or isomorphisms between the structures covered. Of course, it is possible for two structures of distinct cardinality to have the same perfect cover, and so even an isomorphism from one cover onto another (under any reasonable definition) need not yield an embedding for the structures they cover. Possibly this can be rectified by including cardinality considerations and/or the correspondence system in the definition.
8. Since one of the basic results of local computability is that adding the $<$ relation to the field of real numbers destroys all computability, it is much more reasonable to extend our studies of local computability into algebraic topics than into analytic topics. An obvious next step would be the consideration of algebraic groups, over $\mathbb{R}$ or $\mathbb{C}$ or other locally computable fields, since those are defined by polynomial maps, with no use of $<$. Differential algebra over these base fields could also be a fruitful topic of study.
9. A model theorist might make use of Definition 2.9 without the restriction to the countable case. In that situation, the least possible cardinality of an

ω-extensional cover of a structure $\mathcal{S}$ would likely correspond to the size of the type space, with some appropriate adjustment for other levels of extensionality. Proposition 4.4 and Corollary 3.9 both can be adapted to settings where the covers need not be computable or even countable (but must still have correspondence systems). It would also be possible for a pure model theorist to drop the countability restrictions and to consider either covers by uncountably many finitely generated structures, or covers (of a structure of power λ, say) by structures with generating sets of size $< \kappa$, for some fixed $\kappa \leq \lambda$.

10. In the analogy to category theory, the structure $\mathcal{S}$ itself appears as a sort of inverse limit of its perfect cover. ($\mathcal{S}$ actually is the inverse limit of the category **FGSub**($\mathcal{S}$), and its countable simulation is the inverse limit of its perfect cover.) What about direct limits? For example, there are possible ways to view the automorphism group of a countable structure $\mathcal{C}$ as a direct limit of the set of partial automorphisms of $\mathcal{C}$, especially in the case of an algebraic field, and when $\mathcal{C}$ is computable, this may lead to effectiveness notions on the (quite possibly uncountable) automorphism group. Are these dual in some way to local computability?

REFERENCES.

[1] C.J. Ash & J.F. Knight; *Computable Structures and the Hyperarithmetical Hierarchy* (Amsterdam: Elsevier, 2000).

[2] L. Blum, M. Shub, & S. Smale; On a theory of computation and complexity over the real numbers: NP-completeness, recursive functions, and universal machines, *Bulletin of the AMS* **21** (1989) 1–46.

[3] M. Braverman & S. Cook; Computing over the reals: foundations for scientific computing, *Notices of the AMS* **53** (2006) 3, 318–329.

[4] H.M. Edwards, *Galois Theory* (New York: Springer-Verlag, 1984).

[5] J.N.Y. Franklin, A.M. Kach, R. Miller, & R. Solomon; Local computability on ordinals. In *The Nature of Computation: 9th Conference on Computability in Europe, CiE 2013*, P. Bonizzoni, V. Brattka, & B. Löwe (eds). *Lecture Notes in Computer Science* (Berlin: Springer-Verlag, 2013).

[6] V.S. Harizanov; Pure computable model theory. In *Handbook of Recursive Mathematics*, vol. 1, Yu.L. Ershov, S.S. Goncharov, A. Nerode, & J.B. Remmel (eds). (Amsterdam: Elsevier, 1998), 3-114.

[7] D. Hirschfeldt, R. Miller, & S. Podzorov; Order-computable sets, *The Notre Dame Journal of Formal Logic* **48** (2007) 3, 317–347.

[8] W. Hodges; *A Shorter Model Theory* (Cambridge: Cambridge University Press, 1997).

[9] L. Kronecker; Grundzüge einer arithmetischen Theorie der algebraischen Größen, *J. f. Math.* **92** (1882), 1–122.

[10] T.Y. Lam; *Introduction to Quadratic Forms over Fields* Graduate Studies in Mathematics **67** (Providence, RI: American Mathematical Society, 2005).

[11] R.G. Miller; Locally computable structures. in *Computation and Logic in the Real World – Third Conference on Computability in Europe, CiE 2007*, eds. B. Cooper, B. Löwe, & A. Sorbi, *Lecture Notes in Computer Science* **4497** (Springer-Verlag: Berlin, 2007), 575–584.

[12] R.G. Miller & D. Mulcahey; Perfect local computability and computable simulations. In *Logic and Theory of Algorithms, Fourth Conference on Computability in Europe, CiE 2008*, eds. A. Beckmann, C. Dimitracopoulos, & B. Löwe, *Lecture Notes in Computer Science* **5028** (Berlin: Springer-Verlag, 2008), 388–397.

[13] R.I. Soare; *Recursively Enumerable Sets and Degrees* (New York: Springer-Verlag, 1987).

[14] B.L. van der Waerden; *Algebra*, vol. I, 7th ed. (Berlin: Springer-Verlag, 1966). English translation *Algebra*, volume I, trans. F. Blum & J.R. Schulenberger (New York: Springer-Verlag, 1970). Originally published in German 1930–31 as *Moderne Algebra*.

Department of Mathematics,
Queens College, CUNY, NY 11367 USA; *and*
The Graduate Center of CUNY,
New York, NY 10016 USA.
E-mail: Russell.Miller@qc.cuny.edu

BOREL STRUCTURES: A BRIEF SURVEY

ANTONIO MONTALBÁN AND ANDRÉ NIES

Abstract We survey some research aiming at a theory of effective structures of size the continuum. The main notion is the one of a Borel presentation, where the domain, equality and further relations and functions are Borel. We include the case of uncountable languages where the signature is Borel. We discuss the main open questions in the area.

§1. Introduction. When looking at structures of size the continuum from an effective viewpoint, the following definition is a natural generalization of ideas from computable model theory.

DEFINITION 1.1. Let X be either 2^ω, ω^ω or $\mathbb{R}$, and let $\mathcal{C}$ be a (complexity) class of relations on X. A *$\mathcal{C}$-presentation of a structure $\mathcal{A}$* is a tuple of relations $\mathcal{S} = (D, E, R_1, \dots, R_n)$ such that

- All $D, E, R_1, \dots, R_n$ are in $\mathcal{C}$;
- $D \subseteq X$ and E is an equivalence relation on D (D is called the *domain*);
- $R_1, \dots, R_n$ are relations compatible with E.

$\mathcal{S}$ is a $\mathcal{C}$-representation of $\mathcal{A}$ if $\mathcal{A} \cong \mathcal{S}/E$. When E is the identity on D, we say that $\mathcal{S}$ is an *injective* $\mathcal{C}$-presentation of $\mathcal{A}$.

There are various possible choices for $\mathcal{C}$. In this paper we concentrate on the case that $\mathcal{C}$ is the class of Borel relations. Given a topological space X as above, the σ-algebra of *Borel sets* is the smallest σ-algebra containing the open sets. That is, the Borel sets are the ones obtainable from the open sets by closing under complementation, countable unions, and intersections. A structure $\mathcal{A}$ is a *Borel structure* if it has a Borel presentation. The number of classes of a Borel equivalence relation is either countable or $2^{\aleph_0}$ (Silver's Theorem; see [6]). Thus, the same statement holds for the sizes of Borel structures. We give some examples of Borel structures.

1. The fields $(\mathbb{R}, +, \times)$ and $(\mathbb{C}, +, \times)$ are Borel structures.
2. All Büchi automatic structures (see [7]) are Borel structures.
3. The Boolean algebra $\mathcal{B}$ which is $(\mathcal{P}(\mathbb{N}), \subseteq)$ modulo finite differences of sets is a Borel structure.
4. For a countable structure in a countable functional signature, the lattice of substructures, the congruence lattice, and the automorphism group are Borel structures.

[1]Nies was partially supported by the Marsden Fund of New Zealand, grant no. 08-UOA-187. Montalbán was partially supported by NSF grant DMS-0901169 and by the AMS centennial fellowship.

In fact, structures of size at most the continuum one finds in books related to analysis or algebra are usually Borel. In contrast, the well-ordering $(2^{\aleph_0}, \leq)$ is not Borel. For assume it is. Let $\mathcal{S}$ be a Borel presentation. The class $\mathcal{G}$ of linear orderings of $\mathbb{N}$ which embed in $\mathcal{S}$ is Σ^1_1. On the other hand, it is exactly the class of countable well-orderings, and hence Π^1_1 complete (for each Π^1_1 class $\mathcal{C} \subseteq \mathcal{P}(\omega)$, there is a total Turing functional Ψ such that $X \in \mathcal{C} \leftrightarrow \Psi(X) \in \mathcal{G}$). Contradiction. The same argument shows that $(\omega_1, \leq)$ is not Borel.

History. Borel structures were first considered by Friedman in unpublished work dating from the late 1970s; [19] refers to Friedman's unpublished notes [2, 3]. After that, they appear in a few papers till the late 1990s. In the last few years, the authors, together with Hjorth and Khoussainov [7, 8], brought the topic up again, prompted by a question on Büchi automatic structures. After describing some of the earlier work, we survey this more recent research. Many questions, and even whole research directions, remain open.

Friedman [2, 3] studied a logical system where the language is enriched by one of the following quantifiers: "for all but countably many x ... ", "for all x in a co-meager set ... ", or "for all x in a full-measure set ... ". Borel structures are very appropriate to model logics that use these quantifiers. Friedman then studied axiomatizations, completeness, decidability, etc., in these extended languages. A survey including all these results was written by Steinhorn [19]. In [18] he continued to work in this direction.

Further relevant work was on Borel linear orderings. Some very interesting results were obtained. Harrington and Shelah [5] showed that for every Borel linear ordering $\mathcal{A}$ there exists $\xi < \omega_1$ such that $\mathcal{A} \leq 2^{\xi}$, where 2^{ξ} is ordered by the lexicographical ordering and $\leq$ is the embeddability relation. As a corollary, no Borel linear ordering contains a copy of ω_1 or ω_1^*.

Later on, Louveau [16], extending work of Marker, showed the following unexpected result: For every Borel linear ordering $\mathcal{A}$ and $\xi < \omega_1$, either $\mathcal{A} \leq 2^{\omega\cdot\xi}$ or $2^{\omega\cdot\xi+1} \leq \mathcal{A}$. Another surprising result is that under the assumption of hyperprojective determinacy, the Borel suborderings of $\mathbb{R}^{\omega}$ are well-quasi-ordered under $\leq$. This last result, due to Louveau and Saint-Raymond [17], shows how one can obtain interesting properties of a class of structures if one eliminates the pathological cases and restricts oneself to Borel structures.

There was also a considerable amount of work on Borel partial orderings. For example, Harrington, Marker and Shelah [4] showed that every thin Borel partial ordering can be written as the countable unions of Borel chains (to be thin means that there is no uncountable Borel antichain). Kanovei [9] studied under which conditions a Borel partial ordering has a Borel linearization.

The more recent work of Hjorth, Khoussainov, Montalbán and Nies [7] used Borel structures to answer a question on injective presentations that was originally posed only for Büchi presentable structures. The paper of Hjorth and Nies [8] concentrates on theories of Borel structures in uncountable languages.

§2. Effective content of the completeness theorem. The completeness theorem states that each consistent first-order theory T has a model $\mathcal{M}$ no larger than the size of the language. In this section we will study the effective content of this theorem.

Let us first recall the computable case (where the language is countable). Every computable complete theory has a computable model. On the other hand, there is a computable theory without a computable model. Thus, the completeness theorem in the computable setting fails because there are computable theories without computable completions.

We will now look at the completeness theorem in the Borel setting, and still for a countable language. Of course, each countable structure is Borel, so in this case the completeness theorem works for Borel structures. On the other hand, interesting Borel structures have size the continuum. When Friedman introduced Borel structures he obtained the following result.

Theorem 2.1. *[2, 19] Every theory in a countable language with infinite models has an injective Borel model of size $2^{\aleph_0}$. The model can be chosen so that its elementary diagram is Borel.*

Sketch of a proof. Extend T to a complete theory T_1 in some countable language $L_1 \supseteq L$ such that T_1 has Skolem functions. (See [Chang, Keisler; Section 3.3].) Consider constants

$$C = \{c_x : x \in \omega \times \mathbb{R}^{\geq 0}\},$$

where $\mathbb{R}^{\geq 0}$ denotes the non-negative reals. Order these constants as $\mathbb{R}^{\geq 0}$ many copies of ω. Let $\mathcal{U}$ be an L_1-model of T_1 which has C as a set of order indiscernibles and such that every element of $\mathcal{U}$ is a term in the language L_1 using constants from C. Such a model $\mathcal{U}$ is obtained as in [Chang, Keisler; Thm 3.3.11].

Let $\mathcal{U}_0$ be the elementary submodel of $\mathcal{U}$ generated by $C_0 = \{c_x : x \in \omega \times \{0\}\}$. Note that $\mathcal{U}_0$ is countable. Using the theory S of $\mathcal{U}_0$ as a real parameter, we will construct an injective presentation of $\mathcal{U}$ that is $\Delta^1_1(S)$, and hence Borel.

Let $\mathbb{T}$ be the set of all the terms in the language L_1 with constants from C substituted for the free variables. We define an equivalence relation on $\mathbb{T}$ by

$$t_0(\bar{d}_0) \equiv t_1(\bar{d}_1) \iff \mathcal{U} \models t_0(\bar{d}_0) = t_1(\bar{d}_1),$$

where $\bar{d}_0$ and $\bar{d}_1$ are tuples from C. This equivalence relation is $\Delta^1_1(S)$: to tell whether $t_0(\bar{d}_0) \equiv t_1(\bar{d}_1)$, we can consider tuples of constants $\bar{e}_0$ and $\bar{e}_1$ from C_0 which are in the same order as $\bar{d}_0$ and $\bar{d}_1$. Then we have that $t_0(\bar{d}_0) \equiv t_1(\bar{d}_1) \iff \mathcal{U}_0 \models t_0(\bar{e}_0) = t_1(\bar{e}_1)$. In a similar way we can calculate the effect of functions and relations of L_1 over the equivalence classes of terms in $\mathbb{T}$. It is then clear that the Borel presentation with domain $\mathbb{T}/\equiv$ is isomorphic to $\mathcal{U}$.

Now we want to build an injective $\Delta^1_1(S)$ presentation of $\mathcal{U}$. We show that the equivalence relation $\equiv$ on $\mathbb{T}$ has a Borel choice function. Consider some enumeration of the terms in L_1 (without using the constants from C). So, for every $u \in \mathbb{T}$

there is a least term t_0 in this enumeration such that $\mathfrak{U} \models u = t_0(\bar{d})$ for some constants $\bar{d} \in C$. However, there might be many possible choices for $\bar{d}$. Let n be the length of the tuple $\bar{d} = (d_1, \dots, d_n)$. We order C^n lexicographically (so it has order type $(\omega \times \mathbb{R}^{\geq 0})^n$).

Claim 2.2. *For every $u \in \mathbb{T}$ and each term t_0 as above there is a least tuple $\bar{e}$ such that $u \equiv t_0(\bar{e})$. Further, the term $t_0(\bar{e}) \in \mathbb{T}$ can be chosen in a Borel way.*

Without loss of generality, suppose $u = t_0(d_1, \dots, d_n)$ with $d_0 \leq d_1 \leq \cdots \leq d_n \in C$. First we claim that there exists a least e such that $\mathfrak{U} \models u = t_0(e, \bar{c})$ for some $\bar{c} \in C^{n-1}$ where $e \leqslant c_1 \leqslant \cdots \leqslant c_n$. If there is such an e in C_0, there is clearly a least one. Otherwise there is a unique such e: if $\mathfrak{U} \models t(e_1, \bar{c}_1) = t(e_2, \bar{c}_2)$ for $e_1 < e_2$, then, since the constants in C are order indiscernibles, for any $e \in C_0$ with $e \leq e_1$ we have that $\mathfrak{U} \models t(e, \bar{c}) = t(e_2, \bar{c}_2)$ where $\bar{c}$ is obtained from $\bar{c}_1$ by changing the occurrences of e_1 to occurrences of e. In either case, there is a least such e; call it e_0. Now fix e_0. Note that the linear order induced on $\{x \in C : x \geqslant e_0\}$ is isomorphic to C. Hence, by a similar argument as for e_0, there is a least e_1 such that $\mathfrak{U} \models u = t_0(e_0, e_1, \bar{c})$ for some $\bar{c} \in C^{n-2}$ where $e_0 \leqslant e_1 \leqslant c_2 \leqslant \cdots \leqslant c_n$, and so on. In this way we obtain $\bar{e}$ as desired.

To verify the second part of the claim, it suffices to show that the set of $\bar{e} \in C^n$ which are the least ones determining a value of t_0 is Borel. Suppose there exists $\bar{c}$ which is below $\bar{e}$ in C^n such that $\mathfrak{U} \models t_0(\bar{e}) = t_0(\bar{c})$. This can happen if and only if there exist $\bar{e}_1, \bar{e}_2 \in C_0^n$ which are ordered in the same configuration as $\bar{e}, \bar{c}$ such that $\mathfrak{U}_0 \models t_0(\bar{e}_1) = t_0(\bar{e}_2)$. Since we are using the theory S of $\mathfrak{U}_0$ as a parameter, this property is Borel. ⊣

By a *Borel automorphism* of a structure with an injective Borel representation we mean an automorphism of the structure with a Borel pre-image on $D \times D$ where D is the domain of the presentation. For instance, conjugation is a Borel automorphism of the field of complex numbers with the natural presentation. The structure with the Borel presentation obtained in Theorem 2.1 has many Borel automorphisms. Thus we obtain:

Corollary 2.3. *Every theory in a countable language with infinite models has an injective Borel model of size $2^{\aleph_0}$ with $2^{\aleph_0}$ many Borel automorphisms.*

Proof. There are $2^{\aleph_0}$ many Borel automorphisms g of the linear order $\omega \times \mathbb{R}^{\geq 0}$ (obtained by extending automorphisms of the linear order of the positive rationals). Each automorphism g of this linear order extends to an automorphism $\hat{g}$ of the structure $\mathfrak{U}$: $\hat{g}$ is well-defined via the equation

$$\hat{g}(t(\bar{d})) = t(g(\bar{d}))$$

for each term $t \in L_1$ and each tuple $\bar{d}$ from C. If g is Borel then so is $\hat{g}$ for the injective presentation of $\mathfrak{U}$ obtained above. For, in the setting of Claim 2.2 , if t_0 is a term and $\bar{e}$ is the least tuple such that $u \equiv t_0(\bar{e})$, then $g(\bar{e})$ is the least tuple such that $\hat{g}(u) \equiv t_0(g(\bar{e}))$. ⊣

A natural question is what happens to the completeness theorem in the Borel setting with the language the size of the continuum. A little more care is needed here with the basic definitions. We follow [8]. For generality, we allow arbitrary Polish spaces as domains. Thus, *Borel set* will mean a Borel subset of some Polish space. A *Borel signature* is a Borel set $\mathcal{L}$ of function and relation symbols (coded for instance as reals) such that the arity function is Borel.

Using prefix (or Polish) notation one can naturally identify formulas in the resulting first-order language with finite strings in

$$\mathcal{L} \cup \{\neg, \vee, \wedge, \forall, \exists, v_0, v_1, \dots\},$$

where $v_0, v_1, \dots$ are our variable symbols. The collection of well-formed first-order formulas, $\mathcal{L}_{\omega,\omega}$, is a Borel subset of

$$(\mathcal{L} \cup \{\neg, \vee, \wedge, \forall, \exists, v_0, v_1, \dots\})^{<\omega}.$$

Let $\mathcal{L}$ be a Borel signature. Then a *Borel first-order theory in* $\mathcal{L}$ is a Borel subset T of $\mathcal{L}_{\omega,\omega}$. It is not hard to see that the closure under logical inference of a Borel theory is analytical, but may fail to be Borel.

DEFINITION 2.4. Suppose a Borel signature $\mathcal{L}$ has been fixed. Let $\mathcal{M}$ be an $\mathcal{L}$ structure with domain M. We say that $\mathcal{M}$, together with a Polish space X and a Borel equivalence relation $E \subset X \times X$, is a *Borel presentation* if

$$M = X/E = \{[x]_E : x \in X\},$$

and

$$\{(a_0, \dots, a_{n-1}, R) \in X^n \times \mathcal{L} : \\ R \text{ is an } n\text{-ary relation symbol of } \mathcal{L} \;\&\; \mathcal{M} \models R([a_0]_E, \dots, [a_{n-1}]_E)\}$$

is Borel as a subset of $X^n \times \mathcal{L}$; further,

$$\{(a_0, \dots, a_{n-1}, b, f) \in X^{n+1} \times \mathcal{L} : \\ f \text{ is an } n\text{-ary function symbol of } \mathcal{L} \;\&\; \mathcal{M} \models f([a_0]_E, \dots, [a_{n-1}]_E) = [b]_E\}$$

is Borel as a subset $X^{n+1} \times \mathcal{L}$. We say that a structure $\mathcal{N}$ is *Borel* if there is a Borel presentation $\mathcal{M}$ which is isomorphic to $\mathcal{N}$.

We will usually denote presentations as $(X, E; \dots)$ where $\mathcal{M} = X/E$ and the $(\dots)$ refers to the interpretations of the various non-logical symbols of $\mathcal{L}$.

For a structure over a finite language, being Borel in the sense of the present definition is clearly equivalent to being Borel in the sense of Section 1. If the language is uncountable, the present definition is more restrictive than merely requiring that each individual relation or function be Borel: the relations and functions need to be "uniformly Borel". For instance, the fields $\mathbb{R}, \mathbb{C}$ in the extended language with names for all elements and for all continuous functions from the field to itself are Borel structures.

Just as in the computable setting, we can't always find a Borel completion of a Borel theory. This result of [8] first came up in work related to [7].

Theorem 2.5. *There exists a consistent Borel theory with no Borel completion.*

Proof. Let $\mathcal{U}$ be a free ultrafilter on $\mathbb{N}$. We will consider a Borel subtheory of the atomic diagram of the structure $(P(\mathbb{N}), \mathcal{U})$, such that any model of it codes a free ultrafilter on $\mathbb{N}$. The existence of a Borel completion of this theory would contradicts the easy fact that there are no free Borel ultrafilters on $\mathbb{N}$: on the one hand such a filter would have measure 1/2. On the other hand, being closed under finite variants it would have measure 0 or 1 by the 0–1 law. Also see [11, Exercise 8.50].

To give more detail, the signature of our theory contains a unary predicate U, and for each $A \subseteq \mathbb{N}$ a constant symbol c_A. The theory contains the axioms $c_A \neq c_B$, for every $A \neq B \subseteq \mathbb{N}$. Furthermore, it contains axioms saying that U determines a free ultrafilter as far as the elements named by the c_A are concerned. Thus the theory contains the following axioms: $U(c_{\mathbb{N}})$; $U(c_A) \to U(c_B)$, for every pair of sets such that $A \subseteq B \subseteq \mathbb{N}$; $U(c_A) \leftrightarrow \neg U(c_{\mathbb{N}\setminus A})$, for every $A \subseteq \mathbb{N}$; $U(c_A)$ & $U(c_B) \to U(c_{A\cap B})$, for every $A, B \subseteq \mathbb{N}$; $\neg U(c_A)$, for each finite set $A \subseteq \mathbb{N}$.

Clearly, this theory is Borel. The theory is consistent, because it has the model $(P(\mathbb{N}), \mathcal{U})$ extended by constants naming each subset of $\mathbb{N}$. If T is a completion of our theory which is Borel, then

$$\{A \subseteq \mathbb{N} : T \models U(c_A)\}$$

is a Borel free ultrafilter. Contradiction. ⊣

The theory above also does not have any Borel model $\mathcal{X}$: otherwise, $\{A \subseteq \mathbb{N} : c_A^{\mathcal{X}} \in U^{\mathcal{X}}\}$ would be a Borel free ultrafilter on $\mathbb{N}$.

Even if we have a complete theory, the Borel version of the completeness theorem fails. This contrasts with the computable case.

Theorem 2.6 (Hjorth, Nies [8]). *There exists a complete and consistent Borel theory which has no Borel model.*

This theorem relies on a well known fact from descriptive set theory. For a proof see Example 1.6 in [6].

Fact 2.7. *There is no Borel function* $F : \mathcal{P}(\mathbb{N}) \to \mathcal{P}(\mathbb{N})$ *such that*

$$X =^* Y \Leftrightarrow F(X) = F(Y)$$

for each $X, Y \subseteq \mathbb{N}$.

Broadly speaking, to prove Theorem 2.6 one builds a theory such that any Borel model would contain a function contradicting the foregoing fact. Part of the difficulty is that one also has to rule out non-injective Borel presentations.

A *Borel algebraic closure* of a Borel field F of characteristic m is a Borel model of the complete theory which is axiomatized by the atomic diagram of F together with ACF_m. In other words, one embeds F in a Borel way into an algebraically closed Borel field.

The usual construction of an algebraic closure uses a construction of Artin akin to the proof of the completeness theorem (see [14]). In particular, one needs the axiom of choice when the given field is uncountable.

QUESTION 2.8. *Does each Borel field F have a Borel algebraic closure? Note that we ask that the embedding from F to G be Borel as well.*

A negative answer would yield a further example, somewhat more natural than Theorem 2.6, of a complete Borel theory without a Borel model.

§3. Borel presentability. In the computable case there is no need to differentiate between injective and non-injective presentations: if E is a computable equivalence relation, then the quotient under E can be represented computably by taking the first (in $\mathbb{N}$) element of each equivalence class. However, this is not possible in the Borel case.

THEOREM 3.1 (Hjorth, Khoussainov, Montalbán, Nies [7]). *There is a Borel structure in a finite language without an injective Borel presentation.*

They actually build a Büchi automatic structure without an injective Borel presentation. Recall from the introduction that $\mathcal{B}$ denotes the Boolean algebra $\mathcal{P}(\mathbb{N})$ modulo finite differences. The structure built in the proof of Theorem 3.1 is the disjoint union of the Boolean algebras $\mathcal{P}(\mathbb{N})$ and $\mathcal{B}$, together with the canonical projection map $p\colon \mathcal{P}(\mathbb{N}) \to \mathcal{B}$. They apply Fact 2.7 to show that this structure has no injective Borel presentation.

While not very complicated, this example had to be built for the proof; it is not a structure that appears naturally in mathematics. Such a structure would be obtained by answering the following question in the negative.

QUESTION 3.2. *Does the Boolean algebra $\mathcal{P}(\mathbb{N})$ modulo finite differences have an injective Borel presentation?*

§4. Borel dimension. To say that two Borel presentations are equivalent, it is not enough to use the classical notion of isomorphism. We need Borel isomorphism.

DEFINITION 4.1. Two Borel presentations $(X, E; \dots), (Y, F; \dots)$ are said to be *Borel isomorphic* if there is an isomorphism $\Phi : X/E \to Y/F$ such that the preimage on $X \times Y$

$$\hat{\Phi} = \{\langle x, y\rangle\colon\ \Phi([x]_E) = [y]_F\}$$

is Borel.

Borel isomorphism is easily verified to be an equivalence relation on Borel presentations; for transitivity, one uses the Lusin separation theorem to show that the composition of two isomorphisms with Borel preimage also has a Borel preimage.

We could also introduce a slightly stronger notion of Borel isomorphism where we require in 4.1 that both Φ and its inverse are induced by Borel functions on the domains. In many examples of Borel presentations the equivalence relation has only countable classes. In this case, the two definitions are equivalent by the Lusin-Novikov uniformization theorem (see [11]).

A Borel structure $\mathcal{M}$ is *Borel categorical* if any two Borel presentation of it are Borel isomorphic. More generally, one can define the *Borel dimension* of a Borel structure $\mathcal{M}$ to be the number of equivalence classes modulo Borel isomorphism on the set of Borel presentations of $\mathcal{M}$. This is analogous to the notion of computable dimension in the area of recursive model theory. It was suggested by Bakhadyr Khoussainov.

Note that $\mathcal{M}$ is Borel categorical if and only if it has Borel dimension 1. Examples of Borel categorical structures are:

1. The linearly ordered set $(\mathbb{R}, \leq)$.
2. The Boolean algebra $(\mathcal{P}(\mathbb{N}), \subseteq)$.
3. The field $(\mathbb{R}, +, \times)$.

In fact, for these examples, *each* isomorphism between two Borel presentations of the structure has a Borel graph.

An example of a non-Borel categorical structure is the group $(\mathbb{R}, +)$, as shown in [7]. In [8] the stronger result was obtained that its Borel dimension is $2^{\aleph_0}$. To see this, for a real $p > 1$ recall the Banach space

$$\ell^p = \{\vec{x} \in \mathcal{R}^{\mathbb{N}} : \sum_n |x_n|^p < \infty\},$$

where the norm is $|\vec{x}|_p = (\sum_n |x_n|^p)^{1/p}$. Let G_p be (the canonical injective Borel presentation of) the abelian group underlying ℓ_p. Clearly, as abstract groups these are isomorphic for all p, being vector spaces of dimension $2^{\aleph_0}$ over $\mathbb{Q}$. However, they are not Borel isomorphic. For, any isomorphism between the group structure of two Polish groups that is Borel must be a homeomorphism. (See for instance [1, Section 1.2] or [11, Thm. 9.10]; note that each Borel map is Baire measurable.) Hence it would be linear. But for $1 < p < q$ there is no continuous linear bijection between ℓ^p and ℓ^q. See [15, top of pg. 54].

A further example of a non-Borel categorical structure is given by the following result of Nies and Shore.

THEOREM 4.2. *The field* $(\mathbb{C}; +, \times)$ *is not Borel categorical even for injective Borel presentations.*

PROOF. Let $T = ACF_0$ be the theory of algebraically closed fields of characteristic 0. Then T is ω_1-categorical, so every model of size $2^{\aleph_0}$ is classically isomorphic to the field $\mathbb{C}$.

By Corollary 2.3, T has an injective Borel model of size $2^{\aleph_0}$ with $2^{\aleph_0}$ many Borel automorphisms. On the other hand, any Borel automorphism of $\mathbb{C}$ with the natural presentation is continuous by the result mentioned in the foregoing proof.

The only continuous automorphisms of $\mathbb{C}$ are conjugation and identity. Hence the two injective Borel presentations are not Borel isomorphic. ⊣

Nies and Shore actually gave a direct construction of an injective Borel model of ACF_0, of size $2^{\aleph_0}$, and with $2^{\aleph_0}$ many Borel automorphisms. Let B be an uncountable closed set of algebraically independent reals. Then the real closure of B in $\mathbb{R}$ is Borel. The first step is adding the roots of odd-degree polynomials with coefficients in B. One can identify the elements with the polynomials and their roots in order; the roots are computable in the coefficients. Now one iterates the process. The real closed subfield of $\mathbb{R}$ constructed in this way has an injective Borel presentation. Finally one adjoins a solution to $X^2 = -1$ to obtain the required Borel model of ACF_0. There are $2^{\aleph_0}$ many Borel automorphisms for this presentation. They are induced by the Borel permutations of B.

If one chooses B as in the proof of Theorem 5.2 below (namely, reals whose binary presentations are the paths on a perfect tree T such that the effective disjoint union of finitely many paths is arithmetically generic), then by an argument similar to the one given below, the field one obtains is actually Borel as a subfield of $\mathbb{C}$.

The following question remains open.

QUESTION 4.3. *Is there a Borel structure of Borel dimension strictly between 1 and* $2^{\aleph_0}$*?*

§5. Borel models of Peano arithmetic. Recall that a set $\mathcal{S} \subseteq 2^\omega$ is a *Scott set* if $\mathcal{S}$ is closed downwards under Turing reducibility, closed under joins, and each infinite binary tree $T \in \mathcal{S}$ has an infinite path in $\mathcal{S}$. Scott sets occur for instance in reverse mathematics as the ω-models of WKL_0.

Let M be a model of PA. The *standard system* of M consists of the standard parts of M-definable sets. Thus, the standard system is the class

$$\{D \cap \omega \colon D \subseteq M \text{ is parameter definable in } M\}$$

(here we think of M as extending ω).

For $n \in \mathbb{N}$ let p_n denote the nth prime. Let M be a nonstandard model of PA. It is well-known [10] that each set in the standard system has the form $\{n \in \mathbb{N} \colon p_n \mid a\}$ for some $a \in M$.

Scott (see [10, Section 13.1]) showed that the countable Scott sets are precisely the standard systems of countable models of Peano arithmetic. Knight and Nadel [12] proved the analoguous result for the size ω_1. For the size 2^ω, the analogous statement is open. Motivated by this, H. Woodin asked the following effective version of this question.

QUESTION 5.1. *If a Scott set is Borel, is it already the standard system of a Borel model of Peano arithmetic?*

For an upper bound on the complexity, note that the standard system of any Borel model of PA is analytic.

A *jump ideal* is an ideal K in the Turing degrees that is closed under the jump. Note that the sets with degree in K form a Scott set. Thus the following yields an uncountable non-trivial Scott set that is Borel.

THEOREM 5.2 (Slaman, 2010). *There is a proper uncountable jump ideal K in the Turing degrees that is Borel.*

PROOF. K is the jump ideal generated by the degrees of paths on a perfect tree T such that the effective disjoint union of finitely many paths is arithmetically generic. This jump ideal is proper because it does not contain the degree of $\emptyset^{(\omega)}$.

We sketch the argument why K is Borel. For a simple case, consider whether X is recursive in some path G in T via the Turing functional Φ. Let $G \in [T]$. For each n, the value of $\Phi(n, G)$ is determined by a finite initial segment of G, including the value "undefined" since G is generic.

So the set of $G \in [T]$ such that $\Phi(G) = X$ is a $\Pi^0_1(X \oplus T)$ subset of $[T]$. By the compactness of Cantor space, whether this set is nonempty is arithmetic in T and X. The definition of this set only depends on Φ and T. Thus, the set of X such that X is recursive in a path through T by functional Φ is arithmetic in T.

Note that $G^{(n)} \equiv_T G \oplus \emptyset^{(n)}$ for each arithmetically generic G. Similar to the argument above, one can show that it is arithmetic in T whether X is recursive in a sequence of paths of a fixed length and $0^{(n)}$ by functional Φ. The jump ideal generated by $[T]$ is the countable union of these sets, so it is Borel, too. ⊣

A recent preprint of Ali Enayat focuses on Borel models of PA. In particular, he has probed the "Borel content" of a result of Schmerl [13, Theorem 6.4.3] to show that Theorem 5.2 above can be strengthened: there is a proper uncountable Borel jump ideal K in the Turing degrees that can be realized as the standard system of some model of PA.

It would be interesting to determine which Borel upper semilattices with least element and the countable predecessor property are isomorphic to Borel (or analytic) ideals of the Turing degrees.

REFERENCES.

[1] Howard Becker and Alexander S. Kechris. *The Descriptive Set Theory of Polish Group Actions*. Cambridge University Press, 1996.

[2] Harvey Friedman. On the logic of measure and category I. Manuscript, 1978.

[3] Harvey Friedman. Borel structures and mathematics. Manuscript, 1979.

[4] Leo Harrington, David Marker, and Saharon Shelah. Borel orderings. *Trans. Amer. Math. Soc.*, **310**(1), 293–302, 1988.

[5] Leo Harrington and Saharon Shelah. Counting equivalence classes for co-κ-Souslin equivalence relations. In *Logic Colloquium '80 (Prague, 1980)*, D. van Dalen (ed), volume 108 of Stud. Logic Foundations Math., 147–152. North-Holland, 1982.

[6] Greg Hjorth. Borel equivalence relations. In *Handbook on Set Theory, Vol. 1*, Matt Foreman and Aki Kanamori (eds), Springer-Verlag, 2010.

[7] G. Hjorth, Bakh Khoussainov, Antonio Montalbán, and André Nies. From automatic structures to Borel structures. *Proceedings of the Twenty-Third Annual IEEE Symposium on Logic in Computer Science (LICS 2008)*, 431–441, 2008.

[8] Greg Hjorth and André Nies. Borel structures and Borel theories. *J. Symbolic Logic* **76** (2011), 461–476.

[9] Vladimir Kanovei. When a partial Borel order is linearizable. *Fund. Math.*, **155**(3), 301–309, 1998.

[10] R. Kaye. *Models of Peano Arithmetic*, volume 15 of Oxford Logic Guides. Oxford University Press, 1991.

[11] Alexander S. Kechris. *Classical Descriptive Set Theory*, volume 156 of Graduate Texts in Mathematics. Springer-Verlag, 1995.

[12] Julia Knight and Mark Nadel. Models of arithmetic and closed ideals. *J. Symbolic Logic*, **47**(4), 833–840. 1983.

[13] R. Kossak and J.H. Schmerl. *The Structure of Models of Peano Arithmetic*. Oxford Logic Guides, vol. 50. Oxford University Press, 2006.

[14] S. Lang. *Algebra*. Addison-Wesley, 1965.

[15] J. Lindenstrauss and L. Tzafriri. *Classical Banach Spaces I*. Springer, 1996.

[16] Alain Louveau. Two results on Borel orders. *J. Symbolic Logic*, **54**(3) 865–874, 1989.

[17] Alain Louveau and Jean Saint-Raymond. On the quasi-ordering of Borel linear orders under embeddability. *J. Symbolic Logic*, **55**(2) 537–560, 1990.

[18] C.I. Steinhorn. Borel structures and measure and category logics. In *Model-Theoretic Logics*, J. Barwise and S. Feferman (eds), volume 8 of Perspect. Math. Logic, 579–596. Springer, 1985.

[19] C.I. Steinhorn. Borel structures for first-order and extended logics. In *Harvey Friedman's Research on the Foundations of Mathematics*, L.A. Harrington, M.D. Morley, A. Scedrov and S.G. Simpson (eds), volume 117 of Stud. Logic Found. Math., 161–178. North-Holland, 1985.

Department of Mathematics,
University of Chicago,
Chicago, IL, USA.
E-mail: antonio@math.uchicago.edu

Department of Computer Science,
Auckland University,
Auckland, New Zealand.
E-mail: andre@cs.auckland.ac.nz

E-RECURSIVE INTUITIONS

GERALD E. SACKS

In Memoriam
Professor Joseph R. Shoenfield

Abstract An informal sketch (with intermittent details) of parts of E-Recursion theory, mostly old, some new, that stresses intuition. The lack of effective unbounded search is balanced by the availability of divergence witnesses. A set is E-closed iff it is transitive and closed under the application of partial E-recursive functions. Some finite injury, forcing, and model theoretic constructions can be adapted to E-closed sets that are not Σ_1 admissible. Reflection plays a central role.

§1. Initial Intuitions. One of the central intuitions of classical recursion theory is the effectiveness of unbounded search. Let A be a nonempty recursively enumerable set of nonnegative integers. A member of A can be selected by simply enumerating A until some member appears. This procedure, known as unbounded search, consists of following instructions until a termination point is reached. What eventually appears is not merely some number $n \in A$ but a computation that reveals $n \in A$. Unbounded search in its full glory consists of enumerating all computations until a suitable computation, if it exists, is found. It follows there exists a partial recursive function g such that for all e:

$$g(e)\downarrow \longleftrightarrow W_e \neq \emptyset$$
$$g(e)\downarrow \longrightarrow g(e) \in W_e.$$

(Here W_e is the eth recursively enumerable set; $g(e)\downarrow$ means $g(e)$ **converges**, i.e. has a value. The symbol $\uparrow$ indicates **divergence**.)

In E-recursion theory unbounded search is not effective. An E-recursive enumeration of all computations is available, but sifting through them for some desired outcome is demonstrably not effective in most circumstances. The exceptions are called selection principles. Finding them is an important part of the subject. Unbounded search is legal in the setting of α-recursion theory. Both E-recursion and α-recursion, restricted to ω, are equivalent to classical recursion theory. For α-recursion the equivalence is almost immediate, for E-recursion some proof is needed.

---

[1]My thanks to the organizers of "Effective Mathematics of the Uncountable," CUNY Graduate Center, August 2009, for their kind invitation.

[2]Math Subject Codes: 03D30, 03D60, 03E15. Keywords: infinite time Turing machines, infinitary computability, ordinal computation.

In the classical theory unbounded search is enabled by Kleene's least number operator scheme:

(3) $$f(x) \simeq (\text{least } y)(g(x,y) = 0);$$

The symbol $\simeq$ denotes strong equality. Say $f(x)$ is strongly equal to $g(x)$ iff neither is defined or both have the same value. Then in (3) f is partial recursive iff g is. Nothing resembling Kleene's least number operator appears among the Normann [4] schemes[3] for E-recursion. Below x, y, w, z are arbitrary sets, and e, i, j, n are nonnegative integers. The first three are finitary in nature.

(a) projection

$$\{e\}(x_1, \ldots, x_n) = x_i \quad \text{if } e = \langle 1, n, i\rangle.$$

(b) difference

$$\{e\}(x_1, \ldots, x_n) = x_i - x_j \quad \text{if } e = \langle 2, n, i, j\rangle.$$

(c.1) pairing

$$\{e\}(x_1, \ldots, x_n) = \{x_i, x_j\} \quad \text{if } e = \langle 3, 1, n, i, j\rangle.$$

(c.2) union

$$\{e\}(x_1, \ldots, x_n) = \bigcup\{y \mid y \in x_1\} \quad \text{if } e = \langle 3, 2, n\rangle.$$

Scheme (d) below is the only scheme that is potentially infinitary. If x_1 is infinite, then the computation of (d) entails infinitely many subcomputations.

(d) E-recursive bounding

$$\{e\}(x_1, \ldots, x_n) \simeq \{\{c\}(y, x_2, \ldots, x_n) \mid y \in x_1\} \quad \text{if } e = \langle 4, n, c\rangle.$$

The left side of scheme (d) converges iff $\{c\}(y, x_2, \ldots, x_n)\downarrow$ for all $y \in x_1$.

(e) composition

$$\{e\}(x_1, \ldots, x_n) \simeq \{c\}(\{d_1\}(x_1, \ldots, x_n), \ldots, \{d_m\}(x_1 \ldots, x_n))$$
$$\text{if } e = \langle 5, n, m, c, d_1, , , d_m\rangle.$$

(f) enumeration

$$\{e\}(c, x_1, \ldots, x_n, y_1, \ldots, y_m) \simeq \{c\}(x_1, \ldots, x_n) \quad \text{if } e = \langle 6, m, n\rangle.$$

The enumeration scheme leads to Kleene's **fixed point theorem**: let f be a total recursive function; then there exists some e such that

$$\{e\} \simeq \{f(e)\}.$$

(The partial functions $\{e\}$ and $\{f(e)\}$ have the same graph.) The fixed point theorem implies definition by transfinite recursion is effective (Section 3).

[3]Developed independently by Y.N. Moschovakis.

Each Normann scheme is a closure condition in the inductive definition of E, **the class of E-recursive evaluations**. Each member of E is of the form

$$\langle e, \langle x_1, \dots, x_n \rangle, y \rangle.$$

The above tuple is put in E iff the schemes determine a value y for $\{e\}(x_1, \dots, x_n)$. The definition of E is a Σ_1 transfinite recursion on the ordinals σ.

Stage $\sigma = 0$: $\langle e, \langle x_1, \dots, x_n \rangle, x_i \rangle$ is put in E iff $e = \langle 1, n, i \rangle$.

Schemes (b) and (c) are treated similarly.

Stage $\sigma > 0$: $\langle e, \langle x_1, \dots, x_n \rangle, z \rangle$ is put in E iff
it was not put in before stage σ,
$e = \langle 4, n, c \rangle$,
$\forall y_{y \in x_1} \exists w [\langle c, \langle y, x_2, \dots, x_n \rangle, w \rangle$ put in E before stage $\sigma]$,
and $z = \{w \mid \exists y_{y \in x_1} [\langle c, \langle y, x_2, \dots, x_n \rangle, w \rangle$ is already in $E]\}$.

Schemes (e) and (f) are treated similarly.

Define **$\{e\}(x_1, \dots, x_n)$ converges to y** iff

$$\langle e, \langle x_1, \dots, x_n \rangle, y \rangle \in E.$$

A function f is **partial E-recursive** iff there is an e such that

$$f(x_1, \dots, x_n) \simeq \{e\}(x_1, \dots, x_n)$$

for all $x_1, \dots, x_n$. A class of sets is ***E*-recursively enumerable** iff it is the domain of a partial E-recursive function. The graph of a partial E-recursive function is E-recursively enumerable. One consequence of the lack of unbounded search in E-recursion is: a function whose graph is E-recursive need not be E-recursive; an example is $O(x)$, where $x \in L$ and $O(x)$ is the least δ such that x is a first order definable subset of $L(\delta)$. Another consequence is: the range of a E-recursive function on the ordinals need not be E-recursively enumerable, cf. Proposition 3.1.

The class E, of E-recursive evaluations, is E-recursively enumerable thanks to the enumeration scheme.

A set b is ***E*-closed** iff b is transitive and for all $e < \omega$,

$$\{e\}(x_1, \dots, x_n) \in b$$

whenever $x_1, \dots, x_n \in b$ and $\{e\}(x_1, \dots, x_n) \downarrow$.

The above six schemes restricted to the nonnegative integers define the partial recursive functions of classical recursion theory. Recall HF, the set of hereditarily finite sets, defined by

$$x \in HF \leftrightarrow [x \text{ is finite} \wedge \forall y \in x (y \in HF)].$$

A Δ_0 predicate is said to be **lightface** if all its parameters are finite ordinals; Δ_0 means all quantifiers are bounded. The set of nonnegative integers, ω, is a lightface Δ_0-definable subset of HF, hence E-recursive by Proposition 2.5. Let

$f(x_1, \dots, x_n)$ be a partial function from ω^n into ω. Then f is a classical partial recursive function iff it is E-recursive. This follows from Gandy's selection principle, Theorem 4.1, which legitimizes the search for a nonnegative integer in E-recursion theory.

§2. E-Recursively Enumerable Versus Σ_1. Every E-recursively enumerable class is Δ_1 definable; the converse is false. These results can be derived from the notion of computation. A **computation instruction** is an $(n+1)$-tuple $\langle e, x_1, \dots, x_n\rangle$ or more simply $\langle e, \overline{x}\rangle$. Associated with $\langle e, \overline{x}\rangle$ is a tree, $T_{\langle e,\overline{x}\rangle}$. Every node of the tree is a computation instruction; its top node is $\langle e, \overline{x}\rangle$ and it branches downward as determined by the schemes of Section 1.

If e is an index for one of the first three schemes, then $\langle e, \overline{x}\rangle$ is a terminal node.

If e is $\langle 4, n, c\rangle$, then $\langle c, y, x_2, \dots, x_n, \rangle$ is an immediate subcomputation instruction of $\langle e, \overline{x}\rangle$ for each $y \in x_1$.

If e is $\langle 5, n, m, c, d_1, \dots, d_m\rangle$, then $\langle d_j, \overline{x}\rangle$ is an immediate subcomputation instruction of $\langle e, \overline{x}\rangle$ for $1 \le j \le m$; in addition

$$\text{if } \{d_j\}(\overline{x}) \text{ converges to } y_j \text{ for } 1 \le j \le m,$$

then $\langle c, y_1, \dots, y_m\rangle$ is an immediate subcomputation instruction of $\langle e, \overline{x}\rangle$.

If e is not interpretable as a scheme, then $\langle e, \overline{x}\rangle$ has just one immediate subcomputation instruction, a repeat of $\langle e, \overline{x}\rangle$.

Define $b >_U a$ to be a is an immediate subcomputation instruction of b. The predicate $b >_U a$ is E-recursively enumerable. The predicate

$$\exists c[b >_U c >_U a]$$

is not E-recursively enumerable.

Proposition 2.1. *$\{e\}(\overline{x}) \downarrow \longleftrightarrow T_{\langle e,\overline{x}\rangle}$ is wellfounded.*

Both directions of 2.1 are proved by transfinite induction.

Suppose $\{e\}(x) \downarrow$; its **length of computation,** denoted by $| \{e\}(x) |$, is the ordinal height of $T_{\langle e,x\rangle}$. Otherwise its length is ∞.

Proposition 2.2. *If A is E-recursively enumerable, then A is Δ_1 definable.*

Proof. A is Σ_1 because E, the class of E-recursive evaluations, is Σ_1. Suppose for some e,

$$A = \{x \mid \{e\}(x) \downarrow\}.$$

Then for all x,

$$x \notin A \longleftrightarrow T_{\langle e.x\rangle} \text{ is illfounded.}$$

⊣

Proposition 2.3. *There exists a Δ_1 definable class that is not E-recursively enumerable.*

Proposition 2.3 follows from 2.2 and a diagonal argument.

The proof of Proposition 2.2 makes a point whose importance is not readily apparent. Suppose $\{e\}(x) \uparrow$ (diverges). Then the tree $T_{\langle e,x\rangle}$ has an infinite descending path. Any such path witnesses the divergence of $\{e\}(x)$. Moschovakis [2] was the first to realize the importance of **divergence witnesses** and to apply them fruitfully. They are essential ingredients of priority constructions and forcing arguments in E-recursion theory. Their power is ample compensation for the failures of unbounded search.

PROPOSITION 2.4. *If A and $V - A$ are E-recursively enumerable, then A is E-recursive.*

The usual proof of the counterpart of Proposition 2.4 in classical recursion uses unbounded search and so is not applicable in E-recursion; 2.4 follows from Gandy Selection (Theorem 4.1).

A Δ_0 predicate is said to be **lightface** if all its parameters are finite ordinals; Δ_0 means all quantifiers are bounded.

PROPOSITION 2.5. *Every lightface Δ_0 predicate is E-recursive.*

§3. E-Recursion Versus α-Recursion. Recall how transfinite recursion (TR) works in set theory. Let $I : V \longrightarrow V$. Consider the equation

$$f(\gamma) = I(f \restriction \gamma); \tag{4}$$

$f \restriction \gamma$ is the graph of f restricted to γ. There exists a unique f that satisfies (4) for all γ. If I is Σ_1 definable, then f is Σ_1 definable.

Effective Transfinite Recursion (ETR) If I is E-recursive, then the unique f that satisfies (4) is E-recursive.

ETR is a consequence of Kleene's fixed point theorem (Section 1).

Thus $L(\gamma)$, the γth initial segment of Gödel's L, as a function of γ, is E-recursive. Gödel enumerated L one set at a time by means of an E-recursive function.

PROPOSITION 3.1. *$V = L$ iff L is E-recursively enumerable.*

PROOF. Suppose $\forall x(x \in L \longleftrightarrow \{e\}(x) \downarrow)$ and $V \neq L$. Then for some $b \notin L$, $\{e\}(b) \uparrow$. By Proposition 2.2, divergence is Σ_1 definable. Then by Levy–Shoenfield absoluteness, $\{e\}(x) \uparrow$ for some $x \in L$. ⊣

PROPOSITION 3.2. *The predicates, $|\{e\}(x)| < \gamma$ and $|\{e\}(x)| = \gamma$, are E-recursive.*

PROOF. By effective transfinite recursion. ⊣

For any set x, define $E(x)$ to be the least transitive, E-closed set with x as a member. The schemes for E-recursion map $L(tc(\{x\}))$ into $L(tc(\{x\}))$. And

$L(\gamma, tc(\{x\}))$ is an E-recursive function of γ and x. It follows there is an ordinal, κ^x, such that

$$E(x) = L(\kappa^x, tc(\{x\})).$$

Also $\gamma < \kappa^x$ iff $\gamma = \{e\}(x, a_1, \ldots, a_n)$ for some $e < \omega$ and $a_1, \ldots, a_n \in tc(x)$.

For any set x, define $Ad_1(x)$ to be the least Σ_1 admissible set with x as a member. Then

$$Ad_1(x) = L(\alpha^x, tc(\{x\})),$$

where α^x is the least β such that $L(\beta, tc(\{x\}))$ satisfies Σ_1 bounding. Proposition 3.5 implies $\kappa^{\omega_1} < \alpha^{\omega_1}$.

The ordinal κ^x can be regarded as the least β such that $L(\beta, tc(\{x\}))$ satisfies E-recursive bounding. An induction on the length of computations shows

Proposition 3.3. $E(x) \subseteq Ad_1(x)$.

Proposition 3.4. *There exists a partial E-recursive function g such that for all $d < \omega$ and all x;*

(i) $\{d\}(x) \downarrow \longleftrightarrow g(d, x) \downarrow$
(ii) $\{d\}(x) \downarrow \longrightarrow g(d, x) = T_{\langle d,x \rangle}$.

Proof. By effective transfinite recursion. ⊣

Suppose B is E-closed and $x \in B$. If $\{e\}(x) \downarrow$, then $T_{\langle e,x \rangle}$ is wellfounded, hence E-recursive in x and so a member of B. If $\{e\}(x) \uparrow$, then $T_{\langle e,x \rangle}$ is illfounded and might not be in B; nonetheless some infinite descending path through $T_{\langle e,x \rangle}$ might be in B.

Say B **admits divergence witnesses** iff for all e, $x \in B$: if $\{e\}(x)$ diverges, then some witness to the divergence belongs to B.

Proposition 3.5. *$E(\omega_1)$ is not Σ_1 admissible.*

Proof. For some κ, $E(\omega_1) = L(\kappa)$. It suffices to show that $L(\kappa)$ admits divergence witnesses, because then there is a map m from $\omega \times \omega_1$ into $L(\kappa)$ whose graph is a Σ_1 definable subset of $L(\kappa)$ and whose range is unbounded in $L(\kappa)$. The value of $m(e, \beta)$ is either the value of$\{e\}(\omega_1, \beta)$ or the L-least witness to the divergence of $\{e\}(\omega_1, \beta\}$.

The definition of $E(\omega_1)$ implies there is an injective map f of $L(\kappa)$ into ω_1 in L. Suppose $\{e\}(x) \uparrow$ for some $x \in L(\kappa)$; it follows from Proposition 4.4 and Lemma 4.5 that a witness to its divergence is first order definable over $L(\kappa)$. The witness, t, has domain ω. For all n, $t(n) \in L(\kappa)$; f maps the graph of t to a countable subset of ω_1, hence to a member of $L(\omega_1)$. But then $t \in L(\kappa)$. ⊣

Proposition 3.6. *If $b \subseteq \omega$, then $E(b)$ is Σ_1 admissible.*

Proof. If b is finite, then $E(b) = L(\omega)$. Assume b is infinite. Then $E(b) = L(\omega_1^b, b)$, where ω_1^b is the least ordinal not recursive in b. $L(\omega_1^b, b)$ is Σ_1 admissible by a Σ^1_1 bounding argument from hyperarithmetic theory. ($L(\omega_1^b, b) \cap 2^\omega$ is the set of all subsets of ω hyperarithmetic in b.) ⊣

§4. Selection and Reflection. Let $W(e, x)$ be $\{n \mid n \in \omega \wedge \{e\}(n, x) \downarrow\}$. Thus $W(e, x)$ is the eth set (of nonnegative integers) E-recursively enumerable in x. Gandy selection is a uniform method for sifting through computations to find one, if there are any, that puts a non-negative integer into $W(e, x)$; "uniform" means the method is the same for all e and x. The notion of **uniformity** has applications, as in Corollary 4.3.

THEOREM 4.1 (Gandy Selection). *There exists a partial E-recursive function $\phi(e, x)$ such that for all $e > \omega$ and all x:*

(i) $(\exists n < \omega)[\{e\}(n, x) \downarrow] \longrightarrow \phi(e, x) \downarrow$.
(ii) $\phi(e, x) \downarrow \longrightarrow [\phi(e, x) \in \omega \wedge \{e\}(\phi(e, x), x) \downarrow]$.

The proof of Gandy Selection requires a preparatory lemma.

LEMMA 4.2 (Moschovakis). *Suppose $\{d\}(x) \downarrow$ or $\{e\}(y) \downarrow$. Then*

$$\min(|\{d\}(x)|, |\{e\}(y)|)$$

is E-recursive uniformly in d, e, x, y.

PROOF. By effective transfinite recursion on min. A rough approximation of the recursion equation is: $\min(|u|, |v|) =$

$$\max\{\min(|a|, |b|) \mid a <_U u \wedge b <_U v\}$$

where u, v, a, b are computation instructions, and $a <_U b$ means a is an immediate subcomputation instruction of b. The above recursion is not as effective as it might be because if $u \uparrow$, then $\{a \mid a <_U u\}$ may not be E-recursive in u. But enough of the as are explicit to make the recursion (slightly modified) effective, hence successful. ⊣

PROOF OF GANDY SELECTION. For simplicity drop the "x" in the "$\{e\}(n, x)$" of Theorem 4.1. Kleene's fixed point theorem yields a partial recursive function $t(e, n)$ whose definition has two cases.

Case 1: $t(e, n+1) \downarrow$ and $|t(e, n+1)| \leq |\{e\}(n)|$. Then

$$t(e, n) \simeq t(e, n+1) + 1.$$

Case 2: $|\{e\}(n)| < |t(e, n+1)|$. Then $t(e, n) = 0$.

The above split into cases is effective by Lemma 4.2 and Proposition 3.2. Assume $\{e\}(n) \downarrow$ for some n. Then $t(e, 0) \downarrow$ and

$$|\{e\}(n)| < |t(e, n+1)|$$

for some n; let n_0 be the least such. Then $t(e, 0) = n_0$.

Read $w \leq_E z$ as: w is ***E*-recursive in** z. And define it by

$$\exists e[w = \{e\}(z)].$$

There is a $c < \omega$ and a recursive function h such that for all e, w, and z we have $\{h(e)\}(w, z) \downarrow$ iff $[\{e\}(z)$ converges to $w]$ iff $\{c\}\{e, w, z) \downarrow$.

Gandy can select an e, if there is one, uniformly in w and z such that $\{c\}(e, z, w) \downarrow$. It follows that $w \leq_E z$ is an E-recursively enumerable predicate of w and z. ⊣

COROLLARY 4.3. *Suppose P(x,y) is E-recursively enumerable and*

$$\forall_{x \in z} \exists y [y \leq_E x \ \wedge \ P(x, y)]$$

Then there exists a partial E-recursive f such that

$$\forall_{x \in z} [f(x) \downarrow \ \wedge \ P(x, f(x))].$$

The concepts of reflection and divergence are closely linked in E-recursion theory. Define

$$\kappa_0^x = \sup\{\gamma \mid \gamma \leq_E x\}.$$

An ordinal δ is said to be x-reflecting if

$$[L(\delta, tc(\{x\})) \models \mathcal{F}] \rightarrow [L(\kappa_0^x, tc(\{x\})) \models \mathcal{F}]$$

for every Σ_1 sentence $\mathcal{F}$ whose only parameter is x. The predicate, δ is x-reflecting, is E-recursively enumerable in x.

If $\mathcal{F}$ reflects down to κ_0^x, then it reflects down to an ordinal E-recursive in x, because the least member of a nonempty set y of ordinals is E-recursive in x if $y \leq_E x$.

Define

$$\kappa_r^x = \text{ the greatest } x\text{-reflecting ordinal.}$$

Clearly $\kappa_0^x \leq \kappa_r^x$. Define $\kappa_r^{x,a}$ to be $\kappa_r^{\langle x,a \rangle}$.

PROPOSITION 4.4. $\kappa_r^{x,a} \leq \kappa^x$ *for all* $a \in tc(x)$.

PROOF. Suppose not; $E(\langle x, a \rangle) = E(x) = L(\kappa^x, tc(\{x\}))$. Then

$$E(\langle x, a \rangle) \in L(\kappa_r^{x,a}, tc(\{x\})).$$

The latter reflects below $\kappa_0^{x,a}$, because there is a Π_3 sentence $\mathcal{F}$ such that for every transitive class A

$$A \text{ is } E\text{-closed} \longleftrightarrow \langle A, \in \rangle \models \mathcal{F}.$$

Thus $E(\langle x, a \rangle) \in L(\kappa_0^{x,a}, tc(\{x\})$. But then $\kappa^{\langle x,a \rangle} < \kappa_o^{\langle x,a \rangle}$. ⊣

The situation of greatest interest is when $\kappa_0^x < \kappa_r^x < \kappa^\times$. Harrington in his study of $E(2^\omega)$ made a breakthrough when he proved: if $a \in 2^\omega$ and $\{e\}(a, 2^\omega)$ diverges, then some witness to the divergence appears at level $\kappa_r^{2^\omega,a}$ of $E(2^\omega)$. His result inspired

LEMMA 4.5. *Assume some wellordering of* $tc(x)$ *is E-recursive in x. If* $\{e\}(x)$ *diverges, then some witness to the divergence is first order definable over* $L(\kappa_r^x, tc(\{x\}))$.

A witness to divergence is an infinite descending path in $T_{\langle e,x\rangle}$. It has the form $\lambda n \mid t(n)$, where $t(0) = \langle e, x\rangle$ and for each n: $t(n+1)$ is a subcomputation instruction of $t(n)$. In the proof of Lemma 4.5, $t(n)$ is defined by recursion on n and Lemma 4.6 is used to insure that each $t(n) \in L(\kappa_r^x, tc(\{x\}))$.

Lemma 4.6 (Kechris's Basis Theorem). *Suppose $y \leq_E x$ and A is E-recursively enumerable in x. If $y - A$ is nonempty, then*

$$\exists b \left[b \in y - A \ \wedge\ \kappa_r^{x,b} \leq \kappa_r^x \right].$$

Proof. It suffices to find a $b \in y - A$ such that $\kappa_0^{x,b} \leq \kappa_r^x$. Suppose no such b exists. Then

$$y \subseteq A \cup \{b \mid \kappa_r^x < \kappa_0^{x,b}\}.$$

The predicate, $\kappa_r^x < \kappa_0^{x,b}$, is E-recursively enumerable in x, because it is equivalent to

$$\exists \delta[\delta \leq_E x, b \ \wedge\ \delta \text{ is not } x\text{-reflecting}].$$

(The predicate, δ, is not x-reflecting, is E-recursive in x.) Thus y is a subset of a set E-recursively enumerable in x, and so for each $b \in y$, there is a δ_b such that

$$\delta_b \leq_E x, b$$

and (i) or (ii) below holds:

(i) δ_b is the length of a computation that shows $y \in A$;
(ii) δ_b is not x-reflecting.

It follows from Gandy selection, as in Corollary 4.3, that δ_b can be construed as a partial E-recursive function of x, b defined for all $b \in y$. Let

$$\delta^\infty = \sup\{\delta_b \mid b \in y\}.$$

Then $\delta^\infty \leq_E x$ by E-recursive bounding, scheme (d) of Section 1, and so $\delta^\infty < \kappa_r^x$. But then (i) holds for every $b \in y$; hence $y - A$ is empty. ⊣

The Kechris Basis Theorem is analogous to one of Gandy's basis theorems: if D is a nonempty Σ_1^1 (with real parameter x) set of reals, then $\exists b \in D$ such that $\omega_1^{x,b} = \omega_1^x$. In both an element b of a "co-recursively enumerable" set is found such that some ordinal generated by b is minimized. The key step in Kechris's proof is showing (ii) cannot happen (i.e. $\delta_b > \kappa_r^x$) if the desired b does not exist. The corresponding step in Gandy's proof is showing $\omega_1^{x,b} > \omega_1^x$ cannot happen if the desired b does not exist. Analogies can be misleading, but in this case it is reasonable to say that aspects of ω_1^x are shared by κ_0^x together with κ_r^x.

Theorem 4.7. *Let x be a set of ordinals. Then* (i) $\longleftrightarrow$ (ii).

(i) *$E(x)$ is not Σ_1 admissible.*
(ii) *$E(x)$ admits divergence witnesses.*

The proof of (ii)⟶(i) is similar to that of Proposition 3.5.

The proof of ¬(ii)⟶ ¬(i) is more difficult. For some κ, $E(x) = L(x,\kappa)$. Let (I) be the sentence

$$(\exists y_{y\in L(\kappa,x)}[\kappa_r^{x,y} \geq \kappa]. \tag{I}$$

Proof of (I) ⟶ ¬ (i). Suppose $y \in L(\kappa, x)$ and $\kappa_r^{x,y} \geq \kappa$. Assume

$$L(\kappa, x) \models \forall u_{u\in d} \exists v \mathcal{F}(u, v, p),$$

$d, p \in L(\kappa, x)$ and $\mathcal{F}$ is Δ_0. It can be shown that

$$\kappa_r^{x,y,p,b} \geq \kappa_r^{x,y}$$

by a painstaking examination of the details of the proof of Lemma 4.5; in particular there is a preferred divergence witness, namely the "leftmost" infinite branch of $T_{\langle e.x\rangle}$. By reflection

$$\forall b_{bed} \exists c[c \leq_E x, y, p, b \wedge \mathcal{F}(b, c, p)].$$

By Corollary 4.3 c can be construed as a partial E-recursive function of x, y, p, b defined for all $b \in d$. Then E-recursive bounding yields

$$L(\kappa, x) \models \exists v \forall u_{u\in d} \mathcal{F}(u, v, p).$$

⊣

Proof of ¬ (ii)⟶ (I). Assume ¬(I) and obtain (ii) via Lemma 4.5. ⊣

Let $A \subseteq L(\kappa)$. Define A is ***E*-recursively enumerable on** $L(\kappa)$ by

$$\exists e \exists b_{b\in L(\kappa)} \forall y_{y\in L(\kappa)} [y \in A \longleftrightarrow \{e\}(b, y) \downarrow].$$

Define A is ***E*-recursive on** $L(\kappa)$ by $\exists e \exists b_{b\in L(\kappa)}$ such that

$$\forall y_{y\in L(\kappa)} [\{e\}(b, y) \downarrow \ \wedge \ (y \in A \longleftrightarrow \{e\}(b, y) = 1)].$$

The interaction above between Σ_1 admissibility and divergence leads to

The Divergence–Admissibility Split. Every E-recursively closed $L(\kappa)$ belongs to **just one** of two classes.

Class I: $L(\kappa)$ admits divergence witnesses.

Class II: $L(\kappa)$ is Σ_1 admissible; and for all $A \subseteq L(\kappa)$, A is Σ_1 definable over $L(\kappa)$ iff A is E-recursively enumerable on $L(\kappa)$.

§5. Finite Injury Arguments and Post's Problem. The title of this section is misleading. In the setting of E-recursion a classical finite injury argument (or an α-finite injury argument) becomes a wait-and-see argument. The standard approach to Post's Problem seeks to preserve inequalities. Negative requirements arise and are violated for the sake of positive requirements of higher priority. In E-recursion inequalities are still welcome but divergence witnesses are also sought. With their assistance injuries can be avoided when $L(\kappa)$ is E-closed but not Σ_1 admissible.

Let $\mathcal{E}$ be E-closed. Suppose $B \subseteq \mathcal{E}$. The relativisation of E-recursiveness to B is simply a matter of adding a seventh scheme

$$\{c\}^B(x_1, \ldots, x_n) = B \cap x_i \qquad (c = \langle 7, n, i\rangle)$$

to the original six for E-recursion. The additional scheme has the same effect as adding the characteristic function of B to the list of finitary functions (projection, difference etc.). Say f is partial E-recursive relative to B if $f \simeq \{e\}^B$ for some e. And $\mathcal{E}$ is E-closed relative to B iff

$$\{e\}^B(x) \downarrow \longrightarrow \{e\}^B(x) \in \mathcal{E}$$

for all $e < \omega$ and $x \in \mathcal{E}$.

Suppose $L(\kappa)$ is E-closed, but not Σ_1 admissible, in order to guarantee that $L(\kappa)$ admits divergence witnesses. A subset of κ is E-recursively enumerable on $L(\kappa)$ iff it equals

$$\{x \mid x < \kappa \wedge \{e)(x, u) \downarrow\}$$

for some $e < \omega$ and $u < \kappa$; to solve Post's problem, two such subsets, A and B, are constructed. As usual there is a list of requirements. Each requirement is settled before proceeding to the next. In the following A and B ambiguously denote sets and characteristic functions. Let requirement 0 be

$$A(w_0^0) \neq \{e_0\}^B(w_0^0, u_0) \quad (w_0^0, u_0 \in \kappa)$$

At stage 0, $B = \emptyset$. Enumerate all E-recursive computations in $L(\kappa)$. At some stage $\sigma_0 < \kappa$ of the enumeration either (i) or (ii) will happen.

(i) A computation appears that defines a value v for $\{e_0\}^{\emptyset}(w_0^0, u_0)$.The computation is based on positive and negative membership facts that B must now satisfy forever. $A(w_0)$ is set equal to 1 if $v = 0$, and to 0 otherwise. All the witnesses w associated with remaining requirements of the form $B(w) \neq \{e\}^A(w, u)$ are given values large enough to insure they will not injure the negative commitments made for B in requirement 0.

(ii) Computations appear, as in the proof of Lemma 4.5, that establish a divergence witness for $\{e\}^{\emptyset}(w_0, u_0)$. The computations are based on positive and negative membership facts that B must now satisfy forever. And some witness values must be increased as in case (i).

Note that $\sigma_0 \leq \kappa_r^{w_0^0, u_0}$ by Lemma 4.5. Requirements can be indexed by triples of the form $\langle e_\delta, w_\delta^{i^\delta}, u_\delta\rangle$, where $\delta < \kappa$, $e_\delta < \omega$, and $i^\delta \in \{0, 1\}$. With this long indexing, time may run out before all the work is done. There is enough time if

$$\sup\{\kappa_r^{w_\delta^{i^\delta}, u_\delta} \mid \delta < \gamma\} < \kappa$$

for all $\gamma < \kappa$. If not, a shorter indexing of requirements is needed. Define ρ^κ to be the least $\beta \leq \kappa$ such that some f is a *partial* E-recursive on $L(\kappa)$ map of β onto $L(\kappa)$.

Slaman [8] proved splitting and density theorems for sets E-recursively enumerable on $L(\kappa)$ by means of κ-finite injury arguments.

THEOREM 5.1. *If* $\gamma < \rho^\kappa$ *and* $p < \kappa$, *then* $\sup\{\kappa_r^{\delta,p} \mid \delta < \gamma\} < \kappa$.

The proof of Theorem 5.1 is a collapsing (or condensation) argument as in the fine structure of L. A device rightly known as Shore Blocking together with the above theorem buys enough time to satisfy all the requirements.

§6. Forcing: C.C.C. Versus Countably Closed. Suppose $\mathcal{P} \in L(\kappa)$ is a notion of set forcing. (A $\mathcal{P}$-generic G is a subset of some member of $L(\kappa)$.) If $L(\kappa)$ is Σ_1 admissible and G is $\mathcal{P}$-generic, then $L(\kappa, G)$ is Σ_1 admissible. In short, set forcing preserves Σ_1 admissibility.

Now suppose $L(\kappa)$ is not Σ_1 admissible. If $L(\kappa)$ is E-closed and G is $\mathcal{P}$-generic, then $L(\kappa, G)$ need not be E-closed. Set forcing does not in general preserve E-closure.

For example consider $E(\omega_1)$ $(= L(\kappa)$ for some $\kappa)$. Let G be a generic map of ω onto ω_1. Then

$$L(\kappa, G) = L(\kappa, b)$$

for some $b \subseteq \omega$. Suppose $L(\kappa, G)$ is E-closed. Then $L(\kappa, b)$ is $E(b)$. By Proposition 3.6, $E(b)$ is Σ_1 admissible and consequently $L(\kappa)$ is Σ_1 admissible. But Proposition 3.5 says $L(\kappa)$ is not Σ_1 admissible.

An instance of set forcing $\mathcal{P}$ consists of a set P of forcing conditions $p, q, r, \ldots$, and an extension relation $\geq$. If $p \geq q$, then q says as much as, or more than, p says about the generic object. Assume for the rest of this section that $L(\kappa)$ is E-closed but not Σ_1 admissible.

For an arbitrary (not necessarily generic) G, each element of $L(\kappa, G)$ is the value of $\{e\}(a, G)$ for some $a \in L(\kappa)$ and $e < \omega$. To show that $L(\kappa, G)$ is E-closed for a $\mathcal{P}$-generic G, there are two approaches.

1. For all $e < \omega$ and $a \in L(\kappa)$, try to force $\mid \{e\}(a, G) \mid$ to be as small as possible in the hope of forcing a value less than κ. This approach succeeds when $\mathcal{P}$ is c.c.c. (**countable chain condition**); c.c.c means that every antichain is countable (any two distinct elements of an antichain have no common extension).
2. For all $e < \omega$ and $a \in L(\kappa)$, again try to force $\mid \{e\}(a, G) \mid$ to be as small as possible, but allow for the possibility of failure and exploit that possibility to try and force a divergence witness for $\{e\}(a, G)$ into $L(\kappa, G)$. This approach succeeds when $\mathcal{P}$ is countably closed:

$$\forall n(p_n \geq p_{n+1}) \rightarrow \exists q \forall n(p_n \geq q).$$

Success is plausible in this case because a divergence witness is an infinite path through an illfounded computation tree with countably many levels.

Reflection plays a major role in forcing the existence of divergence witnesses with the help of Lemma 4.5 and Kechris's Basis Theorem (Lemma 4.6).

The first approach is conceptually simpler than the second, but more combinatoric. Suppose there exist r and δ such that

$$p \geq r \text{ and } r \Vdash (\| \{e\}(a, G) \models \delta).$$

Then it can be shown that there exist such r and δ E-recursive in p, e, a (and some background parameters such as ω and $\mathcal{P}$). Define

$$\min(p, e, a) \simeq \min_{\delta} \exists r_{p \geq r}(r \Vdash (\| \{e\}(a, G) \models \delta)).$$

An effective transfinite recursion on $\min(p, e, a)$ shows $\min(p, e, a)$ is E-recursive in p, e, a uniformly. The recursion manipulates conditions directly and draws heavily on Gandy selection and the countable chain condition.

§7. Model-Theoretic Completeness and Compactness. Suppose $L(\kappa)$ is E-closed but not Σ_1 admissible. Let $\mathcal{L} \subseteq L(\kappa)$ denote an E-recursive on $L(\kappa)$ set of atomic symbols for a first order language, and let $\mathcal{L}_{\kappa,\omega}$ be the restriction of $\mathcal{L}_{\infty,\omega}$ to $L(\kappa)$. The rules and axiom schemes of $\mathcal{L}_{\infty,\omega}$ are in essence the same as those of infinitary logic with one notable addition: a set containing a deduction of $\mathcal{F}_i$ for each $i \in I$ qualifies as a deduction of the conjunction

$$\wedge\{\mathcal{F}_i \mid i \in I\}.$$

If $L(\kappa)$ is not a union of Σ_1 admissible sets, then there a choice for $\mathcal{L}$ such that some sentence of $\mathcal{L}_{\kappa,\omega}$ has a proof in $\mathcal{L}_{\infty,\omega}$ but not in $\mathcal{L}_{\kappa,\omega}$. Let $\Delta \subseteq \mathcal{L}_{\kappa,\omega}$ denote an E -recursive on $L(\kappa)$ set of sentences throughout the present Section. There exist $\mathcal{L}$ and Δ such that every κ-finite subset of Δ is consistent in the sense of $\mathcal{L}_{\infty,\omega}$ but Δ is not. (A set is κ-finite iff it belongs to $L(\kappa)$.) The lack of Σ_1 admissibility makes life on $L(\kappa)$ difficult for a model theorist But there is hope because some forcing arguments (Section 6) succeed on $L(\kappa)$ despite the lack of Σ_1 admissibility.

Let $\mathcal{F}$ denote a sentence of $\mathcal{L}$; $\mathcal{F}$ is said to be a **logical consequence** of Δ (in symbols $\Delta \vdash \mathcal{F}$) iff $\mathcal{F}$ is deducible from Δ via the axioms and rules of $\mathcal{L}_{\infty,\omega}$.

Define $\Delta \vdash_\kappa \mathcal{F}$ to mean $\Delta \vdash \mathcal{F}$ via a deduction in $L(\kappa)$.

Define $\Delta \vdash_\kappa^E \mathcal{F}$ to mean $\Delta \vdash \mathcal{F}$ via a deduction E-recursive in $\mathcal{F}$.

Say Δ is **κ-consistent** iff no contradiction is deducible from Δ via a deduction in $L(\kappa)$.

Say Δ **admits effectivization of deductions** iff for every sentence $\mathcal{F} \in \mathcal{L}$,

$$\Delta \vdash_\kappa \mathcal{F} \;\rightarrow\; \Delta \vdash_\kappa^E \mathcal{F}.$$

Let $\vee\{\mathcal{F}_i \mid i \in I\}$ be a typical disjunction of $\mathcal{L}_{\kappa,\omega}$; "typical" means $I \in L(\kappa)$ and $\mathcal{F}_i \leq_E i, I$ uniformly in i.

Proposition 7.1. *Suppose Δ admits effectivization of deductions and*

$$\Delta, \vee\{\mathcal{F}_i \mid i \in I\} \text{ is } \kappa\text{-consistent.}$$

then $\Delta, \mathcal{F}_i$ is κ-consistent for some $i \in I$.

Proof. Suppose not. Then $\Delta \vdash_{\kappa}^{E} \neg\mathcal{F}_i$ for all $i \in I$. Thus for each $i \in I$, there is an e such that

$$\{e\}(i, I) \text{ converges to a deduction of } \neg\mathcal{F}_i \text{ from } \Delta.$$

For each i, Gandy selection makes it possible to find such an e effectively; Gandy's method is uniform in i, hence there is one e, call it e_0, that works for all i. (The existence of e_0 via Gandy also needs the fact that the class of all deductions from Δ in $L(\kappa)$ is E-recursively enumerable on $L(\kappa)$; that fact follows from the assumption that Δ is E-recursive on $L(\kappa)$.) The E-recursive bounding scheme implies

$$\{\{e_0\}(i, I) \mid i \in I\} \in L(\kappa).$$

But then $\Delta \vdash_{\kappa} \neg \vee \{\mathcal{F}_i \mid i \in I\}$. ⊣

Lemma 7.2. *Suppose κ is countable, Δ is κ-consistent, and $L(\kappa)$ admits effectivization of deductions. Then Δ has a model.*

The proof of Lemma 7.2 is a standard model theoretic construction with ω steps. Each step adds one sentence and preserves the κ-consistency of the previous step. Proposition 7.1 plays a essential part in the construction.

Lemma 7.3. *If Δ is κ-consistent and admits effectivization of deductions, then Δ is consistent in the sense of $\mathcal{L}_{\infty,\omega}$.*

If Δ is countable, then Lemma 7.3 follows from Lemma 7.2. If Lemma 7.2 failed for some uncountable Δ, then by absoluteness it would fail for some countable Δ.

The above formulation of model theory on a E-closed, but not Σ_1 admissible, $L(\kappa)$ is provisional until some questions are answered.

(Q1) Are there some non-trivial examples of κ-consistent Δs that admit effectivization of deductions?

(Q2) Can some established results such as those in [7], be obtained from Lemma 7.2?

(Q3) Does some form of type omitting make sense in the above formulation?

A partial affirmative answer to (Q1) can be extracted from Section 6. Let $\mathcal{P} \in L(\kappa)$ be a *c.c.c* set forcing relation. There exists a κ-consistent Δ that captures the essential properties of set forcing with $\mathcal{P}$ and that admits effectivization of deductions.

BIBLIOGRAPHY.

[1] Harrington, L. (1973), *Contributions to Recursion Theory in Higher Types*, Ph.D. Thesis, MIT, Cambridge, Massachusetts.[4]

[2] Moschovakis, Y.N. (1967), Hyperanalytic predicates, *Trans. Amer. Math. Soc.*, **129**, 249–282.

[3] Moschovakis, Y.N. (1980), *Descriptive Set Theory*, North-Holland, Amsterdam.

[4] Normann, D. (1978), Set recursion. In *Generalized Recursion Theory II*, North-Holland, Amsterdam, 303–320.

[5] Sacks, G.E. (1971), The k-section of a type n-object, *Amer. J. Math.* **99**, 901–917.

[6] Sacks, G.E. (1990), *Higher Recursion Theory*, Springer-Verlag, Berlin.

[7] Slaman, T.A. (1985), Reflection and forcing in E-recursion theory, *Ann. Pure and Appl. Log.* **29**, 79–106.

[8] Slaman, T.A. (1985a), The E-recursively enumerable degrees are dense. In *Proceedings of Symposia in Pure Math.*, Amer. Math. Soc. **42**, 195–213.

Department of Mathematics,
Harvard University,
Cambridge, MA, USA.
E-mail: Sacks@math.harvard.edu

[4]There is more to E-recursion theory than meets the eye above. To see more consult the References.

REVERSE MATHEMATICS, COUNTABLE AND UNCOUNTABLE

RICHARD A. SHORE

Abstract Reverse mathematics analyzes the complexity of mathematical statements in terms of the strength of axiomatic systems needed to prove them. Its setting is countable mathematics and subsystems of second order arithmetic. We present a similar analysis based on (recursion-theoretic) computational complexity instead. In the countable case, this view is implicit in many of results in the area. By making it explicit and precise, we provide an alternate approach to this type of analysis for countable mathematics. It may be more intelligible to some mathematicians in that it replaces logic and proof systems with relative computability. In the uncountable case, second order arithmetic and its proof theory is insufficient for the desired analysis. Our computational approach, however, supplies a ready made paradigm for similar analyses. It can be implemented with any appropriate notion of computation on uncountable sets.

§1. Introduction. The enterprise of calibrating the strength of theorems of classical mathematics in terms of the (set existence) axioms needed to prove them, was begun by Harvey Friedman in the 1970s (as in [6] and [7]). It is now called Reverse Mathematics as, to prove that some set of axioms is actually necessary to establish a given theorem, one reverses the standard paradigm by proving that the axioms follow from the theorem (in some weak base theory). The original motivations for the subject were foundational and philosophical. It has become a remarkably fruitful and successful endeavor supplying a framework for both the philosophical questions about existence assumptions and foundational or mathematical ones about construction techniques needed to actually produce the objects that the theorems assert exist. The basic text here is [20] to which we refer the reader for background and all unexplained notions.

The proof-theoretical setting for this subject has been second order arithmetic. All systems incorporate some fixed list of elementary axioms for first order arithmetic. (These are the axioms for ordered semirings.) They also have a simple induction axiom: $(0 \in X \ \& \ \forall n(n \in X \rightarrow n+1 \in X) \rightarrow \forall n(n \in X))$. The systems considered are then defined by their comprehension axioms starting with RCA_0 which adds comprehension for Δ^0_1 formulas (and induction for Σ^0_1 formulas) to the

[4]Partially supported by NSF Grants DMS-0554855, DMS–0852811, DMS–1161175, and John Templeton Foundation Grant 13408.

[4]Math Subject Codes: 03B30, 03D45, 03D60, 03D80, 03F35, 03F65. Keywords: reverse mathematics, recursive mathematics, uncountable structures.

basic axioms. They then progress through more complicated existence assumptions. In principle, one can go on through, for example, both arithmetic formulas and then comprehension for Π^1_n formulas for each n. In practice, there are almost no results that extend beyond Π^1_1 comprehension. (See [15] for an example at the level of Π^1_2 and [16] for ones for Π^1_n for all $n \geq 3$.)

In this setting, the mathematics that can naturally be analyzed is that of countable structures both combinatorial and algebraic. Topological and analytic structures can be considered if they have some sort of countable basis, e.g. the reals or any separable metric space, and one can code objects of interest such as continuous functions or Borel sets as countable sets. Such codings can at times seem awkward or unnatural (at least to nonlogicians or even novices in the field). Moreover, the whole approach is tied to second order arithmetic and so cannot in any reasonable way talk about combinatorics, algebra, topology or analysis on structures of arbitrary cardinality.

We present an alternative approach based on classical computability that is implicit in many of the results in the area in the countable case and widely taken for granted. From the viewpoint of traditional logic (model theory or proof theory), it can be seen as restricting attention to the standard models. (A model of second order arithmetic consists of a model $\mathcal{M} = (M, +, \times, <, 0, 1)$ of our basic version of first order arithmetic and a collection $\mathcal{S}$ of subsets of M that serves as the domain of quantification for the second order variables in our language. We say that $(\mathcal{M}, \mathcal{S})$ is *standard* if $M = \mathbb{N}$.) In the countable case, it has the expository advantage of requiring no logical/syntactic or proof-theoretic machinery while still providing a classification of the theorems of mathematics in terms of computational complexity that is quite close to the proof-theoretic ones traditionally used.

Moreover, in most cases our approach serves to answer the classification questions for results and techniques raised in discussion of classical mathematics. To give an example from combinatorics, we point to the analyses of various theorems of matching theory from Hall's to the König Duality theorem (KDT) in [9] and [2]. These basically computational analyses provide an answer to the question raised in [12] (pp. 6, 8) where it is noted that many related theorems of matching theory from Frobenius to König seem to be in some sense equivalent (each is a "special case" of the next in a circular fashion) but nonetheless one thinks that KDT is "the deeper result". The analyses cited in the countable case show that the other "equivalent" results are computationally (and reverse mathematically) equivalent to compactness of Cantor space (WKL_0) or closure under jump (ACA_0) while KDT is equivalent (by [19] as well) to closure under "hyperarithmetic in" or ATR_0. Thus KDT is demonstrably the most complicated of this array of theorems.

Similarly, the combinatorist sees that the proofs of KDT, even in the countable case, do not follow the usual pattern of being deduced from the finite case by some form of compactness (and so WKL or full König's Lemma (KL) which is equivalent to ACA_0). The proofs use transfinite recursion and instances of the axiom of choice. Indeed, these concerns on the part of Aharoni (who proved KDT for graphs of arbitrary cardinality [1]) lead to the analysis in [2] which showed that, in the countable case, compactness (in the form of WKL or KL) do not suffice and transfinite recursion in the form of hyperarithmetic procedures or ATR_0 are actually necessary. In the uncountable case our approach provides a method of tackling the mathematical, foundational and philosophical questions of how to calibrate the strength of the theorems and constructions of mathematics posed by the original subject. In particular, issues about the construction techniques needed to prove theorems such as KDT for uncountable graphs can be addressed.

§2. Computable entailment and equivalence. The theorems analyzed in reverse mathematics are typically Π^1_2 assertions, for every structure of some sort there is a function or relation with some desired property. In this setting, it is easy to think about the analysis as one that calibrates the complexity of constructing the desired function or relation given the initial structure. But, of course, more complicated statements are also analyzed.

As is well known, the standard systems of reverse mathematics, RCA_0, WKL_0, ACA_0, ATR_0 and Π^1_1-CA_0, each correspond to some construction or recursion-theoretic principle. The first is effective mathematics, i.e. closure under relative computability so that (in the Π^1_2 case) the desired function or relation is computable in the given structure. The others correspond to "every infinite binary tree has an infinite path" (or the low basis theorem [10] that every such recursive tree has an infinite path P such that $P' \equiv_T 0'$); König's Lemma (KL) every finitely branching infinite tree has an infinite path or closure under the Turing jump; definition by transfinite recursion or closure under "hyperarithmetic in"; and a version of a comprehension/choice principle that chooses the well orderings from a set of linear orderings or closure under the hyperjump. This correspondence is precise on the standard models, i.e. a standard model $(\mathbb{N}, \mathcal{S})$ is a model of RCA_0 if and only if $\mathcal{S}$ is closed under Turing reducibility and join. Given that it is a model of RCA_0, it is also a model of WKL_0, ACA_0, ATR_0 or Π^1_1-CA_0 if and only if every infinite binary tree coded in $\mathcal{S}$ has a path in $\mathcal{S}$ or $\mathcal{S}$ closed under Turing jump, "hyperarithmetic in" or the hyperjump, respectively.

Quite often analyses in reverse mathematics actually proceed by recursion-(computability-)theoretic methods. In the positive direction for a Π^1_2 assertion, one shows that the desired function or relation is computable in the given structure or computable from some type of jump operator (Turing, or iterations into the transfinite all the way to the hyperjump) applied to it. Such proofs generally provide ones in the analogous axiom system of Reverse Mathematics (at times with more induction needed than the usual minimum of Σ^0_1). In the other direction, one often demonstrates that one principle or Π^1_2 mathematical assertion Φ

does not follow from another Ψ (including, for example, one of the basic systems beyond RCA_0) by providing an ideal in the Turing degrees (i.e. a collection of sets closed under Turing reducibility and join and perhaps the jump operator relevant to the discussion) such that Ψ holds in (the standard model of second order arithmetic corresponding to the sets in) the ideal but Φ does not. This, of course, proves that Ψ does not imply Φ over RCA_0 (or over the system (ACA_0; ATR_0 or Π^1_1-CA_0) corresponding to the jump closure condition). It actually provides a stronger independence result that, for example, applies to the base systems with full induction and more.

We propose a direct formulation of this complexity measure based on the difficulty of computing the desired output (function or relation) from the input (structure) as in the typical case of Π^1_2 theorems. Making this view explicit formalizes the intuition that "being harder to prove" means "harder to compute". It also provides a different expository route into the subject suitable for a mathematical or computer science audience that intuitively understands computability but may find formal proof systems foreign or less appealing. More interestingly, for the practitioners already familiar with this approach, it provides an opportunity to deal with uncountable structures and higher order statements that are out of the reach of standard proof-theoretic methods. The route here is to use one (or more) of the studied definitions of computability on uncountable structures.

Definition 2.1. If $\mathcal{C}$ is a *closed class* of sets, i.e. closed under Turing reducibility and join, we say that $\mathcal{C}$ *computably satisfies* Ψ (a sentence of second order arithmetic) if Ψ is true in the standard model of arithmetic whose second order part consists of the sets in $\mathcal{C}$. We say that Ψ *computably entails* Φ, $\Psi \vDash_c \Phi$, If every closed $\mathcal{C}$ satisfying Ψ also satisfies Φ. We say that Ψ and Φ are *computably equivalent*, $\Psi \equiv_c \Phi$, if each computably entails the other.

One can now express the equivalence of some Ψ with, e.g. ACA_0, ATR_0 or Π^1_1-CA in this way. One can also describe entailment or equivalence over one of these systems by either adding them on to the sentences Ψ and Φ or by requiring that the classes $\mathcal{C}$ be closed under the appropriate operators and reductions (Turing jump, hyperarithmetic in and hyperjump, respectively). More interestingly, one can directly express the relationships between two mathematical statements without going through any formal proof systems.

Turning now to uncountable structures, one can simply interpret computability as some version of generalized computability and then immediately have notions appropriate to uncountable settings. For example, if one is interested in algebraic or combinatorial structures where the usual mathematical setting assumes that an uncountable structure is given with its cardinality, i.e. the underlying set for the structure (vector space, field, graph, etc.) may as well be taken to be a cardinal κ, then a plausible notion of computation is given by α-recursion theory. This seems particularly appropriate when one is willing to also assume that $V = L$ to avoid many purely set-theoretic issues that do not usually affect theorems of classical mathematics. (To be fair, one must also then restrain oneself from using noncomputable combinatorial principles special to L, at least without first providing some appropriate complexity analysis for these principles.) In this setting, one carries out basic computations (including an infinitary sup operation) for α (or in our situation κ) many steps assuming some closure properties such as admissibility on α. (Note that every infinite cardinal is admissible.)

For settings such as analysis where the basic underlying set is the reals $\mathbb{R}$ or the complex numbers $\mathbb{C}$, it seems less natural to assume that one has a well-ordering of the structure and one wants a different model of computation. Natural possibilities include Kleene recursion in higher types, E-recursion (of Normann and Moschovakis) and Blum–Shub–Smale computability. (See for example [17] or [4] for α-recursion theory; [17], [14] or [5] for the various versions of recursion in higher types or E-recursion and [3] for the Blum–Shub–Smale model.)

The general program that we are suggesting consists of the following:

Problem 2.2. Develop a computability-theoretic type of reverse mathematical analyses of mathematical theorems on uncountable structures using whichever generalized notion of computability seems appropriate to the subject being analyzed.

Note that the formulation of the basic yardsticks for this analysis will not, in general be the same as for the countable case. An obvious example is Weak König's Lemma. For uncountable cardinals κ, the assertion that every binary tree of height κ (or even just quite simple ones) has a size κ branch is equivalent to κ being weakly compact. Thus such a principle is not even a candidate yardstick for most cardinals. On the other hand, there are natural candidates for analogs of ACA_0 once one has the right notion for the jump operator or enough closure to make sense of closing under first order definability (as over L_κ). We now consider a few standard examples from reverse mathematics in the setting of α-recursion theory for arbitrary cardinals κ inside L.

§3. WKL and ACA in α-recursion theory. For our α-recursion-theoretic analysis of the computational strength of mathematical theorems and constructions for uncountable structures, we assume that $V = L$ and that ours structures are ones on some uncountable cardinal κ. Thus our models are of the form $(L_\kappa, \mathcal{S}, \in)$ where $\mathcal{S}$ is a collection of subsets of L_κ (or equivalently of κ). We take our notion of relative computability to be "α-recursive in", i.e. in our definition of κ-computable entailment and equivalence, we assume that our classes $\mathcal{S}$ of subsets of L_κ are κ-closed, i.e. closed under an effective join on pairs of sets and, if $A \in \mathcal{S}$ and $B \leq_\kappa A$ then $B \in \mathcal{S}$.

Note that an essential feature of $\leq_\kappa$ is that both input and output information consists of κ-finite sets (i.e. ones in L_κ). More precisely, $B \leq_\kappa A$ if there is a κ-r.e. (i.e. Σ_1 over L_κ) set W_γ such that for all κ-finite sets M and N, $M \subseteq B \Leftrightarrow \exists K, L(\langle M, 1, K, L\rangle \in W_\gamma \mathbin{\&} K \subseteq A \mathbin{\&} L \subseteq \bar{A})$ and $N \subseteq \bar{B} \Leftrightarrow \exists K, L(\langle N, 0, K, L\rangle \in W_\gamma \mathbin{\&} K \subseteq A \mathbin{\&} L \subseteq \bar{A})$ where K and L also range over κ-finite sets. This notion stands in contrast to weak κ-recursiveness, $B \leq_{w\kappa} A$, in which correct decisions are only required for questions about individual membership in B. Our assumption that $V = L$ eliminates worries about nonregularity, i.e. if $A \subseteq \kappa$ and $\gamma < \kappa$ then $A \upharpoonright \gamma$ is κ-finite and so all initial segments of our oracles are available as information for our calculations. On the other hand, it does not eliminate the distinction between weak and ordinary κ-recursiveness when κ is a singular cardinal. In this case, the weak notion is not, in general, transitive. This distinction is a common source of difficulties in α-recursion theory. In our considerations, it first comes to light when we consider the relationship between closure under the jump operator and Σ_1 (or, equivalently, first order) definability. (We refer to [17] or [4] for basic definitions and information about α-recursion theory.) Without going into the competing considerations for defining the jump (see [18] and [17] VII.4.8) we adopt the definition of the κ-jump in [18] and [17] as a universal κ-r.e. set in the following form: $A' = \{\langle \gamma, \delta \rangle \mid (\exists K, L, i \in \{0,1\})(\langle \{\delta\}, i, K, L \rangle \in W_\gamma \ \&\ K \subseteq A \ \&\ L \subseteq \bar{A})\}$. Of course, A' is Σ_1 in A (over L_κ) and so closure under "Σ_1 in" implies closure under the jump. In the other direction, however, while it is easy to see that if B is Σ_1 in A (over $L_\varkappa$) then $B \leq_{w\kappa} A'$ (by regularity) in general it need not be the case that $B \leq_\kappa A'$. Nonetheless, as in the countable case, closure under jump does imply closure under first order definability over L_κ by the following lemma.

Lemma 3.1. *If $X \leq_{w\kappa} Y$ then $X \leq_\kappa Y'$.*

Proof. As $X \leq_{w\kappa} Y$ we have a γ such that, for every δ,

$$\delta \in X \Leftrightarrow \exists K, L(\langle \{\delta\}, 1, K, L \rangle \in W_\gamma \ \&\ K \subseteq Y \ \&\ L \subseteq \bar{Y})$$

and

$$\delta \notin X \Leftrightarrow \exists K, L(\langle \{\delta\}, 0, K, L \rangle \in W_\gamma \ \&\ K \subseteq Y \ \&\ L \subseteq \bar{Y}).$$

So for κ-finite M and N we have $M \subseteq X \Leftrightarrow \neg[\exists \delta \in M \exists K, L(\langle \{\delta\}, 0, K, L \rangle \in W_\gamma \ \&\ K \subseteq Y \ \&\ L \subseteq \bar{Y})] \Leftrightarrow h(M) \notin Y'$ for some κ-recursive function h and $N \subseteq \bar{X} \Leftrightarrow \neg[\exists \delta \in N \exists K, L(\langle \{\delta\}, 1, K, L \rangle \in W_\gamma \ \&\ K \subseteq Y \ \&\ L \subseteq \bar{Y})] \Leftrightarrow g(M) \notin Y')$ for some κ-recursive function g. Thus $X \leq_\kappa Y'$ as required. ⊣

Corollary 3.2. *A κ-closed set $\mathcal{S}$ is closed under κ-jump if and only if it is closed under "Σ_1 in" if and only if it is closed under first order definability over L_κ.*

Next we turn to a classic example from both effective and reverse mathematics of a standard theorem equivalent to closure under the jump or first order definability: the existence of bases for vector spaces. As is often the case, choosing the correct classical construction to generalize is crucial. We adapt an argument from [13].

Theorem 3.3. *The existence of bases for all vector spaces of size κ over fields of size κ is κ-computably equivalent to closure under κ-jump or under first order comprehension (over L_κ).*

Proof. For the classical direction, assume we are given a vector space $V = \{v_\gamma | \gamma < \kappa\}$ over a field $K = \{a_\gamma | \gamma < \kappa\}$ with the usual operations of addition and scalar multiplication. It suffices to define a basis for V that is Π_1 in the given structure. Let $D = \{v_\gamma | \exists n \in \omega \exists \langle \gamma_i | i \leq n \rangle \exists \langle \delta_i | i \leq n \rangle [(\forall i \leq n(\delta_i < \gamma) \; \& \; v_\gamma = \sum_{i \leq n} \alpha_{\gamma_i} v_{\delta_i}]\}$. D is clearly Σ_1 over L_κ in the vector space structure V. We claim that $B = V - D$ is a basis for V. B is an independent set by definition (and the exchange principle for vector spaces). On the other hand, it follows by induction on $\gamma < \kappa$ that each v_γ is either itself a member of B or a linear combination of finitely many v_δ for $\delta < \gamma$ and so (by induction) of $v_\beta \in B$ with $\beta < \gamma$.

For the reversal, we consider any κ-closed class $\mathcal{S}$ of subsets of κ such that any vector space coded by a set in $\mathcal{S}$ has a basis in $\mathcal{S}$. We have to prove that if $X \in \mathcal{S}$ then $X' \in \mathcal{S}$. We begin with the field K generated by κ many variables over the rationals and the vector space V generated over K by independent elements z_γ for $\gamma < \kappa$. Thus V is the set of finite linear combinations of z_γ with coefficients in K. We list V as v_δ so that the set of z_γ appearing in the linear combination that is v_δ is contained in $\{z_\gamma | \gamma \leq \delta\}$. In particular, we take $z_0 = v_0$. We write $(v)_\beta$ for the coefficient of z_β in $v \in V$.

We now define a κ-recursive sequence λ_β of elements of K by recursion. At stage β we check for each $\gamma, \delta < \beta$ if we can find a k and a finite sequence k_j from K and one γ_j and γ from β such that $v_\delta = \sum k_j(z_0 + \lambda_{\beta_j} z_{\gamma_j}) + k(z_0 + \lambda z_\gamma)$ for some $\lambda \in K$ with $\gamma \neq \gamma_i \neq \gamma_j \neq 0$ for each $i \neq j$. Note that by our numbering scheme the only possible γ, γ_j are less than δ and so, by the independence of the z_γ, this is a κ-recursive check. In fact, we claim that for each possible $\delta, \gamma, \langle \gamma_j \rangle$ there can be at most one choice of k, k_j, λ:

First, if $v_\delta = \sum k_j(z_0 + \lambda_{\beta_j} z_{\gamma_j}) + k(z_0 + \lambda z_\gamma)$ then $k_j \lambda_{\beta_j} = (v_\delta)_{\gamma_j}$ and so k_j is uniquely determined. Next, $k + \sum k_j = (v_\delta)_0$ and so k is also uniquely determined. Finally, $k\lambda = (v_\delta)_\gamma$ and so λ is uniquely determined as well.

Thus there are fewer than κ many possible values of such λ over all $\gamma, \delta < \beta$ and so we may κ-recursively choose $\lambda_\beta > \beta$ to be different from all of them.

We now consider any set W_e^X which is κ-r.e. in X and so

$$W_e^X = \{\gamma | \exists K, L(\langle \{\gamma\}, 1, K, L \rangle \in W_e \; \& \; K \subseteq X \; \& \; L \subseteq \bar{X})\}.$$

We consider a subspace $\hat{V}$ of V generated by the set $\hat{B}$ of elements $z_0 + \lambda_\beta z_\gamma$ where γ is *enumerated in* W_e^X *at stage* β, $\gamma \in W_{e,at\ \beta}^X$, i.e.

$$\exists K, L \in L_\beta(\langle \gamma \}, 1, K, L \rangle \in W_\gamma$$

via a witness in L_β with $K \subseteq X \; \& \; L \subseteq \bar{X}$ but there are no such K, L and witness in L_δ for $\delta < \beta$ (and so, in particular, $\gamma < \beta$).

Claim. $\hat{B}, \hat{V} \leq_\kappa X$.

Proof of Claim. As for $\hat{B}$, it is clear that checking membership for any element of the form $z_0 + \lambda_\beta z_\gamma$ only requires information about L_β and $X \restriction \beta$ (a κ-finite set by regularity) and the κ-recursive sequence λ_β and so $\hat{B} \leq_\kappa X$.

As for $\hat{V}$, we claim that $v_\delta \in \hat{V} \Leftrightarrow v_\delta$ is in the span of $\hat{B}\restriction\delta + 1$ as by the definition of the sequence λ_β, it can never be put in by later elements. Thus given the κ-finite information $\hat{B}\restriction\mu + 1$ about $\hat{B}$ we can check κ-recursively for all $\delta \leq \mu$ if $v_\delta = k_0 z_0 + \sum k_j z_{\gamma_j}$ is in the span of $\hat{B}\restriction\mu + 1$ by verifying that the z_{γ_j} are among those appearing in $\hat{B}\restriction\mu + 1$ and, if so, that assuming they appear there as $z_0 + \lambda_{\beta_j} z_{\gamma_j}$ that the coefficients are as required, i.e. there are $\hat{k}_j$ such that $\hat{k}_j \lambda_{\beta_j} = k_j$ and $k_0 = \sum \hat{k}_j$. Thus $\hat{V} \leq_\kappa X$ as required. ⊣

We now define the quotient vector space $V/\hat{V}$ as the set of least elements of the appropriate equivalence classes: $v_\delta \in V/\hat{V} \Leftrightarrow \forall\mu < \delta[(v_\mu - v_\delta) \notin \hat{V}]$ so, in particular, $z_0 \in V/\hat{V}$. We define the natural operations on $V/\hat{V}$ as those inherited from V by taking the least element of the appropriate equivalence class. So, for example, if $v_\gamma, v_\delta \in V/\hat{V}$ then $v_\gamma + v_\delta$ is v_μ where μ is least such that $(v_\mu - (v_\delta + v_\gamma)) \in \hat{V}$. Thus $V/\hat{V} \leq_\kappa X$ and so is coded in $\mathcal{S}$. Our closure assumption now gives us a basis $B \in \mathcal{S}$ for $V/\hat{V}$. Without loss of generality, we may assume that $z_0 \in B$ (otherwise, put z_0 in and perform the standard exchange procedure replacing one of the elements needed to generate it).

We next claim that $W_e^X \leq_{w\kappa} B = \{b_\mu | \mu < \kappa\}$: To see if $\gamma \in W_e^X$ find λ_i and μ_i for $i \leq n$ such that $\sum \lambda_i b_{\mu_i} \equiv_{\hat{V}} z_\gamma$, i.e. the least elements from which they differ by something in $\hat{V}$ are the same. (There must be such as the b_μ form a basis for $V/\hat{V}$.) Now if $n = 1$ and $b_{\mu_0} = z_0$ there is a λ such that $z_0 + \lambda z_\gamma \in \hat{V}$ and so $\gamma \in W_e^X$. On the other hand, if $z_\gamma \not\equiv_{\hat{V}} z_0$, $\gamma \notin W_e^X$ for if it were then there would be a β such that $z_0 + \lambda_\beta z_\gamma \in \hat{V}$.

Now what we really want is that $W_e^X \leq_\kappa B$ as then the proper choice of e would have $X' \leq_\kappa B$ as desired. However, we do not in general know where in B we may need to look to find the components of a representative for every v_δ in some κ-finite M. They could be unbounded if κ is singular. However, the proof of Lemma 3.1 shows that there are κ-recursive functions h and g such that for κ-finite M and N, $M \subseteq W_e^X \Leftrightarrow h(M) \notin B'$ and $N \subseteq \overline{W_e^X} \Leftrightarrow g(N) \notin B'$. If we now repeat our entire argument with B in place of X and B' in place of W_e^X we produce a new vector space such that for any basis C, $B' \leq_{w\kappa} C$. Combining these reductions, we see that $W_e^X \leq_\kappa C$ for any such basis and so, as X' is κ-r.e. in X and $\mathcal{S}$ has the assumed closure properties, $X' \in \mathcal{S}$ as required. ⊣

We now turn to determining the appropriate analog for WKL in this setting. As explained at the end of §2, assuming that every binary branching tree of height κ has a branch of length κ is much too strong. The appropriate tree formulation seems to be the following:

Definition 3.4. A binary tree T on a cardinal κ (i.e. a subset of $2^{<\kappa}$ closed downward under initial segments) is of *finite character* if T is continuous at limit levels, i.e. for any $\gamma \in 2^{<\kappa}$ of length a limit ordinal λ, if $\gamma\restriction\delta \in T$ for every $\delta < \lambda$ then $\gamma \in T$ and for every $\sigma \in T$, if there is a $\gamma > |\sigma|$ such that σ has no successors on

T at level γ then there is a $\tau \subseteq \sigma$ of length a successor ordinal and a $\hat{\gamma}$ such that τ has no successors on T of length $\hat{\gamma}$.

DEFINITION 3.5. The *finite character tree property for a cardinal* κ, $FCTP_\kappa$, says that every binary tree T on κ of finite character has a path of length κ.

We now prove the κ-computable equivalence of $FCTP_\kappa$ with a couple of theorems from logic that, in the countable case, are equivalent to WKL_0.

THEOREM 3.6. *The following are computably equivalent in the sense of* κ*-recursion theory for each cardinal* κ*:*

(1) $FCTP_\kappa$.
(2) *The compactness theorem for first order logic for languages (even propositional ones) and theories of size* κ.
(3) Σ_1*-Separation: for every* X *and pair of disjoint* Σ_1^X *over* L_κ *sets* A *and* B *there is a separating set* C*, i.e.* $A \subseteq C$ *and* $B \cap C = \emptyset$.

PROOF. We prove enough computable entailments among the three conditions to guarantee computable equivalence.

(1) $\vDash_c$ (2): Given a theory T of size κ in a language L of size κ we add Henkin constants c_φ to the language and κ-recursively build a Henkin tree H of sets of sentences T_σ consistent with T for $\sigma \in H \subseteq 2^{<\kappa}$ as usual. At a level corresponding to a sentence φ we split and add on either φ or $\neg\varphi$. If $\varphi = \exists x\psi(x)$ then when we add on φ we also add on $\psi(c_\varphi)$. At limit levels we take unions. We also check at each node if T_σ is inconsistent with T via a proof using axioms from T_σ and $T \cap L_{|\sigma|}$. If so we stop the construction of the tree above σ.

We claim that the tree so constructed is of finite character. Continuity at limit levels is immediate by construction and the finitary nature of proofs. If $\sigma \in H$ with $|\sigma| = \lambda$ a limit ordinal has no successors at level $\gamma > |\sigma|$ then we claim that $T_\sigma \cup (T \cap L_\gamma)$ is inconsistent. If not, then the classical construction of the Henkin tree (in L) would give some path of length γ extending σ that is consistent with T. This path is a constructible subset of γ and so in L_κ contrary to our assumption. Thus there are finitely many sentences in $T_\sigma \cup (T \cap L_\gamma)$ which are inconsistent. All of them from T_σ appear first at some successor level element $\tau \subseteq \sigma$ of H and so τ has no successors at level γ in H as required.

Thus by FCTP_κ, H has a path P of length κ. One can now κ-recursively build a model of $T \cup \{T_\sigma | \sigma \in P\}$ as usual.

(2) $\vDash_c$ (3): Given $X \in \mathcal{S}$, A and B as in the statement of Σ_1-Separation, we define a theory T in the language of propositional logic of size κ. The language consists of the propositional letters p_γ for $\gamma < \kappa$. If β is enumerated in A at stage γ we put $p_{2\gamma} \rightarrow p_{2\beta+1}$ and $\neg p_{2\gamma} \rightarrow p_{2\beta+1}$ into T. If β is enumerated in B at stage γ we put $p_{2\gamma} \rightarrow \neg p_{2\beta+1}$ and $\neg p_{2\gamma} \rightarrow \neg p_{2\beta+1}$ into T. It is clear that $T \leq_\kappa X$ as we can get $T \restriction \gamma$ from κ-finite information about A and B essentially determined by $X \restriction \gamma$. As T is consistent by the disjointness of A and B, by (2) we have a model M of T in $\mathcal{S}$. We can now define the required separator $C \leq_\kappa M$ by $\beta \in C \Leftrightarrow M \vDash p_{2\beta+1}$.

(3) $\models_c$ (1): Let $T \in \mathcal{S}$ be a binary tree on κ of finite character. We define Σ_1^T sets A and B as usual: $A = \{\sigma | \exists\gamma(\sigma\hat{\ }0$ has no successors on T at level γ but $\sigma\hat{\ }1$ has successors on T at every level $\beta < \gamma\}$ and $B = \{\sigma | \exists\gamma(\sigma\hat{\ }1$ has no successors on T at level γ but $\sigma\hat{\ }0$ has successors on T at level $\gamma\}$. It is clear that A and B are Σ_1^T and disjoint. Thus we have a separator $C \in \mathcal{S}$ by (3). We can now κ-recursively in C define a path P in T of length κ as required: We begin with $\emptyset \in P$ and then, recursively, say that if $\sigma \in P$ then $\sigma\hat{\ }1 \in P$ if $\sigma \in C$ and $\sigma\hat{\ }0 \in P$ if $\sigma \notin C$. At limit levels, we take the union σ of the path so far. We argue by induction up to κ that T is unbounded above every $\sigma \in P$ and so P has length κ. At successor levels this is immediate by the definitions of A, B and C and the inductive hypothesis. At limit levels λ, note first that as T is of finite character, the union σ of the path defined so far is at least a node on T. If σ does not have successors at every level of T then as T is of finite character there would be a proper initial segment δ of σ that does not have unboundedly many successors in T as well contradicting our inductive hypothesis. ⊣

We close this section with a sample theorem of classical mathematics that is κ-computably equivalent to FCTP$_\kappa$ (and so to the κ versions of Σ_1-Separation and first order compactness).

Theorem 3.7. *The theorem that every commutative ring of size κ has a prime ideal is κ-computably equivalent to FCTP$_\kappa$.*

Proof. For the classical direction, suppose we have a ring $R = \{a_i | i < \kappa\}$ (with a_0 the 0 of R and a_1 the 1 of R) coded in $\mathcal{S}$. For the purposes of illustration, we offer two proofs. One shows how in this setting we can appeal to classical, non-construcive arguments to get a computably simple solution. The other is more traditionally constructive as one would see in a typical reverse mathematical argument.

For our first proof, we define a binary tree $T \leq_\kappa R$ such that any path P of length κ is (the characteristic function of) a prime ideal I in R. We begin by putting $\emptyset$, $\langle 1\rangle$ and $\langle 1, 0\rangle$ into T and no other strings of length at most 2. We then put every other binary string σ on κ into T unless there are $i_1, \dots, i_n, j_1, \dots, j_n, k_1, \dots, k_m < |\sigma|$ such that $\sigma(j_l) = 1$ for $l \leq n$, $\sigma(k_l) = 0$ for $l \leq m$ but $\sum_{l \leq n} a_{i_l} a_{j_l} = \prod_{l \leq m} a_{k_l}$.

Suppose P is a path in T of length κ. We let $I = \{a_i | P(i) = 1\}$ and note that by our conditions for terminating paths in T, I is a prime ideal of R. (First, $0 \in I$ while $1 \notin I$ by our fixing of the first two entries on T. If $a_i, a_j \in I$ but $a_k = a_i + a_j$ does not, then we would violate the defining condition. Similarly if $a_j \in I$ but $a_k = a_i a_j$ does not or if $a_i, a_j \notin I$ but $a_k = a_i a_j$ does.) It is also clear that T is continuous at limit levels by the finite nature of the condition for termination. Thus, to get a path of length κ in $\mathcal{S}$ we only need to show that if some node σ of limit length λ has no successors at some level $\gamma > \lambda$ then some $\tau \subseteq \sigma$ not of limit length also has no successors at some level $\hat{\gamma}$. (As T will then be of finite character and so have a path of length κ in $\mathcal{S}$.) This however, is immediate from the classical proof. If one begins with σ such that there is no violation of the defining condition with only the j_l and k_l restricted to be less than $|\sigma|$, i.e. no linear combination of elements already in I gives a product of elements already determined to not be in I, then a maximal extension of the set defined by σ with this property is a prime (indeed maximal) ideal and so σ has extensions on T at every level below κ. On the other hand, if σ does violate this unrestricted condition then some finite subset of it contained in some $\tau \subseteq \sigma$ not of limit length also violates it and so there is a level $\hat{\gamma}$ at which the violating combination is found and so τ has no successors on T after $\hat{\gamma}$.

We next provide a similar proof using Σ_1-Separation that does not rely explicitly on the classical argument. We begin with a list Q_δ of all $i_1, \dots, i_n, j_1, \dots, j_n, k_1, \dots, k_m$ such that $\sum_{l\leq n} a_{i_l}a_{j_l} = \prod_{l\leq m} a_{k_l}$. We say that a string σ satisfies Q_δ, $\sigma \vDash Q_\delta$, if $\sigma(j_l) = 1$ for $l \leq n$ and $\sigma(k_l) = 0$ for $l \leq m$. We now define disjoint Σ_1^R sets A and B of binary strings on κ by $A = \{\sigma \supseteq \langle 1,0\rangle \,|\, \exists\delta(\sigma\hat{\ }0 \vDash Q_\delta \ \&\ \forall\gamma \leq \delta(\sigma\hat{\ }1 \nvDash Q_\gamma))\}$ and $B = \{\sigma \supseteq \langle 1,0\rangle \,|\, \exists\delta(\sigma\hat{\ }1 \vDash Q_\delta \ \&\ \forall\gamma < \delta(\sigma\hat{\ }0 \nvDash Q_\gamma))\}$. Now let $C \in \mathcal{S}$ be a separating set for A and B. Define by recursion on κ a characteristic function beginning with $\langle 1,0\rangle$ and continuing at later steps σ by extending to $\sigma\hat{\ }C(\sigma)$. This gives a characteristic function f. We now set $I = \{a_i | f(i) = 1\}$. Clearly $I \leq_\kappa C$ and so is in $\mathcal{S}$. We claim that it is a prime ideal of R. We prove by induction on the length of initial segments σ of f that $\sigma \nvDash Q_\delta$ for every δ. This is clear for $|\sigma| = 2$ and at limit levels. At successor levels the only way it could fail would be for both $\sigma\hat{\ }0$ and $\sigma\hat{\ }1$ to satisfy some Q_{δ_0} and Q_{δ_1}. (If neither do, then there is nothing to prove. If only one does, then the definitions of A and B and the choice of C as a separator guarantees that $\sigma\hat{\ }(1-C(\sigma))$ does and so $\sigma\hat{\ }C(\sigma)$ does not.) So suppose, for the sake of a contradiction, that we have such examples and so, relabeling to reduce the number of subscripts, we have $a_i, c_j \in \sigma$, $b_k, d_l \notin \sigma$ and $a = a_{|\sigma|}$ such that $\sum r_i a_i = a^n \prod b_k$ and $sa + \sum s_j c_j = \prod d_l$ for some $r_i, s_j, s \in R$ and $n \in \mathbb{N}$. Multiplying both sides of the first equation by s^n gives $\sum r_i s^n a_i = (sa)^n \prod b_k$. Solving the second for sa gives $sa = \prod d_l - \sum s_j c_j$. Substituting the second result into the first gives $\sum r_i s^n a_i = \prod b_k(\prod d_l - \sum s_j c_j)^n$. Now expanding out the right hand side of this last equation gives $\prod b_k(\prod d_l)^n$ plus a sum of terms all of which contain at least one c_j, Moving all of these terms over to the left hand side makes that side into a linear combination of elements in σ with coefficients in R while leaving the right hand side a product of elements declared not in σ for the desired contradiction.

We conclude with the reversal by showing the existence of prime ideals implies Σ_1-Separation. Suppose we have $X \in \mathcal{S}$ and disjoint Σ_1^X sets W_e^X and W_i^X (with enumerations of $W_{e,\gamma}^X$ and $W_{i,\gamma}^X$ κ-recursive in X such that at most one element is enumerated at any γ) for which we want to find a separator in $\mathcal{S}$. Note that if either of the enumerations is bounded then the set is κ-finite and separation is immediate so we assume that they are unbounded. We begin with the ring of polynomials over the rationals generated by variables x_β and y_β for $\beta < \kappa$, $R_0 = Q[x_\beta, y_\beta | \beta < \kappa]$. Let $B = \{x_\alpha y_\delta | \alpha \in W_{e,\delta}^X\} \cup \{x_\beta y_\gamma - 1 | \beta \in W_{i,at\ \gamma}^X\}$ and I_0 be the ideal generated in

R_0 by B. It is clear that $B \leq_\kappa X$. We also claim that $I_0 \neq R_0$: The only way that 1 could be in I_0 is if there are α and δ such that $\alpha \in W^X_{e,\delta}$ and $\alpha \in W^X_{i,at\ \delta}$ but this would contradict the disjointness of W^X_e and W^X_i.

We now define a ring R isomorphic to R_0/I_0 by choosing representatives of the equivalence classes and defining the operations accordingly. Every member of R is given initially as a finite sum of terms of the form $q \prod x_{\beta_i} \prod y_{\delta_j}$ for rational q and finite sequences β_i and δ_j. We prescribe a procedure to reduce such terms to a canonical form modulo I_0. Let δ be larger than every β_i and δ_j appearing in this sum. For each pair β_i, δ_j appearing ask if $\beta_i \in W^X_{e,\delta_j}$, if so set all terms containing both x_{β_i} and y_{δ_j} to 0. If not, see if $\beta_i \in W^X_{i,at\ \delta_j}$ and remove x_{β_i} and y_{δ_j} from any term containing them. Note that for each β_i there is at most one δ_j for which $\beta_i \in W^X_{i,at\ \delta_j}$ and vice versa so that this procedure has only one possible outcome. We take the range of this procedure on R_0 as the universe of our desired ring R. The ring operations are defined as expected. We perform the operations in R_0 and the apply the procedure to produce the canonical representative in R. It is routine to check that this endows R with the structure of a ring over Q. Moreover, this ring is κ-recursive in X as the reduction procedure only needs $W^X_{e,\delta}$ and $W^X_{i,\delta}$ where δ is larger than all the subscripts appearing in a given term. Thus $R \in \mathcal{S}$.

Now suppose that $I \in \mathcal{S}$ is a prime ideal in R. If $\beta \in W^X_e$ then $x_\beta y_\delta \in I$ for all sufficiently large δ. On the other hand, as the enumeration of W^X_i is unbounded there are arbitrarily large δ such that, for some ν, $(x_\nu y_\delta - 1) \in B$. No such y_δ can be in I as if it were, then $x_\nu y_\delta$ and hence 1 would also be in I. Thus $x_\beta \in I$. On the other hand, if $\beta \in W^X_i$ then for some δ, $\beta \in W^X_{i,at\ \delta}$ and so $(x_\beta y_\delta - 1) \in B$ and so again as $1 \notin I$, $x_\beta \notin I$. Thus $\{\beta | x_\beta \in I\}$ is a separating set for W^X_e and W^X_i which is κ-recursive in I and so in $\mathcal{S}$ as required. ⊣

An obvious issue here is whether FCTP_κ and closure under κ-jump are different conditions. They are, of course, for $\kappa = \omega$ and we believe they are for certain uncountable κ of countable cofinality but, in general, it seems an interesting question.

Problem 3.8. Is FCTP_κ a strictly weaker condition on κ-closed sets than closure under κ-jump for every uncountable κ.

§4. ATR and Π^1_1-CA in α-recursion theory. A natural candidate for the analog of Π^1_1-CA_0 is, of course, closure under definability by formulas with a single second order quantifier. We have not yet looked for any equivalences at this level.

Problem 4.1. What mathematical theorems are computably equivalent (in the sense of α-recursion theory) to closure under definability over L_κ by formulas with a single second order quantifier.

It is not clear what the appropriate basic yardstick corresponding to ATR_0 should be. A candidate for analysis here is König's Duality Theorem (KDT) which is equivalent to ATR_0 in the countable case [2],[19]. The arguments of [2] show that KDT is strictly stronger than closure under first order definability for every κ, i.e. it is not computably entailed by closure under the κ-jump. We do not know what reasonable closure notion (if any) might κ-computably entail KDT for uncountable κ but one possibility is suggested by the analysis of KDT in [2] is a direct analog of the definition of the H sets.

Definition 4.2. We say that a κ-closed set $\mathcal{S}$ is *κ-hyperarithmetically closed* if it is closed under projections (i.e. if $S \in S$ then $\{x | \exists \beta(\langle x, \beta\rangle \in S\} \in \mathcal{S}$ and under κ-recursive unions (i.e. if there is a sequence $\left\langle S_\beta | \beta < \lambda \right\rangle$ for some $\lambda < \kappa$ of sets given uniformly κ-recursively as projections at successor levels and as effective unions at limit levels, then the effective join, $\{\langle \beta, x\rangle \,| \beta < \lambda \;\&\; x \in S_\beta\}$, is also in $\mathcal{S}$.

The analysis of KDT in [2] shows that it κ-computably entails hyperarithmetic closure. It is far from clear if any reversal is possible.

Problem 4.3. Is hyperarithmetic closure equivalent to KDT for every κ? What about comparability of well orderings (of subsets of κ)? (Note that if $cf(\kappa) > \omega$ then being well-founded is a co-κ-r.e. relation so the situation in general for well-orderings is quite different than in the countable case.) If this is not the "right" analog for ATR_0 in α-recursion theory, then what is?

§5. E-recursion and related systems. Turning to analysis and related subjects about $\mathbb{R}$, we just note that an old result of Grilliot [8] can be seen from our point of view as showing that the existence of a noncontinuous functional is computably equivalent to the existence of ${}^2E$. In this setting there are also proof-theoretic approaches that correspond to Kleene recursion in higher types as the classical proof-theoretic systems do to Turing computability (see [11] where other examples from E-recursion are analyzed proof-theoretically).

Problem 5.1. Analyze the classical theorems of analysis in terms of computable entailment and equivalence with computation taken to be Kleene Recursion in higher types, E-recursion or Blum–Shub–Smale computability.

REFERENCES.

[1] Aharoni, R., König's duality theorem for infinite bipartite graphs, *J. Lon. Math. Soc. (3)* **29** (1984), 1–12.

[2] Aharoni, R., Magidor, M., and Shore, R.A., On the strength of König's duality theorem for infinite bipartite graphs, *J. Combin. Theory Ser. B* **54** (1992) 257–290.

[3] Blum, Lenore, Cucker, Felipe, Shub, Michael and Smale, Steve, *Complexity and Real Computation*, Springer-Verlag, New York, 1998.

[4] Chong, C.T. *Techniques of Admissible Recursion Theory*, LNM **1106**, Springer-Verlag, Berlin, 1984.

[5] Fenstad, Jens Erik, *General Recursion Theory: An Axiomatic Approach*, Perspectives in Mathematical Logic, Springer-Verlag, Berlin–New York, 1980.

[6] Friedman, H., Higher set theory and mathematical practice, *Ann. Math. Logic* **2** (1971), 326–357.

[7] Friedman, H., Some systems of second order arithmetic and their use, *Proc. Intl. Cong. Math.*, Vancouver 1974, vol. 1, Can. Math. Congress, 1975, 235–242.

[8] Grilliot, T.J. On effectively discontinuous type-2 objects. *J. Symbolic Logic* **36** (1971) 245–248.

[9] Hirst, J., Marriage theorems and reverse mathematics. In *Logic and Computation* (Pittsburgh, PA, 1987), *Contemporary Mathematics* **106**, Amer. Math. Soc., Providence, RI, 1990, 181–196.

[10] Jockusch, C.G. jr. and Soare, R.I., Π^0_1 classes and the degrees of theories, *Tran. AMS* **173** (1972), 33–56.

[11] Kohlenbach, U., Higher order reverse mathematics. In *Reverse Mathematics 2001*, S. Simpson ed., 281–295, Lect. Notes Log. **21**, Association for Symbolic Logic and A.K. Peters, Wellesley MA 2005.

[12] Lovasz, L. and Plummer, M. D., *Matching Theory*, Ann. Discrete Math. **29**, North-Holland, Amsterdam, 1986.

[13] Metakides, G. and Nerode, A., Recursively enumerable vector spaces, *Ann. Math. Logic* **11** (1977), 147–171.

[14] Moldestad, J., *Computations in Higher Types*, LNM **574**, Springer-Verlag, Berlin, 1977.

[15] Mummert, C. and Simpson, S.G., Reverse Mathematics and Π^1_2 comprehension, *Bulletin of Symbolic Logic* **11** (2005) 526–533.

[16] Montalbán, A. and Shore, R.A., The limits of determinacy in second order arithmetic, *Proceedings of the London Mathematical Society* **104**(3) (2012), 223–252.

[17] Sacks, Gerald E., *Higher Recursion Theory*, Perspectives in Mathematical Logic, Springer-Verlag, Berlin, 1990.

[18] Shore, R.A., On the jump of an α-recursively enumerable set, *Trans. AMS* **217** (1976), 351–364.

[19] Simpson, S., On the strength of König's duality theorem for countable bipartite graphs, *J. Symbolic Logic* **59** (1994) 113–123.

[20] Simpson, S., *Subsystems of Second Order Arithmetic*, 2nd edition, Perspectives in Logic, Association for Symbolic Logic and Cambridge University Press, New York, 2009.

Department of Mathematics,
Cornell University,
Ithaca, NY 14853, USA.
E-mail: shore@math.cornell.edu

EFFECTIVE MODEL THEORY: AN APPROACH VIA Σ-DEFINABILITY

ALEXEY STUKACHEV

Abstract We present an approach to abstract computability based on the notion of Σ-definability, and survey some results on effective model theory which are obtained within this framework.

§1. Introduction. We look at computability over abstract structures via the formalism based on the notion of Σ-definability in admissible sets and, in particular, in hereditarily finite (HF) superstructures. HF-computability, as well as some other different (yet equivalent) approaches to computability over abstract structures provide a natural framework for considering "computability" ("effectiveness") relative to this structure. A survey of results on HF-computability in general can be found in [15]. Here we present some of the results focused (mostly) on effective model theory.

Considering the effectiveness of model-theoretical constructions, we first look at the notion of interpretation of one structure in another, which seems to be much older than model theory and probably as old as mathematics itself (for examples, one should think of numeral systems, geometries, fields, etc.). Classical computable model theory studies interpretations and presentations of countable structures in the standard model of arithmetics or, equivalently, in the least admissible set $\mathbb{HF}(\varnothing)$. So the notion of Σ-definability of a structure in HF-superstructure over another structure is, on the one hand, an effectivization of the notion of interpretability from model theory, and, on the other hand, a generalization of the notion of constructivizability from computable (or constructive) model theory. Moreover, the notion of Σ-degree of a structure turns out to be closely related to the notion of degree (in the sense of Turing or enumeration reducibility) from classical computability theory, as well as with the notion of degree spectra of a (countable) structure. So, the notion of Σ-degree, which has the advantage of being defined for structures of arbitrary cardinality, is a natural tool for measuring the (relative) complexity of a structure.

[1]This research was partially supported by the Ministry of Education and Science of Russian Federation (project 8227), the Russian Foundation for Basic Research (grants 11–01-=00688-a, 13–01–91001-ANF-a), and the State Maintenance Program for the Leading Scientific Schools of the Russian Federation (grants N.Sh.-3606.2010.1, N.Sh.-276.2012.1).

It turns out that many of the important model-theoretical constructions are natural tools for studying effective model theory. We will mention, in addition to interpretations, some other typical examples, including Skolem expansions, Marker's extensions, Fraïssé limits, indiscernibles, etc.

Since we are concerned with our particular approach based on Σ-definability in HF-superstructures, we just mention some closely related approaches, such as search computability [64], Montague computability [53], and BSS-computability [5, 2]. It should be noted that the notion of search computability, as well as the notion of abstract computability in the sense of Montague, are equivalent (in accordance with [18]) to the notion of HF-computability.

The author would like to thank the organizers of the Effective Mathematics of the Uncountable (EMU) series of workshops for the opportunity to present a talk on the topics surveyed in the present paper.

§2. Basic Definitions and Facts. By ω we denote the set of natural numbers. For an arbitrary set M, we construct the collection $HF(M)$ of hereditarily finite sets over M as follows:

$HF_0(M) = \varnothing$;
$HF_{n+1}(M) = \mathcal{P}_\omega(M \cup HF_n(M)),\ n < \omega$
(here $\mathcal{P}_\omega(X)$ is the collection of all finite subsets of X);
$HF(M) = \bigcup_{n<\omega} HF_n(M)$.

For simplicity, we consider structures of relational signatures only, identifying functions with their graphs. We consider at most countable signatures, usually finite but always computable. We assume that each signature is equipped with some fixed Gödel numbering of its (first-order) formulas.

If $\mathfrak{M}$ is a structure of some relation signature σ then one can define on $M \cup HF(M)$ a structure $\mathbb{HF}(\mathfrak{M})$ of signature $\sigma' = \sigma \cup \{U, \varnothing, \in\}$ ($U, \varnothing, \in$ are some symbols not in σ) so that

$U^{\mathbb{HF}(\mathfrak{M})} = M$;
$P^{\mathbb{HF}(\mathfrak{M})} = P^{\mathfrak{M}},\ P \in \sigma$;
$\varnothing^{\mathbb{HF}(\mathfrak{M})} = \varnothing \in HF_0(M)$;
$\in^{\mathbb{HF}(\mathfrak{M})} = \in \cap ((M \cup HF(M)) \times HF(M))$.

We will also assume that the signature of $\mathbb{HF}(\mathfrak{M})$ contains a binary relational symbol Sat^2 interpreted as the satisfiability relation for the atomic formulas of $\mathfrak{M}$, with respect to the fixed Gödel numbering. In the case of finite signatures this additional assumption is not essential.

A class of Δ_0-*formulas* of signature σ' is the least one containing atomic formulas which is closed under $\vee$, $\wedge$, $\rightarrow$, $\neg$ and restricted quantifiers $\forall x \in y$ and $\exists x \in y$ ($\forall x \in y\varphi$ and $\exists x \in y\varphi$ are abbreviations for $\forall x(x \in y \rightarrow \varphi)$ and $\exists x(x \in y \wedge \varphi)$ respectively).

A class of *Σ-formulas* of signature σ' is the least one containing Δ_0-formulas and closed under $\vee$, $\wedge$, restricted quantifiers $\forall x \in y$, $\exists x \in y$, and $\exists x$.

A *Σ_1-formula* is a formula of the kind $\exists u \varphi_0$ where φ_0 is Δ_0-formula. It is known that any Σ-formula is equivalent in the theory KPU (see [4]) to some Σ_1-formula.

A *Σ-predicate* is a relation definable by some Σ-formula (possibly with parameters). A *Δ-predicate* is a Σ-predicate whose complement is also Σ. A partial operation is called a *(partial) Σ-function* if its graph is Σ.

There is a useful result which presents the relationship between Σ-definability and definability by infinitary computable formulas.

THEOREM 2.1 ([99]). *A predicate $P \subseteq M^n$ is Σ-definable in $\mathbb{HF}(\mathfrak{M})$ iff there is a computable family $\varphi_s(\bar{x}, \bar{y})$, $s \in \omega$ ($\bar{x} = (x_0, \dots, x_k)$, $\bar{y} = (y_0, \dots, y_{n-1})$), of $\exists$-formulas and a k-tuple $\bar{a} \in M^k$ such that, for any $\bar{b} \in M^n$,*

$$\bar{b} \in P \iff \mathfrak{M} \models \bigvee_{s \in \omega} \varphi_s(\bar{a}, \bar{b})$$

It is usually convenient to use in constructions not only the elements from $\mathbb{HF}(\varnothing) \subseteq \mathbb{HF}(\mathfrak{M})$, but also the elements from $\mathbb{HF}(\mathcal{N})$, where $\mathcal{N}$ is isomorphic to the standard model of arithmetic. To avoid confusion with ordinals from $\mathbb{HF}(\varnothing)$, we denote the domain of $\mathcal{N}$ and its elements as $\underline{\omega}$ and $\underline{n}$, $n \in \omega$, respectively. Since $\mathbb{HF}(\mathcal{N})$ is constructivizable, it can be effectively defined in any hereditarily finite superstructure.

In what follows, we use definitions and constructions from [10]. For all $n \in \omega$, $\varkappa \in \mathrm{HF}(\underline{n})$ ($\underline{n} = \{\underline{0}, \underline{1}, \dots, \underline{n-1}\}$), and $\bar{x} \in M^n$, we define an element $\varkappa(\bar{x}) \in \mathrm{HF}(\mathfrak{M})$ as follows. Define a mapping $\lambda_{\bar{x}} : n \to M$ as $\lambda_{\bar{x}}(i) = x_i$, where $\bar{x} = \langle x_0, \dots, x_{n-1} \rangle$. The mapping $\lambda_{\bar{x}}$ can be uniquely extended to $\lambda_{\bar{x}}^{\omega} : \mathrm{HF}(\underline{n}) \to \mathrm{HF}(\mathfrak{M})$ so that $\lambda_{\bar{x}}^{\omega}(a_0, \dots, a_k) \rightleftharpoons \{\lambda_{\bar{x}}^{\omega}(a_0), \dots, \lambda_{\bar{x}}^{\omega}(a_k)\}$ for each set $\{a_0, \dots, a_k\} \in \mathrm{HF}(\underline{n})$. Then we put $\varkappa(\bar{x}) \rightleftharpoons \lambda_{\bar{x}}^{\omega}(\varkappa)$.

For every $\varkappa \in \mathrm{HF}(\underline{n})$, we can effectively define a term $t_\varkappa(x_0, \dots, x_{n-1})$ of signature $\langle \{\}, \cup, \varnothing \rangle$ in such a way that, for all elements $x_0^0, \dots, x_{n-1}^0 \in M$, the equality $t_\varkappa(x_0^0, \dots, x_{n-1}^0) = \varkappa(\bar{x}^0)$ is valid.

Hereditarily finite superstructures are the "simplest" admissible sets, from the set-theoretical point of view. In addition to this, Σ-definability in hereditarily finite superstructures is one of the natural approaches for generalizing classical computability theory on natural numbers to the case of computability over arbitrary structures. For results on computability on admissible sets in general, we refer the reader to [4] and [10].

§3. Semilattices of Σ-Degrees of Structures. The theory of constructive (computable) models is one of the key research areas of classical computability theory, and also of model theory. Because of evident cardinality limitations, in classical computable model theory only countable structures are considered. An approach regarding generalized computability as Σ-definability in admissible sets lets us consider structures with arbitrary cardinality.

Hence, for a structure $\mathfrak{M}$ the following problems naturally arises: describe

- those structures Σ-definable in $\mathbb{HF}(\mathfrak{M})$;
- those structures such that $\mathfrak{M}$ is Σ-definable in their HF-superstructures.

Let us formalize those problems. Let $\mathfrak{M}$ be a structure of a finite predicate signature $\langle P_1^{n_1}, \dots, P_k^{n_k} \rangle$ and let $\mathbb{A}$ be an admissible set. The following notion is an effectivization of the model-theoretical notion of interpretability of one structure in another, and also a natural generalization of the notion of constructivizability of a (countable) structure on natural numbers.

Definition 3.1 ([8]). $\mathfrak{M}$ is *Σ-definable in* $\mathbb{A}$ if there exist Σ-formulas

$$\Phi(x_0, y), \Psi(x_0, x_1, y), \Psi^*(x_0, x_1, y), \Phi_1(x_0, \dots, x_{n_1-1}, y),$$

$$\Phi_1^*(x_0, \dots, x_{n_1-1}, y), \dots, \Phi_k(x_0, \dots, x_{n_k-1}, y), \Phi_k^*(x_0, \dots, x_{n_k-1}, y),$$

such that for some parameter $a \in A$, and letting

$$M_0 \leftrightharpoons \Phi^{\mathbb{A}}(x_0, a), \quad \eta \leftrightharpoons \Psi^{\mathbb{A}}(x_0, x_1, a) \cap M_0^2,$$

one has that $M_0 \neq \varnothing$ and η is a congruence relation on the structure

$$\mathfrak{M}_0 \leftrightharpoons \langle M_0, P_1^{\mathfrak{M}_0}, \dots, P_k^{\mathfrak{M}_0} \rangle,$$

where $P_i^{\mathfrak{M}_0} \leftrightharpoons \Phi_i^{\mathbb{A}}(x_0, \dots, x_{n_i-1}) \cap M_0^{n_i}$ for all $1 \leqslant i < k$,

$$\Psi^{*\mathbb{A}}(x_0, x_1, a) \cap M_0^2 = M_0^2 \setminus \Psi^{\mathbb{A}}(x_0, x_1, a),$$

$$\Phi_i^{*\mathbb{A}}(x_0, \dots, x_{n_i-1}, a) \cap M_0^{n_i} = M_0^{n_i} \setminus \Phi_i^{\mathbb{A}}(x_0, \dots, x_{n_i-1})$$

for all $1 \leqslant i < k$, and the structure $\mathfrak{M}$ is isomorphic to the quotient structure $\mathfrak{M}_0/\eta$.

Remark 3.2. This definition can be naturally generalized to the case of structures with infinite computable signatures. That is, a structure $\mathfrak{M}$ with a computable predicate signature $\langle P_0^{n_0}, P_1^{n_1}, \dots \rangle$ is called *Σ-definable in* $\mathbb{A}$ if there exists a computable sequence $\Phi(x_0, y)$, $\Psi(x_0, x_1, y)$, $\Psi^*(x_0, x_1, y)$, $\Phi_0(x_0, \dots, x_{n_0-1}, y)$, $\Phi_0^*(x_0, \dots, x_{n_0-1}, y)$, $\dots$, $\Phi_k(x_0, \dots, x_{n_k-1}, y)$, $\Phi_k^*(x_0, \dots, x_{n_k-1}, y), \dots$ of Σ-formulas and a parameter $a \in A$, which forms a Σ-definition of $\mathfrak{M}$ in $\mathbb{A}$, in the sense of Definition 3.1.

Also, for a structure with an infinite computable signature, we assume that some Gödel numbering of formulas of this signature is fixed. Σ-reducibility $\leqslant_\Sigma$ is defined as follows: for structures $\mathfrak{A}$ and $\mathfrak{B}$, $\mathfrak{A} \leqslant_\Sigma \mathfrak{B}$ if $\mathfrak{A}$ is Σ-definable in $\mathbb{HF}(\mathfrak{B})$. We assume that the signature of $\mathbb{HF}(\mathfrak{B})$ contains a predicate symbol Sat^2 interpreted by the satisfiability relation for atomic formulas in $\mathfrak{B}$, with respect to a fixed Gödel numbering. In the case of structures with a finite signature this assumption is not essential.

Since, for the set of natural numbers, Σ-definability in $\mathbb{HF}(\varnothing)$ is equivalent to classical computability, we get the following fact.

PROPOSITION 3.3. *Let $\mathfrak{M}$ be a countable structure. The following are equivalent:*

1. *$\mathfrak{M}$ is constructivizable;*
2. *$\mathfrak{M}$ is Σ-definable in $\mathbb{HF}(\varnothing)$.*

For structures $\mathfrak{M}$ and $\mathfrak{N}$, we denote by $\mathfrak{M} \leqslant_\Sigma \mathfrak{N}$ the fact that $\mathfrak{M}$ is Σ-definable in $\mathbb{HF}(\mathfrak{N})$. From the definition it follows that the relation $\leqslant_\Sigma$ is reflexive and transitive. We now look at the general properties of this relation, regarding it as a kind of effective reducibility on structures.

For any infinite cardinal α, we denote by $\mathcal{K}_\alpha$ the class of structures with cardinality less than or equal to α.

As usual, preordering $\leqslant_\Sigma$ generates on $\mathcal{K}_\alpha$ a relation of Σ-equivalence: $\mathfrak{A} \equiv_\Sigma \mathfrak{B}$ if $\mathfrak{A} \leqslant_\Sigma \mathfrak{B}$ and $\mathfrak{B} \leqslant_\Sigma \mathfrak{A}$. Classes of Σ-equivalence are called *degrees of Σ-definability*, or *Σ-degrees*. The poset

$$\mathcal{S}_\Sigma(\alpha) = \langle \mathcal{K}_\alpha / \equiv_\Sigma, \leqslant_\Sigma \rangle$$

is an upper semilattice with least element, which is the degree consisting of computable structures. We denote the Σ-degree of a structure $\mathfrak{A}$ by $[\mathfrak{A}]_\Sigma$. The notion of Σ-degree of a structure is independent of the choice of a semilattice $\mathcal{S}_\Sigma(\alpha)$, because all infinite structures of the same Σ-degree have the same cardinality. For any structures $\mathfrak{A}, \mathfrak{B} \in \mathcal{K}_\alpha$, $[\mathfrak{A}]_\Sigma \vee [\mathfrak{B}]_\Sigma = [(\mathfrak{A}, \mathfrak{B})]_\Sigma$, where $(\mathfrak{A}, \mathfrak{B})$ is the pair of $\mathfrak{A}$ and $\mathfrak{B}$ in model-theoretical sense.

For a structure $\mathfrak{A} \in \mathcal{K}_\alpha$ and infinite cardinals $\beta \leqslant \alpha$, $\gamma \geqslant \alpha$, the sets

$$I_\beta(\mathfrak{A}) = \{[\mathfrak{B}]_\Sigma \mid \mathfrak{B} \in \mathcal{K}_\beta,\ \mathfrak{B} \leqslant_\Sigma \mathfrak{A}\}, \quad F_\gamma(\mathfrak{A}) = \{[\mathfrak{B}]_\Sigma \mid \mathfrak{B} \in \mathcal{K}_\gamma,\ \mathfrak{A} \leqslant_\Sigma \mathfrak{B}\}$$

are, correspondingly, an ideal in $\mathcal{S}_\Sigma(\beta)$ (principal for $\beta = \alpha$) and a filter in $\mathcal{S}_\Sigma(\gamma)$ (principal for any $\gamma \geqslant \alpha$). The sets $F_\gamma(\mathfrak{A})$ in semilattices $\mathcal{S}_\Sigma(\gamma)$ are natural analogues of the *spectrum* of a structure $\mathfrak{A}$. The sets $I_\beta(\mathfrak{A})$ in semilattices $\mathcal{S}_\Sigma(\beta)$ consist of Σ-degrees of structures Σ-presentable over $\mathfrak{A}$.

A *presentation* of a structure $\mathfrak{M}$ in an admissible set $\mathbb{A}$ is any structure $\mathfrak{C}$ which is isomorphic to $\mathfrak{M}$ and whose domain C is a subset of A (the relation $=$ is treated as a congruence relation on $\mathfrak{C}$, and it may differ from the standard equality relation on $\mathfrak{C}$). In what follows, we will identify the presentation $\mathfrak{C}$ (more precisely, its atomic diagram) with some subset of A, fixing a Gödel numbering of atomic formulas of the signature $\sigma_{\mathfrak{M}}$.

DEFINITION 3.4. A *problem of presentability* of a structure $\mathfrak{M}$ in $\mathbb{A}$ is the set $\mathrm{Pr}_{\mathbb{A}}(\mathfrak{M})$ consisting of all possible presentations of $\mathfrak{M}$ in $\mathbb{A}$.

Denote by $\underline{\mathfrak{M}}$ the set $\mathrm{Pr}(\mathfrak{M}, \mathbb{HF}(\varnothing))$ of presentations of $\mathfrak{M}$ in the least admissible set.

There exist natural embeddings of the semilattice $\mathcal{D}$ of Turing degrees and the semilattice $\mathcal{D}_e$ of degrees of enumerability of sets of natural numbers into semilattice $\mathcal{S}_\Sigma(\omega)$ (and hence into any semilattice $\mathcal{S}_\Sigma(\alpha)$) via the mappings $i : \mathcal{D} \to \mathcal{S}_\Sigma(\omega)$ and $j : \mathcal{D}_e \to \mathcal{S}_\Sigma(\omega)$ defined below. These definitions show that the notion of Σ-degree of a structure, which is total, i.e. defined for any structure, no matter countable or not, is a natural generalization of the (partial) notion of a degree of a countable structure, introduced in [72]. Hence, we get that semilattices $\mathcal{S}_\Sigma(\alpha)$ extend in a natural way semilattices $\mathcal{D}$ and $\mathcal{D}_e$.

Definition 3.5 ([72, 93]). Let $\mathfrak{M}$ be a countable structure. We say that $\mathfrak{M}$ has a *degree (e-degree)* if there exists the least degree in the set of T-degrees (e-degrees) of all possible presentations of $\mathfrak{M}$ on natural numbers.

Using the equivalence of "∀-recursiveness" and "∃-definability", in the sense of [50], [63], see also [1], [6], we get following result.

Theorem 3.6 ([93]). *For a countable structure $\mathfrak{M}$, the following are equivalent:*

1. *$\mathfrak{M}$ has a degree (e-degree);*
2. *there exists a presentation $\mathfrak{C} \in \underline{\mathfrak{M}}$ which is a Δ-subset (Σ-subset) of $\mathbb{HF}(\mathfrak{M})$.*

We define mappings $i : \mathcal{D} \to \mathcal{S}_\Sigma(\omega)$ and $j : \mathcal{D}_e \to \mathcal{S}_\Sigma(\omega)$ in the following way: for every degree $\mathbf{a} \in \mathcal{D}$, put

$$i(\mathbf{a}) = [\mathfrak{M}_{\mathbf{a}}]_\Sigma, \text{ where } \mathfrak{M}_{\mathbf{a}} \text{ is any structure having degree } \mathbf{a}.$$

Similarly, for every e-degree $\mathbf{b} \in \mathcal{D}_e$, put

$$j(\mathbf{b}) = [\mathfrak{M}_{\mathbf{b}}]_\Sigma, \text{ where } \mathfrak{M}_{\mathbf{b}} \text{ is any structure having } e\text{-degree } \mathbf{b}.$$

Lemma 3.7. *The mappings i and j are well defined: for any (e-)degree* **a** *there are structures having (e-)degree* **a**. *Moreover, for any countable structures $\mathfrak{M}$ and $\mathfrak{N}$;*

1. *if $\mathfrak{M}$ has (e-)degree* **a** *and $\mathfrak{M} \equiv_\Sigma \mathfrak{N}$, then $\mathfrak{N}$ also has (e-)degree* **a**;
2. *if $\mathfrak{M}$ and $\mathfrak{N}$ have the same (e-)degree then $\mathfrak{M} \equiv_\Sigma \mathfrak{N}$.*

(Notice, however, that the property of having a (e-)degree is not closed downwards with respect to $\leqslant_\Sigma$.) Usually, we just write **a** instead of $i(\mathbf{a})$.

For a structure $\mathfrak{A}$, a *Σ-jump* of $\mathfrak{A}$ is the structure

$$\mathfrak{A}' = (\mathbb{HF}(\mathfrak{A}), \Sigma\text{-Sat}^{\mathbb{HF}(\mathfrak{A})}),$$

where $\Sigma\text{-Sat}^{\mathbb{HF}(\mathfrak{A})}$ denotes the satisfiability relation for the set of Σ-formulas in $\mathbb{HF}(\mathfrak{A})$. The definition of Σ-jump is persistent with respect to the Σ-equivalence: for any structures $\mathfrak{A}$ and $\mathfrak{B}$, from $\mathfrak{A} \equiv_\Sigma \mathfrak{B}$ it follows that $\mathfrak{A}' \equiv_\Sigma \mathfrak{B}'$. Hence, we may assume that semilattices $\mathcal{S}_\Sigma(\alpha)$ are equipped with the jump operation.

Remark 3.8. In a similar way the jump operation was introduced in [3] for the semilattice of s-degrees of countable structures. Also, in the same way, a notion of the jump of an admissible set with respect to various effective reducibilities was introduced in [57, 71].

The operation of Σ-jump agrees with the jump operations for Turing and enumeration degrees with respect to the natural embeddings i and j: if a structure $\mathfrak{A}$ has a (e-)degree $\mathbf{a}$, then the structure $\mathfrak{A}'$ has (e-)degree $\mathbf{a}'$.

PROPOSITION 3.9. *The mappings* $i : \mathcal{D} \to \mathcal{S}_\Sigma$ *and* $j : \mathcal{D}_e \to \mathcal{S}_\Sigma$ *are embeddings preserving* 0, $\vee$ *and the jump operation.*

The existence of an embedding of $\mathcal{D}$ in $\mathcal{S}_\Sigma$ was first noted in [30].

The jump inversion theorem from classical computability theory can also be generalized to the case of semilattices of Σ-degrees of structures.

THEOREM 3.10 ([95, 96]). *Let* $\mathfrak{A}$ *be a structure such that* $\mathbf{0}' \leqslant_\Sigma \mathfrak{A}$. *Then there exists a structure* $\mathfrak{B}$ *such that*

$$\mathfrak{B}' \equiv_\Sigma \mathfrak{A}.$$

The proof uses Marker's extensions in form similar to that proposed in [17] and used in [83, 84]. It should be noted that the relation of Σ-reducibility, since it is defined on structures of arbitrary cardinality, can be viewed, in the case of countable structures, as the strongest reducibility in the hierarchy of effective reducibilities on structures [93], [94] (see Section 5). One of the weak reducibilities in this hierarchy is Muchnik reducibility. In [83, 84], the jump inversion theorem is proved for semilattices of degrees of presentability of countable structures with respect to Muchnik reducibility. As a corollary of Theorem 3.10, we get the jump inversion theorem for all known effective reducibilities on countable structures.

It follows from the definition of Σ-jump that $\mathfrak{A} \vee \mathbf{0}' \leqslant_\Sigma \mathfrak{A}'$ for any structure $\mathfrak{A}$. We say that a structure $\mathfrak{A}$ is *generalized low* with respect to the Σ-jump (in short, *generalized* Σ-*low*) if

$$\mathfrak{A}' \equiv_\Sigma \mathfrak{A} \vee \mathbf{0}'.$$

It turns out that the class of structures with a *c-simple* theory (i.e., model complete, ω-categorical, decidable, and with a decidable set of atoms) constitutes a series of examples (in arbitrary cardinality) of generalized Σ-low structures.

THEOREM 3.11 ([98]). *For a structure* $\mathfrak{A}$, *if* Th($\mathfrak{A}$) *is c-simple, then* $\mathfrak{A}' \equiv_\Sigma \mathfrak{A} \vee \mathbf{0}'$.

So, a jump of a structure $\mathfrak{A}$ in the sense of [54] is any expansion $(\mathfrak{A}, P_0, P_1, \dots)$, where $P_0, P_1, \dots$ is a complete set of Π^c_1-relations on $\mathfrak{A}$. As an example, one can always choose an expansion

$$\mathfrak{A}^* = (\mathfrak{A}, P_1, P_2, \dots),$$

where $P_1, P_2, \dots$ is the list of all relations on $\mathfrak{A}$ which are Π-definable in $\mathbb{HF}(\mathfrak{A})$, corresponding to some computable numbering $\Phi_1, \Phi_2, \dots$ of all Π-formulas of signature $\sigma'_{\mathfrak{A}}$. As follows from the results on the equivalence of computable infinitary formulas and Π-formulas, [99], $\mathfrak{A}^*$ is a jump of this structure in the sense of [54]. However, for a given structure there can be many different jumps in this sense: for example, any dense linear order is a jump of itself, see [54]. We propose calling a *weak jump* of a structure any jump of that structure in the sense of [54]. The terminology we use is justified by the relationship between a weak jump of a structure $\mathfrak{A}$ and the Σ-jump of $\mathfrak{A}$. More precisely, we have

Proposition 3.12. *For any structure $\mathfrak{A}$ and any of its weak jumps, $\mathfrak{A}^*$, it follows that*

$$\mathfrak{A}' \equiv_\Sigma \mathfrak{A}^* \vee \mathbf{0}'.$$

Proof. Reducibility $\mathfrak{A}^* \vee \mathbf{0}' \leqslant_\Sigma \mathfrak{A}'$ is evident from the equivalence of computable infinitary formulas and Π-formulas [99]. To prove the reverse reducibility, we use notations from [10]. Any Π-subset $P \subseteq \mathbb{HF}(\mathfrak{A})$ can be represented in the form $P = \bigcup_{\varkappa \in \mathrm{HF}(\omega)} P_\varkappa$, where, for any $\varkappa \in \mathrm{HF}(\omega)$, we have $P_\varkappa = \{\varkappa(\overline{a}) \mid \mathfrak{A} \models \Phi_\varkappa(\overline{a})\}$, $\Phi_\varkappa$ is a computable conjunction of ∀-formulas of signature σ (with parameters $\bar{c} \in A^{<\omega}$), and $\{\Phi_\varkappa \mid \varkappa \in \mathrm{HF}(\omega)\}$ is a computable family.

Consider an element $\varkappa(\overline{a}) \in \mathbb{HF}(\mathfrak{A})$, where $\varkappa \in \mathrm{HF}(\omega)$ and $\overline{a} \in A^{<\omega}$ are urelements. We have $\varkappa(\bar{a}) \in P$ if and only if $\mathfrak{A} \models \bigwedge_{n\in\omega} \forall \bar{y}_n \varphi_{\varkappa,n}(\bar{c}, \bar{y}_n, \bar{a})$. Since in $\mathbb{HF}(\mathfrak{A}^*)$ one can effectively (with $\mathbf{0}'$) find a Σ-formula equivalent to the Π_1^c-formula described above, we get that the relation $\Sigma\text{-Sat}^{\mathbb{HF}(\mathfrak{A})}$ is Δ-definable in $\mathbb{HF}(\mathfrak{A}^*)$ with $\mathbf{0}'$. ⊣

As was already mentioned, the original definition from [54] is not quite precise, but it is precise in the sense that any two jumps of a given structure are Σ-equivalent over $\mathbf{0}'$. In fact, directly from the definition we get

Corollary 3.13. *If $\mathfrak{B}_1, \mathfrak{B}_2$ are the weak jumps of a structure $\mathfrak{A}$ then*

$$\mathfrak{B}_1 \vee \mathbf{0}' \equiv_\Sigma \mathfrak{B}_2 \vee \mathbf{0}'.$$

We mention the results on "Σ-universality" (i.e., universality with respect to the Σ-equivalence) of two classes of posets – graphs and lattices.

Proposition 3.14 ([98]). *For any structure $\mathfrak{A}$, there is an irreflexive graph $G_\mathfrak{A}$ such that $\mathfrak{A} \equiv_\Sigma G_\mathfrak{A}$.*

Proposition 3.15 ([98]). *For any structure $\mathfrak{A}$, there is a lattice $L_\mathfrak{A}$ such that $\mathfrak{A} \equiv_\Sigma L_\mathfrak{A}$.*

It should be noted that the class of linear orders is not Σ-universal (see the results on local constructivizability below).

§4. Effective Self-Presentations of Admissible Sets. In Theorem 3.6, we have already seen examples of non-trivial effective self-presentations of admissible sets of the kind $\mathbb{HF}(\mathfrak{M})$, where $\mathfrak{M}$ is a countable structure. Namely, $\mathbb{HF}(\mathfrak{M})$ has a 'pure' copy (i.e., a copy which is a subset of $\mathbb{HF}(\varnothing)$) as a Δ-subset (resp., Σ-subset) of $\mathbb{HF}(\mathfrak{M})$ if and only if $\mathfrak{M}$ have a degree (resp., an *e*-degree). Now let us consider effective self-presentations of admissible sets which are, in some sense, as close to the ground structure as possible.

In some cases, for structures $\mathfrak{A}$ and $\mathfrak{B}$ one can do more than just state the fact that $\mathfrak{A} \leqslant_\Sigma \mathfrak{B}$. For example, it is obvious that $\mathbb{HF}(\mathfrak{A}) \leqslant_\Sigma \mathfrak{A}$ for any $\mathfrak{A}$, but, in the case of the standard model of arithmetics $\mathbb{N}$, a much stronger result holds: $\mathbb{HF}(\mathbb{N})$ is Σ-definable within $\mathbb{N}$, without using the elements of the superstructure.

In particular, a natural additional restriction on Σ-definability of structures in admissible sets is the restriction on the rank of elements used in this process. To describe the situation formally, we now give some definitions.

Fix some signature σ, and let P be an unary predicate symbol not in σ. For any formula Φ of the signature $\sigma \cup \{\in\}$, with the bounded quantifiers of the form $\forall x \in t$ and $\exists x \in t$, we define by induction the *relativization* Φ^P of Φ by P:

— if Φ is an atomic formula, put $\Phi^P = \Phi$;
— if $\Phi = (\Phi_1 * \Phi_2)$, $* \in \{\wedge, \vee, \to\}$, put $\Phi^P = (\Phi_1^P * \Phi_2^P)$;
— if $\Phi = \neg\Psi$, put $\Phi^P = \neg\Phi^P$;
— if $\Phi = (Qx \in y)\Psi$, $Q \in \{\forall, \exists\}$, put $\Psi^P = (Qx \in y)\Psi^P$;
— if $\Phi = \exists x\Psi$, put $\Phi^P = \exists x(P(x) \wedge \Psi^P)$;
— if $\Phi = \forall x\Psi$, put $\Phi^P = \forall x(P(x) \to \Psi^P)$.

Let now $\mathbb{A}$ be an admissible set, $B \subseteq A$ be some transitive subset of $\mathbb{A}$, and $\Phi(x_0, \dots, x_{n-1})$ be a formula of the signature $\sigma_{\mathbb{A}}$. Define the set

$$(\Phi(x_0, \dots, x_{n-1}))^B = \{\langle a_0, \dots, a_{n-1}\rangle \in A^n \mid \langle \mathbb{A}, B\rangle \models \Phi^P(a_0, \dots, a_{n-1})\}.$$

Definition 4.1 ([90]). Let $\mathbb{A}$ be an admissible set, $B \subseteq A$ be some transitive subset of $\mathbb{A}$. A structure of a computable predicate signature $\langle P_0^{n_0}, P_1^{n_1}, \dots\rangle$ is called *Σ-definable in $\mathbb{A}$ inside B* if there exist a computable sequence

$$\Phi(x_0, y), \Psi(x_0, x_1, y), \Psi^*(x_0, x_1, y), \Phi_0(x_0, \dots, x_{n_0-1}, y),$$

$$\Phi_0^*(x_0, \dots, x_{n_0-1}, y), \dots, \Phi_k(x_0, \dots, x_{n_k-1}, y), \Phi_k^*(x_0, \dots, x_{n_k-1}, y), \dots$$

of Σ-formulas of $\sigma_{\mathbb{A}}$, and a parameter $b \in B$, such that, for the sets

$$M_0 \leftrightharpoons \Phi^B(x_0, b),\ M_0 \subseteq B,\ \ \eta \leftrightharpoons \Psi^B(x_0, x_1, b) \cap M_0^2,$$

the following holds: $M_0 \neq \varnothing$, η is a congruence relation on the structure

$$\mathfrak{M}_0 \leftrightharpoons \langle M_0, P_0^{\mathfrak{M}_0}, \dots, P_k^{\mathfrak{M}_0}, \dots\rangle,$$

where $P_k^{\mathfrak{M}_0} \leftrightharpoons (\Phi_k(x_0, \dots, x_{n_k-1}))^B \cap M_0^{n_k}$, $k \in \omega$,

$$(\Psi^*(x_0, x_1, a))^B \cap M_0^2 = M_0^2 \setminus (\Psi(x_0, x_1, a))^B,$$

$$(\Phi_k^*(x_0, \dots, x_{n_k-1}, a))^B \cap M_0^{n_k} = M_0^{n_k} \setminus (\Phi_k(x_0, \dots, x_{n_k-1}))^B$$

for any $k \in \omega$, and the quotient structure $\mathfrak{M}$ is isomorphic to $\mathfrak{M}_0/\eta$.

For an admissible set $\mathbb{A}$ and a subset $B \subseteq A$, define the ordinal $\mathrm{rnk}(B)$ as follows:

$$\mathrm{rnk}(B) = \sup\{\mathrm{rnk}(b) | b \in B\},$$

where $\mathrm{rnk}(b)$ is the rank of the set b in $\mathbb{A}$, see [4].

Definition 4.2 ([90]). The *rank of inner constructivizability* of an admissible set $\mathbb{A}$ is the ordinal

$$\mathrm{cr}(\mathbb{A}) = \inf\{\mathrm{rnk}(B) \mid \mathbb{A} \text{ is } \Sigma\text{-definable in } \mathbb{A} \text{ inside } B\}.$$

The next theorem gives the precise estimate for the rank of inner constructivizability of hereditarily finite superstructures. It can be viewed as an effective analogue of some results from [53] on definability in higher-order languages.

Theorem 4.3 ([90]). *Let $\mathfrak{M}$ be a structure of a computable signature.*
(1) If $\mathfrak{M}$ is finite then $\mathrm{cr}(\mathbb{HF}(\mathfrak{M})) = \omega$.
(2) If $\mathfrak{M}$ is infinite then $\mathrm{cr}(\mathbb{HF}(\mathfrak{M})) \leqslant 2$.

As a corollary of Theorem 4.3 we get the following. For structures $\mathfrak{M}$, $\mathfrak{N}$, and a natural number $n \in \omega$, we denote by $\mathfrak{M} \leqslant^n_\Sigma \mathfrak{N}$ the fact that $\mathfrak{M}$ is Σ-definable in $\mathbb{HF}(\mathfrak{N})$ inside the subset consisting of all elements with rank less than or equal to n. If $\mathfrak{N}$ is an infinite structure then

$$\mathfrak{M} \leqslant^n_\Sigma \mathfrak{N} \text{ if and only if } \mathfrak{M} \leqslant_\Sigma \mathfrak{N}$$

for any $\mathfrak{M}$ and any $n \geqslant 2$.

Typical examples of structures $\mathfrak{M}$ with $\mathrm{cr}(\mathbb{HF}(\mathfrak{M})) = 2$ are infinite structures with empty signature, dense linear orders, and, more interestingly, the structure $\langle \omega, s\rangle$ of natural numbers with successor function. This fact follows from the next proposition, taking into account the decidability of $\mathrm{Th}_{\mathrm{WM}}(\langle \omega, s\rangle)$, where $\mathrm{Th}_{\mathrm{WM}}(\mathfrak{M})$ is the weak monadic second-order theory of $\mathfrak{M}$.

Proposition 4.4 ([90]). *If* $\mathrm{Th}_{\mathrm{WM}}(\mathfrak{M})$ *is decidable then*

$$\mathrm{cr}(\mathbb{HF}(\mathfrak{M})) = 2.$$

An example of a structure $\mathfrak{M}$ with $\mathrm{cr}(\mathbb{HF}(\mathfrak{M})) = 0$ is, obviously, the standard model of arithmetic. An example of a structure with the rank of the inner constructivizability of the hereditarily finite superstructure that is equal to 1 is the field $\mathbb{R}$ of real numbers.

Proposition 4.5 ([90]). $\mathrm{cr}(\mathbb{HF}(\mathbb{R})) = 1$.

Another natural special type of a Σ-presentation of a structure $\mathfrak{M}$ in an admissible set $\mathbb{A}$, such that $M \subseteq U(\mathbb{A})$, is a Σ-presentation that preserves the domain of a structure. For a signature σ and an ordinal $n \leqslant \omega$, we denote by $\mathrm{Form}_n(\sigma)$ the set of (finite first-order) formulas of the signature σ, which have a prenex normal form with no more than n alterating groups of quantifiers.

We assume that, for any signature considered, some Gödel numbering $\ulcorner\cdot\urcorner$ of its terms and formulas is fixed.

Definition 4.6. Let $\mathfrak{M}$ be a structure of a finite signature σ, $\mathbb{A}$ an admissible set, and let $M \subseteq U(A)$. The structure $\mathfrak{M}$ is *n-decidable in $\mathbb{A}$* ($n \leqslant \omega$) if

$$\{\langle \ulcorner\varphi\urcorner, \overline{m}\rangle \mid \varphi \in \mathrm{Form}_n(\sigma), \overline{m} \in M^{<\omega}, \mathfrak{M} \models \varphi(\overline{m})\}$$

is Δ-definable in $\mathbb{A}$.

A structure $\mathfrak{M}$ is *computable in* $\mathbb{A}$ if $\mathfrak{M}$ is 0-decidable in $\mathbb{A}$, and *decidable in* $\mathbb{A}$ if $\mathfrak{M}$ is ω-decidable in $\mathbb{A}$. It is easy to prove that if $\mathrm{Th}(\mathfrak{M})$ is regular, then $\mathfrak{M}$ is decidable in $\mathbb{HF}(\mathfrak{M})$.

The decidability is rather a strong condition. For example, we have

PROPOSITION 4.7. *A linear order* $\mathfrak{L}$ *is* 1*-decidable in* $\mathbb{HF}(\mathfrak{L})$ *if and only if* $\mathfrak{L}$ *is a sum of a finite number of dense linear orders and points.*

A structure $\mathfrak{M}$ of signature σ is *n-complete* [13] ($n \leqslant \omega$) if for any formula $\varphi(\overline{x}) \in \mathrm{Form}_n(\sigma)$ and for any $\overline{m} \in M^{<\omega}$ such that $\mathfrak{M} \models \varphi(\overline{m})$ there exists a $\exists$-formula $\psi(\overline{x})$ such that $\mathfrak{M} \models \psi(\overline{m})$ and $\mathfrak{M} \models \forall\overline{x}(\psi(\overline{x}) \to \varphi(\overline{x}))$. The following proposition follows immediately from the definitions.

PROPOSITION 4.8.

1. *Suppose* $\mathfrak{M}$ *is n-decidable in* $\mathbb{HF}(\mathfrak{M})$ ($n \leqslant \omega$)*. Then* $\mathfrak{M}$ *is n-complete in some expansion of* $\mathfrak{M}$ *by a finite number of constants.*
2. *Suppose* $\mathfrak{M}$ *is n-complete and* $\mathrm{Th}(\mathfrak{M})$ *is decidable. Then* $\mathfrak{M}$ *is n-decidable in* $\mathbb{HF}(\mathfrak{M})$.

Suppose $\mathfrak{M}$ is 1-decidable in $\mathbb{HF}(\mathfrak{M})$. Then $\mathbb{HF}(\mathfrak{M})$ has a universal Σ-function and the reduction property, but not necessarily the uniformization property (see [10]).

A structure $\mathfrak{A}$ is *s*Σ*-definable* [95, 96] in $\mathbb{HF}(\mathfrak{B})$ (denoted by $\mathfrak{A} \leqslant_{s\Sigma} \mathfrak{B}$) if $A \subseteq \mathrm{HF}(B)$ is a Σ-subset in $\mathbb{HF}(\mathfrak{B})$, and all the signature relations and functions of $\mathfrak{A}$ are Δ-definable in $\mathbb{HF}(\mathfrak{B})$. The relation $\leqslant_{s\Sigma}$ is reflexive and transitive, under some additional assumptions on the structures being considered [98]. For countable structures, in a slightly different form, $s\Sigma$-reducibility was introduced in [3].

We write $\mathfrak{A} <_{s\Sigma} \mathfrak{B}$ to denote the fact that $\mathfrak{A} \leqslant_{s\Sigma} \mathfrak{B}$ and $\mathfrak{B} \nleqslant_{s\Sigma} \mathfrak{A}$.

In [95, 96] it was noted that

$$\mathfrak{A} <_{s\Sigma} \mathfrak{A}'$$

for any structure $\mathfrak{A}$, whether countable or not. This means that the operation of Σ-jump has no fixed points with respect to $s\Sigma$-reducibility, one of the strongest effective reducibilities on countable structures (see [93, 94]). However, the relation of $s\Sigma$-reducibility is not persistent relative to the isomorphism, and the corresponding relation of $s\Sigma$-equivalence is stronger than the isomorphism relation. Also, the embeddings of the semilattices of Turing and enumeration degrees into the semilattice of $s\Sigma$-degrees are not as natural as the corresponding embeddings in the case of the semilattices of Σ-degrees [93, 94].

However, $s\Sigma$-reducibility is useful when studying generalized computability. Actually, it was implicitly used in [86] to formulate a criterion of the uniformization property.

Recall that an admissible set $\mathbb{A}$ is said to satisfy

- *reduction* if, for any Σ-subsets B_0 and B_1, there are disjoint Σ-subsets $C_0 \subseteq B_0$ and $C_1 \subseteq B_1$ such that $C_0 \cup C_1 = B_0 \cup B_1$.

- *uniformization* if, for any binary Σ-predicate R on $\mathbb{A}$, there is a partial Σ-function $\varphi(x)$ with $\delta\varphi = \mathrm{Pr}_1(R)$ and $\Gamma_\varphi \subseteq R$.

Suppose $\mathfrak{M}$ has a *regular* (i.e., model complete and decidable [10]) first-order theory. Then $\mathfrak{M}$ is 1-decidable in $\mathbb{HF}(\mathfrak{M})$, and hence has a universal (partial, single-valued) Σ-function and satisfies reduction, but not necessarily uniformization. It turns out that the relation of Σ-reducibility (in fact, of $s\Sigma$-equivalence) can be used to provide a criterion.

Recall that a theory T of signature σ is said to be a *theory with definable Skolem functions* [101], provided that, for each formula $\varphi(x_0, \dots, x_n)$ of signature σ, there exists a formula $\psi(x_0, \dots, x_n)$ of the same signature such that

$$T \vdash \forall x_1 \dots \forall x_n\Big[\exists x_0\, \varphi(x_0, \dots, x_n) \to \exists\,!\, x_0\, (\varphi(x_0, \dots, x_n) \wedge \psi(x_0, \dots, x_n))\Big].$$

Actually, the requirement of definability of Skolem functions is too stringent. Let $\mathfrak{M}$ be a structure of signature σ, and let σ' denotes the signature of $\mathbb{HF}(\mathfrak{M})$. In formulas of signature σ', we conventially distinguish between variables with values in the set of urelements and general variables, i.e. variables whose values may be arbitrary elements of an admissible set. In what follows, given a formula of signature σ, we assume all its variables, free or bounded, to be variables for urelements.

A structure $\mathfrak{M}$ is said to have *Σ-definable Skolem functions* [86], provided that, given any formula $\varphi(x_0, \dots, x_n)$ of signature σ, we can effectively find a Σ-formula $\psi(x_0, \dots, x_n)$ of signature σ' such that

$$\mathbb{HF}(\mathfrak{M}) \models \forall x_1 \dots \forall x_n\Big[\exists x_0 \varphi(x_0, \dots, x_n) \to \exists!\, x_0(\varphi(x_0, \dots, x_n) \wedge \psi(x_0, \dots, x_n))\Big]$$

(recall that $x_0, \dots, x_n$ together with all bounded variables in φ are variables for urelements).

The above definition can be easily expressed in terms of $s\Sigma$-equivalence. Let $\mathfrak{M}$ be a structure of signature σ and let the signature σ_{Skolem} consist of all symbols of σ and new functional symbols $f_\varphi(x_1, \dots, x_n)$ for all formulas $\varphi(x_0, x_1, \dots, x_n)$ of signature σ. The structure $\mathfrak{M}^S$ of signature σ_{Skolem} is called a (non-iterated) *Skolem expansion of* $\mathfrak{M}$ if $M^S = M$, $\mathfrak{M} \restriction_\sigma = \mathfrak{M}^S \restriction_\sigma$, and, for any formula $\varphi(x_0, x_1, \dots, x_n)$ of signature σ,

$$\mathfrak{M}^S \models \forall x_1 \dots \forall x_n(\exists x \varphi(x, x_1, \dots, x_n) \to \varphi(f_\varphi(x_1, \dots, x_n), x_1, \dots, x_n)).$$

It is easy to see that, if $\mathfrak{M}$ is a structure with a regular theory, then $\mathfrak{M}$ is a structure with Σ-definable Skolem functions if and only if, for some Skolem expansion $\mathfrak{M}^S$ of $\mathfrak{M}$ we have

$$\mathfrak{M}^S \equiv_{s\Sigma} \mathfrak{M}.$$

The Skolem expansion $\mathfrak{M}^S$ of a structure $\mathfrak{M}$ is *well defined* if for every formula $\varphi(x_0, x_1, \dots, x_n)$ of signature σ, every $\overline{m} \in M^n$, and every permutation ρ of the set $\{1, \dots, n\}$,

$$\mathfrak{M} \models (\varphi(x_0, \overline{m}) \leftrightarrow \varphi(x_0, \rho(\overline{m}))) \text{ implies } \mathfrak{M}^S \models (f_\varphi(\overline{m}) = f_\varphi(\rho(\overline{m}))),$$

where $\rho(\overline{m}) = \langle m_{\rho(1)}, \dots, m_{\rho(n)} \rangle$.

Recall that $\mathbb{HF}(\mathfrak{M})$ has the *uniformization property* if for every Σ-predicate $E \subseteq \mathrm{HF}(M) \times \mathrm{HF}(M)$ there exists a Σ-function F such that the following assertions are valid:

1. $\mathrm{dom}(F) = \mathrm{Pr}_1(E)$,
2. $\mathrm{graph}(F) \subseteq E$,

where

$$\mathrm{dom}(F) = \{x \mid F(x)\downarrow\}, \qquad \mathrm{graph}(F) = \{\langle x, y\rangle \bigm| F(x) = y\},$$
$$\mathrm{Pr}_1(E) = \left\{x \bigm| \exists y(\langle x, y\rangle \in E)\right\}.$$

The next theorem is a reformulation (and correction) of the main result from [86]. (Unfortunately, the property of well-definedness for Skolem expansions was not explicitly stated there, yet it was implicitly used in the text.) This theorem gives a natural (and useful) example of using $s\Sigma$-reducibility on structures, and Proposition 4.13 can be viewed as a natural (and non-trivial) example of $s\Sigma$-equivalence.

THEOREM 4.9. *Let $\mathfrak{M}$ be a structure with a regular theory. Then $\mathbb{HF}(\mathfrak{M})$ satisfies the uniformization property if and only if, for some well-defined Skolem expansion $\mathfrak{M}^S$ of $\mathfrak{M}$, we have*

$$\mathfrak{M}^S \equiv_{s\Sigma} \mathfrak{M}.$$

PROOF. Fix a Gödel numbering of formulas of signature σ' which distinguishes variables for urelements. The Gödel number of a formula φ is denoted by $[\varphi]$. Note that if T is a regular theory then, by model completeness, each formula of its signature is T-equivalent to some $\exists$-formula and, moreover, by decidability, this formula can be found effectively (henceforth, by effectiveness we mean existence of an appropriate computable function on the set of Gödel numbers).

Let us first prove the necessity. Suppose that $\mathbb{HF}(\mathfrak{M})$ satisfies the uniformization property. Let $\varphi(x_0, x_1, \dots, x_n)$ be an arbitrary formula of signature σ, and let $\psi(x_0, x_1, \dots, x_n)$ be some $\exists$-formula of signature σ which is $\mathrm{Th}(\mathfrak{M})$-equivalent to φ (such a formula exists and can be found effectively since $\mathrm{Th}(\mathfrak{M})$ is regular). First define a binary predicate G_0 as follows:

$$G_0 = \{\langle a, b\rangle | a = \langle [\varphi], m_1, \dots, m_{n-1}\rangle, b = \{\langle [\varphi], m_{\rho(1)}, \dots, m_{\rho(n-1)}\rangle |$$
$$| \rho \in S_{n-1} \text{ is a permutation such that}$$
$$\mathfrak{M} \models (\varphi(x_0, m_1, \dots, m_{n-1}) \leftrightarrow \varphi(x_0, m_{\rho(1)}, \dots, m_{\rho(n-1)}))\}\}.$$

From the regularity it follows that G_0 is a Σ-predicate on $\mathbb{HF}(\mathfrak{M})$. Let $F_0(x)$ be a Σ-function which uniformizes G_0. Second, define a binary Σ-predicate G_1 on $\mathbb{HF}(\mathfrak{M})$ as follows:

$$\langle a, m\rangle \in G_1 \iff \Big(a = \langle [\varphi], m_1, \dots, m_{n-1}\rangle\Big) \wedge$$
$$\wedge (\psi \text{ is an } \exists\text{-formula equivalent to } \varphi) \wedge \Sigma\text{-Sat}([\psi], \langle m, m_1, \dots, m_{n-1}\rangle).$$

Again, there exists a Σ-function F_1 that uniformizes G_1. The function

$$f_\varphi(x_1,\dots,x_{n-1}) = \lambda x_1.\ \dots\ .\lambda x_{n-1}.\, F_1\big(F_0(\langle[\varphi], x_1,\dots,x_{n-1}\rangle)\big)$$

is a Skolem function for the formula φ, which is well defined by the construction.

Let us now prove the sufficiency. Hereinafter, let T be a regular theory of signature σ and let $\mathfrak{M} = \langle M, \sigma^{\mathfrak{M}}\rangle$ be a model of T with a well-definable Skolem expansion $\mathfrak{M}^S$ such that $\mathfrak{M}^S \equiv_{s\Sigma} \mathfrak{M}$.

Lemma 4.10. *Suppose that P is a definable n-ary predicate over $\mathfrak{M}$. Then each formula defining P can be effectively transformed into a Σ-formula, with parameters used in the definition of P, and in the definition of $\mathfrak{M}^S$, that defines a predicate Q on $\mathbb{HF}(\mathfrak{M})$ such that*

1. *if $P = \varnothing$ then $Q = \varnothing$,*
2. *if $P \neq \varnothing$ then $Q = \{\bar{x}\}$, $\bar{x} \in P$.*

Proof. The case $n = 1$ is evident from the existence of a Skolem expansion $\mathfrak{M}^S$ with the property $\mathfrak{M}^S \equiv_{s\Sigma} \mathfrak{M}$; so, assume that $n > 1$ and that the statement is true for all $k < n$. Suppose that the predicate P is defined by a formula $\varphi(x_0,\dots,x_{n-1},\bar{y})$ of signature σ with parameters $\overline{m}$. Given the predicate

$$X = \{x_0 \mid \mathfrak{M} \models \exists x_1 \dots \exists x_{n-1}\, \varphi(x_0,\dots,x_{n-1},\overline{m})\},$$

we can effectively find by induction a Σ-formula $\Phi(x,\bar{y})$ that defines a single element in X (we can also assume that all the parameters used are urelements). Consider the predicate

$$Y = \Big\{\langle x_1,\dots,x_{n-1}\rangle \in M^{n-1} \;\Big|\; \mathbb{HF}(\mathfrak{M}) \models \exists x_0(\varphi(\bar{x},\overline{m}) \wedge \Phi(x_0,\overline{m}))\Big\}.$$

By Lemma 4.11, $\mathbb{HF}(\mathfrak{M}) \models \Phi(x_0,\overline{m}) \iff \mathfrak{M} \models \bigvee_{i\in\omega} \varphi_i(x_0,\overline{m})$, where φ_i are formulas of signature σ and the set $\{[\varphi_i] \mid i \in \omega\}$ is computably enumerable. Whence

$$Y = \Big\{\langle x_1,\dots,x_{n-1}\rangle \in M^{n-1} \;\Big|\; \mathfrak{M} \models \bigvee_{i\in\omega} \exists x_0(\varphi(\bar{x},\overline{m}) \wedge \varphi_i(x_0,\overline{m}))\Big\}.$$

Assume that $i_0 = \mu i\big(\mathfrak{M} \models \exists \bar{x}(\varphi(\bar{x},\overline{m}) \wedge \varphi_i(x_0,\overline{m}))\big)$; by regularity, i_0 is defined by a Σ-formula in $\mathbb{HF}(\mathfrak{M})$. Since the formula $\Phi(x_0,\overline{m})$ is true for at most one element, we have

$$Y = \Big\{\langle x_1,\dots,x_{n-1}\rangle \in M^{n-1} \;\Big|\; \mathfrak{M} \models \exists x_0(\varphi(\bar{x},\overline{m}) \wedge \varphi_{i_0}(x_0,\overline{m}))\Big\}.$$

We find by induction a Σ-formula $\Psi(x_1,\dots,x_{n-1},\bar{y})$ that defines a single element in Y, with the same restrictions on the parameters. The required predicate Q is defined by the Σ-formula $\Phi(x_0,\bar{y}) \wedge \Psi(x_1,\dots,x_{n-1},\bar{y})$. The lemma is proven. ⊣

Define a function $h : \omega \to \mathrm{HF}(\underline{\omega})$. For each $n \in \omega$, we put

(5)
$$h(n) = \begin{cases} \underline{n_1}, & \text{if } n = c(0, n_1), \\ \{h(n_1)\}, & \text{if } n = c(1, n_1), \\ h(n_1) \cup h(n_2), & \text{if } n = c(2, c(n_1, n_2)), l(n_1) \cdot l(n_2) > 0, \text{ and } n_1 < n_2, \\ \varnothing, & \text{otherwise,} \end{cases}$$

where $c(n, m) = \frac{(n+m)^2+3n+m}{2}$ is Cantor's bijection, and $l(n)$ and $r(n)$ are the left and right projections. It is easy to see from the definition that h is a numbering of $\mathrm{HF}(\underline{\omega})$, and since ω is a Δ-subset in $\mathrm{HF}(\mathfrak{M})$, we conclude that, in terms of [10], h is an $\mathbb{HF}(\mathfrak{M})$-constructivization of $\mathrm{HF}(\underline{\omega})$. Thus, $\mathrm{HF}(\underline{\omega})$ can be effectively defined in each superstructure.

LEMMA 4.11. *Suppose that $\varphi(x)$ is a Δ_0-formula of signature σ' and let $\varkappa \in \mathrm{HF}(n)$. Then we can effectively find a formula $\varphi^*(x_0, \dots, x_{n-1})$ of signature σ so that, for each valuation $\gamma : \{x_0, \dots, x_{n-1}\} \to M$,*

$$\mathbb{HF}(\mathfrak{M}) \models \varphi(x)^x_{t_\varkappa(\bar{x})}[\gamma] \iff \mathfrak{M} \models \varphi^*(x_0, \dots, x_{n-1})[\gamma].$$

PROOF. Given a formula $\varphi(x)$ and an element $\varkappa \in \mathrm{HF}(n)$, we construct a formula $\varphi^x_\varkappa(x_0, \dots, x_{n-1})$ of signature $\sigma' \cup \{\varnothing, \{\}, \cup\}$ as follows:

1. if $\varphi = \varphi_1 \, q \, \varphi_2$, $q \in \{\vee, \wedge, \to\}$, then $\varphi^x_\varkappa \rightleftharpoons (\varphi_1)^x_\varkappa \, q \, (\varphi_2)^x_\varkappa$;
2. if $\varphi = \neg\varphi_1$ then $\varphi^x_\varkappa \rightleftharpoons \neg(\varphi_1)^x_\varkappa$;
3. if $\varphi = (t_1 \, p \, t_2)$, $p \in \{\in, =\}$, then $\varphi^x_\varkappa \rightleftharpoons (t_1 \, p \, t_2)^x_{t_\varkappa(\bar{x})}$;
4. if $\varphi = \exists y \in x(\varphi_1)$ then $\varphi^x_\varkappa \rightleftharpoons \bigvee_{\varkappa' \in \varkappa} ((\varphi_1)^y_{\varkappa'})^x_\varkappa$;
5. if $\varphi = \forall y \in x(\varphi_1)$ then $\varphi^x_\varkappa \rightleftharpoons \bigwedge_{\varkappa' \in \varkappa} ((\varphi_1)^y_{\varkappa'})^x_\varkappa$;
6. if $\varphi = U(x)$ then $\varphi^x_\varkappa \rightleftharpoons \begin{cases} \tau, & \text{if } \varkappa \in n \\ \neg\tau, & \text{otherwise;} \end{cases}$
7. if $\varphi = P(t_0, \dots , t_k)$, $P \in \sigma$, then
$$\varphi^x_\varkappa \rightleftharpoons \begin{cases} P(t_0, \dots t_k)^x_{t_\varkappa(\bar{x})}, & \text{if } \varkappa \in n \\ \neg\tau, & \text{otherwise,} \end{cases}$$
where τ denotes the statement $\exists x(x = x)$ (without loss of generality we may assume that σ does not contain functional symbols).

Next, for any pair of terms t_0, t_1 of signature $\langle \varnothing, \{\}, \cup \rangle$ over variables for urelements $x_0, \dots, x_{n-1}$, we can effectively define formulas Φ_{t_0,t_1} and Ψ_{t_0,t_1} of empty signature so that $\mathrm{FV}(\Phi_{t_0,t_1}) = \mathrm{FV}(\Psi_{t_0,t_1}) = \mathrm{FV}(t_0) \cup \mathrm{FV}(t_1)$ and, for each valuation $\gamma : \mathrm{FV}(t_0 = t_1) \to M$, the following statements are true:

$$t_0^{\langle \mathbb{HF}(\mathfrak{M}), \{\}, \cup \rangle}[\gamma] \in t_1^{\langle \mathbb{HF}(\mathfrak{M}), \{\}, \cup \rangle}[\gamma] \iff \mathfrak{M} \models \Phi_{t_0,t_1}[\gamma]$$
$$t_0^{\langle \mathbb{HF}(\mathfrak{M}), \{\}, \cup \rangle}[\gamma] \subseteq t_1^{\langle \mathbb{HF}(\mathfrak{M}), \{\}, \cup \rangle}[\gamma] \iff \mathfrak{M} \models \Psi_{t_0,t_1}[\gamma]$$

(see [10] for a proof). The formula $\varphi^*(\bar{x})$ is obtained from $\varphi^x_\varkappa(\bar{x})$ by replacing the subformulas of kind $(t_0 \in t_1)$ by Φ_{t_0,t_1} and the subformulas of kind $(t_0 = t_1)$ by $(\Psi_{t_0,t_1} \wedge \Psi_{t_1,t_0})$. The lemma is proven. ⊣

Lemma 4.11 can be easily extended to formulas with several variables. This lemma also implies that we can restrict our consideration to formulas with parameters in M only.

Assume that $\Phi(x, \overline{m})$ is a Δ_0-formula of signature σ' with parameters $\overline{m}$ in M. For each $n \in \omega$, we define the set

$$\mathrm{H}_n \rightleftharpoons \{\varkappa \in \mathrm{HF}(n) \mid \mathrm{HF}(\mathfrak{M}) \models \exists x_0 \dots \exists x_{n-1}(\Phi(x, \overline{m}))^x_{t_\varkappa(\bar{x})}\}$$

and put $\mathrm{H} \rightleftharpoons \bigcup_{n\in\omega} \mathrm{H}_n$. The following lemma is valid:

Lemma 4.12. *The set* H *is a* Δ*-subset in* $\mathbb{HF}(\mathfrak{M})$.

Proof. Let $\overline{\mathrm{H}}_n \rightleftharpoons \mathrm{HF}(n) \setminus \mathrm{H}_n$, $\overline{\mathrm{H}} \rightleftharpoons \mathrm{HF}(\omega) \setminus \mathrm{H}$; then $\overline{\mathrm{H}} = \bigcup_{n\in\omega} \overline{\mathrm{H}}_n$. So, it suffices to prove that H_n is a Δ-subset in $\mathbb{HF}(\mathfrak{M})$.

Making use of Lemma 4.11, given a formula Φ and an element $\varkappa$, we effectively find a formula $\Psi_\varkappa(\bar{x}, \overline{m})$ of signature σ such that

$$\varkappa \in \mathrm{H}_n \iff \mathfrak{M} \models \exists x_0 \dots \exists x_{n-1}\, \Psi_\varkappa(\bar{x}, \overline{m}).$$

By regularity, given the formula $\exists \bar{x}\, \Psi_\varkappa(\bar{x}, \bar{y})$, we can effectively find an $\exists$-formula $\Theta_\varkappa(\bar{y})$ equivalent to it. Thus,

$$\varkappa \in \mathrm{H}_n \iff \mathbb{HF}(\mathfrak{M}) \models \Sigma\text{-Sat}([\Theta_\varkappa], \overline{m}).$$

The case $\varkappa \in \overline{\mathrm{H}}_n$ is handled similarly. The lemma is proven. ⊣

So, let $E \subseteq \mathrm{HF}(M) \times \mathrm{HF}(M)$ be an arbitrary Σ-predicate. Without loss of generality, we may assume that the predicate $E(x, y)$ is defined by a formula $\exists z\, \Phi(x, y, z, \overline{m})$, where $\Phi(x, y, z, \overline{m})$ is a Δ_0-formula with parameters $\overline{m}$ in M.

It is evident that $\mathrm{Pr}_1(E)$ is a Σ-predicate. Indeed, consider the Δ_0-formula

$$\Psi(x, t, \overline{m}) \rightleftharpoons \exists u \in t\ \exists v \in t\ \exists y \in u\ \exists z \in v\ (t = \langle y, z\rangle \wedge \Phi(x, y, z, \overline{m})).$$

It is clear that $x \in \mathrm{Pr}_1(E) \iff \mathbb{HF}(\mathfrak{M}) \models \exists t\, \Psi(x, t, \overline{m})$.

For each $a \in \mathrm{HF}(M)$, there exist $n \in \omega$, $\varkappa \in \mathrm{HF}(\underline{n})$, and $a_0, \dots, a_{n-1} \in M$ such that $a = \varkappa(\bar{a})$. (Here is the point where we assume the tuple $\bar{a}$ being ordered in some way, but the ordering does not matter, because the Skolem expansion is well defined.) Let $x^* \in \mathrm{HF}(\mathfrak{M})$, $x^* = \varkappa_0(\bar{x})$, where $\varkappa_0 \in \mathrm{HF}(\underline{l})$, $\bar{x} = \langle x_0, \dots, x_{l-1}\rangle \in M^l$. In the same way as for Lemma 4.11, we define the sets

$$H_n \rightleftharpoons \{\, \varkappa \in \mathrm{HF}(\underline{n}) \mid \mathbb{HF}(\mathfrak{M}) \models \exists t_0 \dots \exists t_{n-1}(\Psi(x^*, t, \overline{m}))^t_{t_\varkappa(\bar{t})}\}$$

for all $n \in \omega$ and put $\mathrm{H} \rightleftharpoons \bigcup_{n\in\omega} \mathrm{H}_n$.

If $x^* \in \mathrm{Pr}_1(E)$ then the set $\{t \mid \mathbb{HF}(\mathfrak{M}) \models \Psi(x^*, t, \overline{m})\}$ is nonempty; hence, the set H is nonempty too. In this case, the element $\varkappa_1 \in \mathrm{H}$ with the least number in the enumeration h is uniquely defined. In other words, $\varkappa_1$ is taken so as to satisfy the following conditions:

$$\exists k\big((k \in \omega) \wedge (\varkappa_1 = h(k)) \wedge (\varkappa_1 \in \mathrm{H}) \wedge \forall k' < k(h(k') \notin \mathrm{H})\big).$$

By virtue of Lemma 4.12, this condition is expressed in $\mathbb{HF}(\mathfrak{M})$ by some Σ-formula $\Psi_1(\varkappa_1, x^*, \overline{m})$.

Suppose that $\varkappa_1 \in \mathrm{HF}(\underline{n})$. Consider the set

$$T = \left\{\langle t_0, \dots, t_{n-1}\rangle \in M^n \;\middle|\; \mathbb{HF}(\mathfrak{M}) \models \Psi(x^*, t, \overline{m})^{t}_{\varkappa_1(\bar{t})}\right\}.$$

By Lemma 4.11 we can effectively construct a formula $\Theta(\bar{x}, \bar{t}, \bar{y})$ of signature σ so that, for each valuation $\gamma\ \{\bar{x}, \bar{t}\} \to M$, the following is true:

$$\mathbb{HF}(\mathfrak{M}) \models \Psi(x, t, \overline{m})^{x,\ \ \ \ \ \ t}_{\varkappa_0(\bar{x}),\ \varkappa_1(\bar{t})}[\gamma] \iff \mathfrak{M} \models \Theta(\bar{x}, \bar{t}, \overline{m})[\gamma].$$

By Lemma 4.10, given the formula Θ, we can effectively find an $\exists$-formula $\Theta^*(\bar{x}, \bar{t}, \overline{m})$ that defines a unique element $\overline{t^*}$ in the set T.

The element $t^* \rightleftharpoons \varkappa_1(\overline{t^*})$ satisfies the formula $\Psi(x^*, t^*, \overline{m})$; hence, it has the form $t^* = \langle y^*, z^*\rangle$ and, moreover, $\langle x^*, y^*\rangle \in E$. We put $F(x^*) \rightleftharpoons y^*$ by definition.

The required Σ-function F is defined as follows: let $F(x^*) = y^*$ if

$$\mathbb{HF}(\mathfrak{M}) \models \exists t^* \exists z^* \exists \varkappa_0 \exists \varkappa_1 ((t^* = \langle y^*, z^*\rangle) \wedge \Psi_1(\varkappa_1, x^*, \overline{m}) \wedge$$
$$\wedge \Sigma\text{-}Sat([\exists x_0 \dots \exists x_{l-1} \exists t_0 \dots \exists t_{n-1} (x^* = \varkappa_0(\bar{x}) \wedge$$
$$\wedge t^* = \varkappa_1(\bar{t}) \wedge \Theta^*(\bar{x}, \bar{t}, \bar{y}))], \overline{m})).$$

⊣

One of the important corollaries of this criterion follows from the next result.

Proposition 4.13 ([86]). *There exist well-defined Skolem expansions* $\mathbb{R}^S$ *and* $(\mathbb{Q}_p)^S$, *of the fields* $\mathbb{R}$ *and* $\mathbb{Q}_p$, *such that* $\mathbb{R}^S \equiv_{s\Sigma} \mathbb{R}$ *and* $(\mathbb{Q}_p)^S \equiv_{s\Sigma} \mathbb{Q}_p$.

Corollary 4.14 ([86]). *The structures* $\mathbb{HF}(\mathbb{R})$ *and* $\mathbb{HF}(\mathbb{Q}_p)$ *satisfy uniformization and have a universal* Σ*-function.*

For $\mathbb{HF}(\mathbb{R})$, the uniformization property and existence of a universal Σ-function was independently proved by the author in [85] and by M.V. Korovina in [34]. In [10], a general sufficient condition for an admissible set to have a universal Σ-function was found, which implies the existence of a universal Σ-function for $\mathbb{HF}(\mathfrak{M})$, where $\mathrm{Th}(\mathfrak{M})$ is regular.

The role of parameters in the Σ-definition of a structure is rather important. For example, as is easy to see, any countable structure is Σ-definable in $\mathbb{HF}(\mathbb{R})$, where $\mathbb{R}$ is the field of real numbers. The case of Σ-definability without parameters turned out to be quite interesting, as it was shown recently in [60].

Theorem 4.15 ([60]). *Suppose a countable structure* $\mathfrak{M}$ *is* Σ*-definable in* $\mathbb{HF}(\mathbb{R})$ *without parameters. Then* $\mathfrak{M}$ *has a hyperarithmetic presentation.*

This estimate is precise, as follows from the next theorem.

THEOREM 4.16 ([60]). *For any $\delta < \omega_1^{CK}$ there is a countable structure $\mathfrak{M}$ such that:*

1. *$\mathfrak{M}$ is Σ-definable in $\mathbb{HF}(\mathbb{R})$ without parameters;*
2. *for any $H \subseteq \omega$ such that $\mathfrak{M}$ has an H-computable presentation, we have $0^{(\delta)} \leqslant_T H$.*

In the case where we fix some restrictions on the cardinality of the congruence classes, the estimate of complexity becomes much lower.

THEOREM 4.17 ([60]). *Let $\mathfrak{M}$ be a countable structure with a finite signature. The following are equivalent:*

1. *$\mathfrak{M}$ is Σ-definable without parameters in $\mathbb{HF}(\mathbb{R})$, and all congruence classes are at least countable;*
2. *$\mathfrak{M}$ is computable.*

§5. Degrees of Presentability of Structures in Admissible Sets. The relation $\leqslant_\Sigma$ of Σ-reducibility, which is defined on structures of arbitrary cardinality, can, in the case of countable structures, be viewed as the strongest reducibility in the hierarchy of effective reducibilities on structures, as was shown in [93, 94]. We overview some of the results in this field.

Let $\mathbb{A}$ be an admissible set. A mapping $F : P(A)^n \to P(A)$ $(n \in \omega)$ is called a *Σ-operator* [10] if there exists a Σ-formula $\Phi(x_0, \dots, x_{n-1}, y)$ of signature $\sigma_\mathbb{A}$ such that, for any $S_0, \dots, S_{n-1} \in P(A)$,

$$F(S_0, \dots, S_{n-1}) = \left\{ a \mid \exists a_0, \dots, a_{n-1} \in A \left(\bigwedge_{i<n} a_i \subseteq S_i \wedge \mathbb{A} \models \Phi(a_0, \dots, a_{n-1}, a) \right) \right\}.$$

The next condition is necessary for the transitiveness of the reducibilities defined below. An operator $F : P(A) \to P(A)$ is called *strongly continuous in $S \in P(A)$* if, for any $a \subseteq F(S)$, $a \in A$, there exists an $a' \subseteq S$, $a' \in A$, such that $a \subseteq F(a')$ (this definition can be easily generalized for the case of operators with the number of arguments more than 1).

For an operator $F : P(A)^n \to P(A)$, we denote by $\delta_c(F)$ the set of elements of $P(A)^n$ in which F is strongly continuous. A set $S \in P(A)^n$ is called a *Σ_*-set* if $S \in \delta_c(F)$ for any Σ-operator $F : P(A)^n \to P(A)$. It is easy to verify that, in admissible sets of the kind $\mathbb{HF}(\mathfrak{M})$, any subset is a Σ_*-set. However, in general this is not so: for example, in [88] were studied Σ_*-sets in $\mathbb{HYP}(\mathbb{L})$, where $\mathbb{L}$ is a dense linear order. Even in this simplest case the class of Σ_*-sets is non-trivial.

Let $B, C \subseteq A$. We give below reducibilities that are direct generalizations of *e*- and *T*-reducibilities on natural numbers:

1. $B \leqslant_{e\Sigma} C$ if there is a unary Σ-operator F for which $C \in \delta_c(F)$ and $B = F(C)$;
2. $B \leqslant_{T\Sigma} C$ if there are binary Σ-operators F_0 and F_1 such that $\langle C, A \setminus C \rangle \in \delta_c(F_0) \cap \delta_c(F_1)$ and $B = F_0(C, A \setminus C)$, $A \setminus B = F_1(C, A \setminus C)$.

Let $\mathbb{A}$ be an admissible set. We define uniform reducibilities on subsets of A, which are direct generalizations of Medvedev, Muchnik, and Dyment reducibilities on mass problems [81]. Let $\mathcal{X}, \mathcal{Y} \subseteq P(A)$. Then:

1. $\mathcal{X}$ is *Medvedev reducible* to $\mathcal{Y}$ (i.e. $\mathcal{X} \leqslant_s \mathcal{Y}$) if there are binary Σ-operators F_0 and F_1 such that for all $Y \in \mathcal{Y}$, $\langle Y, A \setminus Y\rangle \in \delta_c(F_0) \cap \delta_c(F_1)$, and for some $X \in \mathcal{X}$, $X = F_0(Y, A \setminus Y)$ and $A \setminus X = F_1(Y, A \setminus Y)$;
2. $\mathcal{X}$ is *Dyment reducible* to $\mathcal{Y}$ (i.e. $\mathcal{X} \leqslant_e \mathcal{Y}$) if there is a unary Σ-operator F such that, for all $Y \in \mathcal{Y}$, $Y \in \delta_c(F)$ and $F(\mathcal{Y}) \subseteq \mathcal{X}$;
3. $\mathcal{X}$ is *Muchnik reducible* to $\mathcal{Y}$ (i.e. $\mathcal{X} \leqslant_w \mathcal{Y}$) if for any $Y \in \mathcal{Y}$ there are binary Σ-operators F_0 and F_1 such that $\langle Y, A \setminus Y\rangle \in \delta_c(F_0) \cap \delta_c(F_1)$ and, for some $X \in \mathcal{X}$, $X = F_0(Y, A \setminus Y)$ and $A \setminus X = F_1(Y, A \setminus Y)$;
4. $\mathcal{X}$ is *weakly Dyment reducible* to $\mathcal{Y}$ (i.e. $\mathcal{X} \leqslant_{ew} \mathcal{Y}$) if for any $Y \in \mathcal{Y}$ there is a unary Σ-operator F such that $Y \in \delta_c(F)$ and $F(\mathcal{Y}) \subseteq \mathcal{X}$.

For an admissible set $\mathbb{A}$ and a symbol $r \in \{e, s, w, ew\}$, we denote by $\mathcal{M}_r(\mathbb{A})$ the degree structure $\langle P(P(A))/\equiv_r, \leqslant_r\rangle$. For simplicity, we use the notation $\mathcal{M}_r$ instead of $\mathcal{M}_r(\mathbb{HF}(\varnothing))$. All structures of the kind $\mathcal{M}_r(\mathbb{A})$ are lattices with 0 and 1, moreover, $\mathcal{M}, \mathcal{M}_e, \mathcal{M}_w$ are isomorphic to the Medvedev, Dyment, and Muchnik lattices, respectively.

Proposition 5.1. *For any admissible set* $\mathbb{A}$ *and any reducibility symbol* $r \in \{e, s, w, ew\}$*, the structure*

$$\mathcal{M}_r(\mathbb{A}) = \langle P(P(A))/\equiv_r, \leqslant_r\rangle$$

is a distributive lattice with 0 *and* 1.

Proof. Fix some admissible set $\mathbb{A}$, together with some reducibility symbol $r \in \{e, s, w, ew\}$. For arbitrary $a \in A$ and $X \subseteq A$, we denote by $a * X$ the set $\{\langle a, x\rangle | x \in X\}$. For any $\mathcal{X}, \mathcal{Y} \subseteq P(A)$, define, as in the classical case,

$$\mathcal{X} \vee \mathcal{Y} = \{X \oplus Y | X \in \mathcal{X}, Y \in \mathcal{Y}\}, \ \mathcal{X} \wedge \mathcal{Y} = 0 * \mathcal{X} \cup 1 * \mathcal{Y}.$$

It is easy to check that $\mathcal{X} \vee \mathcal{Y}$ and $\mathcal{X} \wedge \mathcal{Y}$ are the l.u.b. and the g.l.b. for $\mathcal{X}$ and $\mathcal{Y}$, correspondingly. We define $[\mathcal{X}]_r \vee [\mathcal{Y}]_r = [\mathcal{X} \vee \mathcal{Y}]_r$ and $[\mathcal{X}]_r \wedge [\mathcal{Y}]_r = [\mathcal{X} \wedge \mathcal{Y}]_r$. To prove distributivity, it is enough to check it straightforwardly as in the classical case. As usual, $\mathbf{1} = [\varnothing]_r$ and $\mathbf{0} = [\{\varnothing\}]_r$. ⊣

Recall, [93], that the problem of presentability of a structure $\mathfrak{M}$ in an admissible set $\mathbb{A}$ is the mass problem

$$\mathrm{pr}_{\mathbb{A}}(\mathfrak{M}) = \{\mathcal{C} \mid \mathcal{C} \subseteq A, \ \mathcal{C} \simeq \mathfrak{M}\}.$$

Here, we identify presentations of structures with the atomic diagrams of these presentations and assume that some Gödel numbering for the signature symbols is fixed. We denote by $\mathfrak{M} \leqslant_r^{\mathbb{A}} \mathfrak{N}$ the fact that $\mathrm{pr}_{\mathbb{A}}(\mathfrak{M}) \leqslant_r \mathrm{pr}_{\mathbb{A}}(\mathfrak{N})$. For an admissible set $\mathbb{A}$ and a reducibility symbol $r \in \{e, s, w, ew\}$, preorder $\leqslant_r^{\mathbb{A}}$ generates the semilattice $\mathcal{S}_r(\mathbb{A})$ of degrees of presentability of structures (with cardinality $\leqslant \mathrm{card}(\mathbb{A})$) in $\mathbb{A}$ with respect to reducibility r.

Recall that we denote by $\underline{\mathfrak{M}}$ the set $\mathrm{Pr}(\mathfrak{M}, \mathbb{HF}(\varnothing))$ of presentations of $\mathfrak{M}$ in the least admissible set $\mathbb{HF}(\varnothing)$.

For a countable structure $\mathfrak{M}$, we consider the following classes consisting of structures that are effectively reducible to $\mathfrak{M}$:

$\mathcal{K}_\Sigma(\mathfrak{M}) = \{\mathfrak{N} \mid \mathfrak{N} \leqslant_\Sigma \mathfrak{M}\}$,
$\mathcal{K}_e(\mathfrak{M}) = \{\mathfrak{N} \mid \underline{\mathfrak{N}} \leqslant_e \underline{(\mathfrak{M}, \bar{m})}$ for some $\bar{m} \in M^{<\omega}\}$,
$\mathcal{K}_s(\mathfrak{M}) = \{\mathfrak{N} \mid \underline{\mathfrak{N}} \leqslant_s \underline{(\mathfrak{M}, \bar{m})}$ for some $\bar{m} \in M^{<\omega}\}$,
$\mathcal{K}_{ew}(\mathfrak{M}) = \{\mathfrak{N} \mid \underline{\mathfrak{N}} \leqslant_{ew} \underline{\mathfrak{M}}\}$,
$\mathcal{K}_w(\mathfrak{M}) = \{\mathfrak{N} \mid \underline{\mathfrak{N}} \leqslant_w \underline{\mathfrak{M}}\}$.

It is known, [94], that for any structure $\mathfrak{M}$, the following inclusions hold:

$$\mathcal{K}_\Sigma(\mathfrak{M}) \subseteq \mathcal{K}_e(\mathfrak{M}) \subseteq \mathcal{K}_s(\mathfrak{M}) \subseteq \mathcal{K}_w(\mathfrak{M}),$$

and

$$\mathcal{K}_e(\mathfrak{M}) \subseteq \mathcal{K}_{ew}(\mathfrak{M}) \subseteq \mathcal{K}_w(\mathfrak{M}).$$

In general, all these inclusions are proper: [20, 21].

For any $r \in \{e, s, w, ew\}$, we define a relation $\leqslant_r$ on $\mathcal{K}_\omega$ by setting $\mathfrak{M} \leqslant_r \mathfrak{N}$ iff $\mathcal{K}_r(\mathfrak{M}) \subseteq \mathcal{K}_r(\mathfrak{N})$ and letting $\mathcal{S}_r = \langle \mathcal{K}_\omega / \equiv_r, \leqslant_r \rangle$ be the structure of degrees of presentability corresponding to this relation.

THEOREM 5.2 ([93]). *For any $r \in \{e, s, w, ew\}$, the structure $\mathcal{S}_r$ is an upper semilattice with* 0, *and the following embeddings ($\hookrightarrow$) and homomorphisms ($\to$) hold:*

$$\mathcal{D} \hookrightarrow \mathcal{D}_e \hookrightarrow \mathcal{S}_\Sigma \to \mathcal{S}_e \to \mathcal{S} \hookrightarrow \mathcal{M}.$$

The next theorem clarifies the role of Σ-reducibility as the strongest possible effective reducibility on structures. Moreover, it gives an "existential" characterization (item (1) below) of a "universal" sentence (item (2)).

THEOREM 5.3 ([98]). *For any structures $\mathfrak{M}$, $\mathfrak{N}$, and any reducibility symbol $r \in \{e, ew\}$, the following are equivalent:*

(1) $\mathfrak{M} \leqslant_\Sigma \mathfrak{N}$;
(2) *for any admissible set $\mathbb{A}$, we have $\mathfrak{M} \leqslant_r^{\mathbb{A}} \mathfrak{N}$.*

REMARK. The cardinalities of $\mathfrak{M}$, $\mathfrak{N}$, and $\mathbb{A}$ in this theorem are arbitrary: $\mathrm{pr}_{\mathbb{A}}(\mathfrak{M}) = \varnothing$ (i.e., the "hardest" problem) if $\mathrm{card}(\mathbb{A}) < \mathrm{card}(\mathfrak{M})$.

As a corollary of the Jump Inversion Theorem for the semilattices of Σ-degrees and Theorem 5.3, we get

PROPOSITION 5.4 ([98]). *Suppose $\mathbb{A}$ is an admissible set, and r is a reducibility symbol, $r \in \{e, ew\}$. Let $\mathfrak{A}$ be a structure such that $\mathbf{0}' \leqslant_\Sigma \mathfrak{A}$. Then there exists a structure $\mathfrak{B}$ for which*

$$\mathfrak{B}' \equiv_r^{\mathbb{A}} \mathfrak{A}.$$

Proposition 5.5 ([98]). *Let $\mathbb{A}$ be an admissible set with $\mathbf{0}' \in \Delta(\mathbb{A})$. Then, for any generalized Σ-low structure $\mathfrak{A}$ and for any effective reducibility $r \in \{e, ew\}$,*

$$\mathfrak{A}' \equiv_r^{\mathbb{A}} \mathfrak{A}.$$

Proof. It is sufficient to consider the case of $\mathfrak{A}$ being a generalized Σ-low structure with $\mathrm{card}(\mathfrak{A}) \leqslant \mathrm{card}(\mathbb{A})$. Since there is a natural homomorphism from $\mathcal{S}_\Sigma(\mathrm{card}(\mathfrak{A}))$ into $\mathcal{M}_r^{\mathbb{A}}$, we get the desired statement. ⊣

Remark 5.6. The existence of fixed points for the operation of Σ-jump with respect to Muchnik reducibility (i.e., the existence of structures $\mathfrak{A}$ with the property $\mathfrak{A} \equiv_w \mathfrak{A}'$) was announced by the author in his talk at the CiE2009 Conference. However, the proof turned to be more complicated than he assumed. The corrected proof, valid for all abstract effective reducibilities, will appear in [98].

Remark 5.7. If an admissible set $\mathbb{A}$ is recursively listed, [4], then Theorems 5.3, 5.4, and Proposition 5.5 are true also for reducibilities $r \in \{s, w\}$, since in this case an analogue of Theorem 2 from [93] is valid.

For arbitrary structures $\mathfrak{M}$ and $\mathfrak{M}'$ with the same signature and any $n \in \omega$, we denote by $\mathfrak{M} \equiv_n^{\mathrm{HF}} \mathfrak{M}'$ the fact that $\mathbb{HF}(\mathfrak{M}) \equiv_n \mathbb{HF}(\mathfrak{M}')$, and by $\mathfrak{M} \preccurlyeq_n^{\mathrm{HF}} \mathfrak{M}'$ the fact that $\mathbb{HF}(\mathfrak{M}) \preccurlyeq_n \mathbb{HF}(\mathfrak{M}')$. It is easy to check that, for $n < 2$, $\mathfrak{M} \equiv_n^{\mathrm{HF}} \mathfrak{M}'$ if and only if $\mathfrak{M} \equiv_n \mathfrak{M}'$. In case $n = 2$, $\mathfrak{M} \equiv_2^{\mathrm{HF}} \mathfrak{M}'$ if and only if, for any computable sequence $\{\varphi_{mn}(\overline{x}_m, \overline{y}_n) | m, n \in \omega\}$ of quantifier-free formulas of signature $\sigma_{\mathfrak{M}}$,

$$\mathfrak{M}' \models \bigvee_{m\in\omega} \exists \overline{x}_m \bigwedge_{n\in\omega} \forall \overline{y}_n \varphi_{mn}(\overline{x}_m, \overline{y}_n)$$

if and only if the same sentence is true in $\mathfrak{M}$.

For arbitrary structures $\mathfrak{M}$ and $\mathfrak{N}$, we denote by $\mathfrak{M} \leqslant_\exists \mathfrak{N}$ the fact that, for any tuple $\overline{m} \in M^{<\omega}$, there exists a tuple $\overline{n} \in N^{<\omega}$ such that $\mathrm{Th}_\exists(\mathfrak{M}, \overline{m}) \leqslant_e \mathrm{Th}_\exists(\mathfrak{N}, \overline{n})$. In particular, if $\mathfrak{M}$ is locally constructivizable then $\mathfrak{M} \leqslant_\exists \mathfrak{N}$ for any structure $\mathfrak{N}$. As was noted in [10], if $\mathfrak{M} \leqslant_\Sigma \mathfrak{N}$ and $\mathfrak{N}$ is locally constructivizable then $\mathfrak{M}$ is also locally constructivizable. A straightforward generalization of this fact is as follows: $\mathfrak{M} \leqslant_\Sigma \mathfrak{N}$ implies $\mathfrak{M} \leqslant_\exists \mathfrak{N}$.

Definition 5.8 ([93]). A structure $\mathfrak{M}$ is *uniformly locally constructivizable of level n* $(1 < n \leqslant \omega)$ if there exists a constructivizable structure $\mathfrak{N}$ for which $\mathfrak{M} \preccurlyeq_n^{\mathrm{HF}} \mathfrak{N}$.

For instance, the structure $\langle \omega_1^{CK}, \leqslant \rangle$ is uniformly locally constructivizable of level ω since $\langle \omega_1^{CK}, \leqslant \rangle \preccurlyeq^{\mathrm{HF}} \langle \omega_1^{CK}(1+\eta), \leqslant \rangle$, where the last ordering (known as the *Harrison ordering*) is constructivizable.

Proposition 5.9 ([94]). *If $\mathfrak{M} \leqslant_\Sigma \mathfrak{N}$ and a structure $\mathfrak{N}$ is (uniformly) locally constructivizable of level n $(1 < n \leqslant \omega)$, then $\mathfrak{M}$ is also (uniformly) locally constructivizable of level n.*

The next proposition shows that a class of locally constructivizable (of level 1) countable structures is closed downward with respect to $\leqslant_w$, which is weakest among the reducibilities under consideration.

PROPOSITION 5.10 ([94]). *Let $\mathfrak{M}$ and $\mathfrak{N}$ be structures. Then $\mathfrak{N} \leqslant_\exists \mathfrak{M}$ if $\mathfrak{N} \in \mathcal{K}_w(\mathfrak{M})$. In particular, if $\mathfrak{M}$ is locally constructivizable, then every structure $\mathfrak{N} \in \mathcal{K}_w(\mathfrak{M})$ is also locally constructivizable.*

A pair $(\mathfrak{M}, \mathfrak{N})$ is locally constructivizable iff $\mathfrak{M}$ and $\mathfrak{N}$ are also; therefore, a set of degrees generated by locally constructivizable structures is an ideal in semilattices $\mathcal{S}_r$, $r \in \{\Sigma, e, s, w, ew\}$. Classes of locally constructivizable structures of level n, $n > 1$, however, are downward closed with respect to $\leqslant_\Sigma$ only (so they form initial segments in $\mathcal{S}_\Sigma$). For weaker reducibilities, this is not the case.

THEOREM 5.11 ([94]). *There exists a countable structure $\mathfrak{M}_0$ which is locally constructivizable of level 1 exactly and is such that $\mathfrak{M}_0 \leqslant \mathfrak{M}$ for every nonconstructivizable countable structure $\mathfrak{M}$. If $\mathfrak{M}$ is locally constructivizable of level $n > 1$ but is not constructivizable, then $\mathcal{K}_\Sigma(\mathfrak{M}) \subsetneq \mathcal{K}(\mathfrak{M})$.*

The proof makes use of the result (obtained by T. Slaman [80], and, independently, S. Wehner [102]) which states that there exists a structure whose problem of presentability belongs to the least nonzero degree of the Medvedev lattice (which, in particular, means that a semilattice $\mathcal{S}$ of degrees of presentability has a least nonzero element). Every such structure is locally constructivizable. In particular, the following fact was proved in [93].

THEOREM 5.12. *There exist a countable structure $\mathfrak{M}$ and a unary relation $P \subseteq M$ for which $\underline{(\mathfrak{M}, P)} \equiv_s \underline{\mathfrak{M}}$ but $(\mathfrak{M}, P) \nleqslant_\Sigma \mathfrak{M}$.*

This theorem is of interest in connection with the following result from [1, 6]: for any countable structure $\mathfrak{M}$, a relation $P \subseteq M^n$, $n \in \omega$, is Σ-definable in $\mathbb{HF}(\mathfrak{M})$ iff $P^{\mathfrak{C}}$ is $\mathfrak{C} \restriction \sigma_{\mathfrak{M}}$-c.e. for every $\mathfrak{C} \in \underline{(\mathfrak{M}, P)}$.

The next result from [93] gives some sufficient conditions for the equality of the principal ideals generated by a structure $\mathfrak{M}$ with respect to different effective reducibilities.

THEOREM 5.13. *If $\mathfrak{M}$ has a degree then $\mathcal{K}_\Sigma(\mathfrak{M}) = \mathcal{K}_e(\mathfrak{M}) = \mathcal{K}_s(\mathfrak{M}) = \mathcal{K}_w(\mathfrak{M})$. If $\mathfrak{M}$ has an e-degree then $\mathcal{K}_\Sigma(\mathfrak{M}) = \mathcal{K}_e(\mathfrak{M}) = \mathcal{K}_{ew}(\mathfrak{M})$.*

For an admissible set $\mathbb{A}$ and a structure $\mathfrak{M}$, consider the class

$$\mathcal{K}_s^{\mathbb{A}}(\mathfrak{M}) = \{\mathfrak{M}' \mid \mathrm{Pr}_{\mathbb{A}}(\mathfrak{M}') \leqslant_s \mathrm{Pr}_{\mathbb{A}}(\mathfrak{M}, \bar{m}) \text{ for some } \bar{m} \in M^{<\omega}\}.$$

The classes $\mathcal{K}_e^{\mathbb{A}}(\mathfrak{M})$, $\mathcal{K}_w^{\mathbb{A}}(\mathfrak{M})$, and $\mathcal{K}_{ew}^{\mathbb{A}}(\mathfrak{M})$ are defined similarly.

PROPOSITION 5.14 ([93]). *Let $\mathfrak{M}$ and $\mathfrak{N}$ be countable structures and let $\mathfrak{N}$ be a structure of the empty signature, or dense linear order. Then $\mathcal{K}_\Sigma(\mathfrak{M}) = \mathcal{K}_e^{\mathbb{HF}(\mathfrak{N})}(\mathfrak{M}) = \mathcal{K}_s^{\mathbb{HF}(\mathfrak{N})}(\mathfrak{M})$.*

As a consequence, there exist natural isomorphisms between a semilattice $\mathcal{S}_\Sigma$ of degrees of Σ-definability and semilattices $\mathcal{S}_r^{\mathbb{HF}(\mathfrak{N})}$ of degrees of presentability, where $\mathfrak{N}$ is a countable structure of empty signature, or dense linear order.

We mention one more result using the equivalence of "$\forall$-recursiveness" and "$\exists$-definability", based on results from [50], [63], and [1, 6].

THEOREM 5.15 ([93]). *For any countable structures* $\mathfrak{M}$ *and* $\mathfrak{N}$ *and any relation* $R \subseteq \mathbb{HF}(\mathfrak{N})$, *the following conditions are equivalent:*

1. $R \leqslant_{e\Sigma} \mathcal{C}$ *for every presentation* $\mathcal{C}$ *of* $\mathfrak{M}$ *in the admissible set* $\mathbb{HF}(\mathfrak{N})$;
2. R *is* Σ*-definable in* $\mathbb{HF}(\mathfrak{M}, \mathfrak{N})$, *where* $(\mathfrak{M}, \mathfrak{N})$ *is the pair of* $\mathfrak{M}$ *and* $\mathfrak{N}$).

DEFINITION 5.16. Let $\mathfrak{M}$ and $\mathfrak{N}$ be countable structures. The structure $\mathfrak{M}$ *has a degree* (an *e-degree*) *over the structure* $\mathfrak{N}$ if there exists a least degree among all $T\Sigma$-degrees ($e\Sigma$-degrees) of all possible presentations of $\mathfrak{M}$ in $\mathbb{HF}(\mathfrak{N})$.

An immediate consequence of Theorem 5.15 is the following generalization of item 3.6:

THEOREM 5.17. *Let* $\mathfrak{M}$ *and* $\mathfrak{N}$ *be countable structures. Then the following conditions below are equivalent:*

1. $\mathfrak{M}$ *has a degree (an e-degree) over* $\mathfrak{N}$;
2. *some presentation* $\mathcal{C} \subseteq HF(N)$ *of* $\mathfrak{M}$ *is* Δ*-subset (*Σ*-subset) in* $\mathbb{HF}(\mathfrak{M}, \mathfrak{N})$.

As a corollary, for $\mathfrak{M} \leqslant_\exists \mathfrak{N}$, the structure $\mathfrak{M}$ has a degree (an e-degree) over $\mathfrak{N}$ iff $\mathfrak{M} \leqslant_\Sigma \mathfrak{N}$. It is also true that if $\mathfrak{M}$ has a degree (an e-degree) over $\mathfrak{N}$, and $\mathfrak{N} \leqslant_\Sigma \mathfrak{N}'$, then $\mathfrak{M}$ has a degree (an e-degree) over $\mathfrak{N}'$. Furthermore, we have, for any countable structure $\mathfrak{A}$, that there exists a structure $\mathfrak{M}$ which has a degree but is not Σ-definable in $\mathbb{HF}(\mathfrak{A})$.

As in the nonrelativized case, we have the following result:

THEOREM 5.18. *Let* $\mathfrak{M}$ *and* $\mathfrak{N}$ *be countable structures. If* $\mathfrak{M}$ *has a degree over* $\mathfrak{N}$, *then* $\mathcal{K}_\Sigma(\mathfrak{M}) = \mathcal{K}_e^{\mathbb{HF}(\mathfrak{N})}(\mathfrak{M}) = \mathcal{K}_s^{\mathbb{HF}(\mathfrak{N})}(\mathfrak{M})$. *If* $\mathfrak{M}$ *has an e-degree over* $\mathfrak{N}$, *then* $\mathcal{K}_\Sigma(\mathfrak{M}) = \mathcal{K}_e^{\mathbb{HF}(\mathfrak{N})}(\mathfrak{M})$.

§6. Effective Presentations of Special Structures. As already mentioned, cardinality boundaries are unavoidable in the classical theory of computability (CTC). Numberings permit us to use CTC for countable objects. Admissible sets of the form $\mathbb{HF}(\mathfrak{M})$ can have an arbitrary cardinality. Hence, the following question naturally arises: does there exist a "reasonably good" theory T such that the class of admissible sets of the form $\mathbb{HF}(\mathfrak{M})$, with $\mathfrak{M} \models T$, allows us to extend, in some natural way, the classical theory CTC to the case of objects with an arbitrary cardinality?

Recall that a theory T of a finite signature is called *regular*, [10], if it is decidable and model complete.

REMARK 6.1. Let T be a regular theory. Then, for any formula $\Phi(\overline{x})$ of the signature of T, there exists an $\exists$-formula $\Psi(\overline{x})$ which is equivalent (with respect to the theory T) to $\Phi(\overline{x})$. Moreover, $\Psi(\overline{x})$ can be found effectively from $\Phi(\overline{x})$.

Recall that a theory T is called *c-simple* (constructively simple), [10], if it is regular, ω-categorical, and has a decidable set of the complete formulas.

REMARK 6.2. In [10] such theories were called simple, but this terminology was simultaneously used in model theory for a different notion.

In the definition of a c-simple theory, ω-categoricity gives the uniqueness, up to an isomorphism, of a countable model of such a theory. Model completeness, decidability of a theory, and decidability of the set of its complete formulas, guarantee the autostability of every constructivization of this countable theory, i.e., the uniqueness of the "computability" on its countable models.

Furthermore, if T is a c-simple theory, $\mathfrak{M}_0$ and $\mathfrak{M}_1$ are any models of T (i.e. $\mathfrak{M}_i \models T$, $i = 0, 1$), then $\mathbb{HF}(\mathfrak{M}_0) \equiv \mathbb{HF}(\mathfrak{M}_1)$, since the models of ω-categorical theories are saturated enough (see [10]).

Henceforth, for a c-simple theory T, the class of admissible sets of the form $\mathbb{HF}(\mathfrak{M})$, $\mathfrak{M} \models T$, extends "uniformly" the classical theory of computability for arbitrary infinite cardinalities.

An example of a c-simple theory is the theory T_{E} of infinite structures with the empty signature. But this theory is too "weak", if we regard a theory T as being "strong" whenever there are enough uncountable structures Σ-definable in $\mathbb{HF}(\mathfrak{M})$, $\mathfrak{M} \models T$. The reason for this "weakness" of T_{E} is the following property: for an arbitrary set X and arbitrary permutation f on X, we can f extend (in a unique way) to an automorphism f^* of $\mathbb{HF}(X)$.

Following [8, 9, 10], we present a characterization of the theories having uncountable models which are Σ-definable in $\mathbb{HF}(\mathfrak{L})$ for $\mathfrak{L} \models T_{\mathrm{DLO}}$.

The category ${}^*\omega$ is defined as follows. Its objects are sets of the form $[\mathbf{n}] \rightleftharpoons \{0, 1, \dots, n-1\}$, $n \in \omega$ ($[\mathbf{0}] \rightleftharpoons \varnothing$), and its morphisms are order-preserving embeddings. It should be noted that there is a unique morphism from $[\mathbf{0}]$ into $[\mathbf{n}]$ for any $n \in \omega$.

DEFINITION 6.3. By a *${}^*\omega$-spectrum* we mean any functor S from the category ${}^*\omega$ into the category Mod^*_σ of structures (of some fixed signature σ), whose morphisms are all possible embeddings.

To define a ${}^*\omega$-spectrum S, it is necessary to give an infinite sequence $\mathfrak{M}_0$, $\mathfrak{M}_1$, $\dots$, $\mathfrak{M}_n$, $\dots$, $n \in \omega$, of structures of signature σ, and associate with each order-preserving embeddings $\mu : [\mathbf{n}] \to [\mathbf{m}]$ an embedding $\mu_* : \mathfrak{M}_n \to \mathfrak{M}_m$ such that, if $\mu_0 : [\mathbf{n}] \to [\mathbf{m}]$ and $\mu_1 : [\mathbf{m}] \to [\mathbf{k}]$, $n \leqslant m \leqslant k \in \omega$, are morphisms of the category ${}^*\omega$, then $(\mu_0\mu_1)_* = \mu_{1*}\mu_{0*}$, and if $\mu : [\mathbf{n}] \to [\mathbf{n}]$ is the unique morphism from $[\mathbf{n}]$ into $[\mathbf{n}]$ (which is $\mathrm{id}_{[\mathbf{n}]}$), then $\mu_* = \mathrm{id}_{\mathfrak{M}_n} : \mathfrak{M}_n \to \mathfrak{M}_n$, $n \in \omega$.

If the ${}^*\omega$-spectrum $S = \{\mathfrak{M}_n, \mu_* | n \in \omega, \mu \in \mathrm{Mor}^*\omega\}$ has been defined, then for any linearly ordered set L, it is possible to define the structure $\mathfrak{M}_L(\mathfrak{M}^S_L)$ as a direct limit $\lim_{\vec{L_0}} \mathfrak{M}'_{L_0}$ of the spectrum

$$\{\mathfrak{M}'_{L_0}, \varphi_{L_0,L_1} \mid L_0 \subseteq L_1 \subseteq L,\ L_1 \text{ is finite}\},$$

where $\mathfrak{M}'_{L_0} \rightleftharpoons \mathfrak{M}_n$, if $L_0 \subseteq L$ is finite and $|L_0| = n$, and the embedding φ_{L_0,L_1} : $\mathfrak{M}'_{L_0} \to \mathfrak{M}'_{L_1}$ is defined for finite $L_0 \subseteq L_1 (\subseteq L)$ as follows: if $L_1 = \{l_0 < l_1 < \cdots < l_{m-1}\}$ and $L_0 = \{l_{i_0} < l_{i_1} < \cdots < l_{i_{n-1}}\}$ (in which case $0 \leqslant i_0 < i_1 < \cdots < i_{n-1} \leqslant m$) and $\mu : [\mathbf{n}] \to [\mathbf{m}]$ is defined as $\mu(j) \rightleftharpoons i_j$, $j < n$, then

$$\varphi_{L_0,L_1} \rightleftharpoons \mu_* : \mathfrak{M}'_{L_0} = \mathfrak{M}_n \to \mathfrak{M}_m = \mathfrak{M}'_{L_1}.$$

If $L \subseteq L'$ are linearly ordered sets, then the structure $\mathfrak{M}_L$ can be identified with a substructure of $\mathfrak{M}_{L'}$ in a natural way.

Any isomorphism between linearly ordered sets L and L' induces an isomorphism between $\mathfrak{M}_L$ and $\mathfrak{M}_{L'}$. Also if $L \subseteq L'$ are dense linear orders without endpoints, then $\mathfrak{M}_L \preccurlyeq \mathfrak{M}_{L'}$. As a corollary, if L and L' are dense linear orders without endpoints, then $\mathfrak{M}_L \equiv \mathfrak{M}_{L'}$.

Let μ_0 and μ_1 be morphisms from $[\mathbf{1}]$ into $[\mathbf{2}]$ such that $\mu_0(0) = 0$ and $\mu_1(0) = 1$. The condition

$$\mu_{0*} \neq \mu_{1*} \tag{$*$}$$

is sufficient for $|\mathfrak{M}^S_L| \geqslant |L|$ to hold for any linearly ordered set L.

Definition 6.4. A system of numberings $\nu_n : \omega \to M_n$, $n \in \omega$, is called a *computable sequence of constructivization*

$$(\mathfrak{M}_0, \nu_0), (\mathfrak{M}_1, \nu_1), \ldots, (\mathfrak{M}_n, \nu_n), \ldots, \quad n \in \omega,$$

if the following conditions hold (we assume that the signature σ of the structures $\mathfrak{M}_0, \mathfrak{M}_1, \ldots$ is finite and without function symbols):

1. $E \rightleftharpoons \{\langle n, m_0, m_1\rangle | n, m_0, m_1 \in \omega, \nu_n(m_0) = \nu_n(m_1)\}$ is a Δ-predicate on ω;
2. $N_P \rightleftharpoons \{\bar{n} = \langle n_0, n_1, \ldots, n_k\rangle | \bar{n} \in \omega^{k+1}, \langle \nu_{n_0}(n_1), \ldots, \nu_{n_0}(n_k)\rangle \in P^{\mathfrak{M}_{n_0}}\}$ is a Δ-predicate on ω for any (k-ary) predicate symbol $P \in \sigma$;
3. for any constant symbol $c \in \sigma$ there exists a Σ-function $f_c : \omega \to \omega$ such that $c^{\mathfrak{M}_n} = \nu_n f_c(n)$.

Every morphism μ : $[\mathbf{n}] \to [\mathbf{m}]$ of the category is uniquely defined by the number m and the subset $\mu([\mathbf{n}]) \subseteq [\mathbf{m}]$. This remark allows one to define a one-to-one correspondence $\mu^* : \Delta \to \mathrm{Mor}^*\omega$ between the subset $\Delta \rightleftharpoons \{n | n \in \omega, r(n) < 2^{l(n)}\} \subseteq \omega$ and the set $\mathrm{Mor}^*\omega$, provided that $n \in \Delta$ is assumed to code the morphism $\mu : [\mathbf{k}] \to [\mathbf{l}]$ such that $l = l(n)$ and $r(n)$ is the number of the subset $\mu([\mathbf{k}]) \subseteq [\mathbf{l}] = [\mathbf{l(n)}]$ in some standard listing of the finite subsets of ω (here, $l(n)$ and $r(n)$ are the left and right projections). It is evident that Δ is a Δ-subset of ω.

Definition 6.5. Let $S = \{\mathfrak{M}_n, \mu_* | n \in \omega, \mu \in \mathrm{Mor}^*\omega\}$ be a ${}^*\omega$-spectrum. By a *constructivization* of S we mean any computable sequence of constructivizations

$$(\mathfrak{M}_0, \nu_0), (\mathfrak{M}_1, \nu_1), \ldots, (\mathfrak{M}_n, \nu_n), \ldots, \quad n \in \omega,$$

together with a Σ-function $f : \Delta \times \omega \to \omega$ such that, for any $n, m, k \in \omega$ and $\mu : [\mathbf{n}] \to [\mathbf{m}] \in \mathrm{Mor}^*\omega$, if $n^* \in \Delta$ is such that $\mu^*(n^*) = \mu$, then $\mu_* \nu_n(k) = \nu_m f(n^*, k)$.

A ${}^*\omega$-spectrum S is called *constructivizable* if there exists a constructivization for it.

THEOREM 6.6 ([10]). *Let L be a dense linear order without endpoints. A theory T has an uncountable model Σ-definable in $\mathbb{HF}(L)$ if and only if there exists a constructivizable ${}^*\omega$-spectrum S, satisfying condition $(*)$, and such that $\mathfrak{M}_L^S \models T$.*

One of the important corollaries of this theorem is the first part of the following result, showing that the field $\mathbb{C}$ of complex numbers is rather "simple". The second part shows that $\mathbb{C}$ is not "too simple".

THEOREM 6.7 ([10]).

1. *$\mathbb{C}$ is Σ-definable in $\mathbb{HF}(\mathfrak{L})$ for any dense linear order $\mathfrak{L}$ with the same cardinality;*
2. *$\mathbb{C}$ is not Σ-definable in $\mathbb{HF}(\mathfrak{S})$ for any structure $\mathfrak{S}$ with empty signature.*

Another example of a c-simple theory is the theory T_{DLO} of dense linear orders (without endpoints). This theory seems to be quite reasonable candidate for a "correct extension of CTC for arbitrary cardinalities". Below we present two different characterizations of those theories having uncountable models which are Σ-definable in $\mathbb{HF}(\mathfrak{L})$ $\mathfrak{L} \models T_{\mathrm{DLO}}$.

We now formalize a desired property of T_{DLO} to be the "strongest" in the class of c-simple theories.

CONJECTURE 6.8 ([11]). *Suppose a theory T has an uncountable model which is Σ-definable in $\mathbb{HF}(\mathfrak{M})$, for some structure $\mathfrak{M}$ with a c-simple theory. Then T has an uncountable model which is Σ-definable in $\mathbb{HF}(\mathfrak{L})$ for some $\mathfrak{L} \models T_{\mathrm{DLO}}$.*

It is an open question whether this conjecture is equivalent to the following one (which is its formal consequence).

CONJECTURE 6.9. *Any c-simple theory has an uncountable model which is Σ-definable in $\mathbb{HF}(\mathfrak{L})$ for some $\mathfrak{L} \models T_{\mathrm{DLO}}$.*

It turns out that there are counterexamples to Conjectures 6.8 and 6.9. The next definition is a generalization of the model-theoretical notions of order and total indiscernibility.

DEFINITION 6.10. For structures $\mathfrak{A}$, $\mathfrak{B}$ and some $k > 0$, a set $I \subseteq A^k \cap B$ is called a *set of $\mathfrak{A}$-indiscernibles in $\mathfrak{B}$* (with dimension k) if for any pair of tuples $\bar{i}, \bar{i}' \in I^{<\omega}$ with the same length,

$$\langle \mathfrak{A}, \bar{i} \rangle \equiv \langle \mathfrak{A}, \bar{i}' \rangle \text{ implies } \langle \mathfrak{B}, \bar{i} \rangle \equiv \langle \mathfrak{B}, \bar{i}' \rangle.$$

PROPOSITION 6.11 ([96]). *Suppose $\mathfrak{A}$ is an uncountable structure, a structure $\mathfrak{B}$ is saturated enough and locally constructivizable of level ω, and let $\mathfrak{A} \leqslant_\Sigma \mathfrak{B}$. There exist computable structures $\mathfrak{A}_0$ and $\mathfrak{B}_0$ such that $\mathfrak{A}_0 \equiv \mathfrak{A}$, $\mathfrak{B}_0 \equiv \mathfrak{B}$, and* *there is an infinite computable set of* $(\mathfrak{B}_0, \overline{b}_0)$*-indiscernibles in* $\mathfrak{A}_0$ *with dimension* k*, for some* $k > 0$ *and* $\overline{b}_0 \in (B_0)^{<\omega}$.

For certain c-simple theories this necessary condition of Σ-definability of uncountable models can be simplified (by assuming the dimension to be equal to 1). Namely, for the theory T_{DLO} of dense linear orders without endpoints, and the theory T_{E} of infinite structures with empty signature, there is the following theorem which is a correct part of the corresponding incorrect statement from [89] (the proof can be found in [97]).

THEOREM 6.12. *Let* T *be a c-simple theory, and let* $\mathfrak{A}$ *be any computable model of* T*. Then*

1. *if there exists an uncountable* $\mathfrak{M} \models T$ *such that* $\mathfrak{M} \leqslant_\Sigma \mathfrak{L}$, $\mathfrak{L} \models T_{\mathrm{DLO}}$, *then there exists an infinite computable set of order indiscernibles in* $\mathfrak{A}$ *(with dimension* 1*);*
2. *if there exists an uncountable* $\mathfrak{M} \models T$ *such that* $\mathfrak{M} \leqslant_\Sigma \mathfrak{S}$, $\mathfrak{S} \models T_{\mathrm{E}}$, *then there exists an infinite computable set of total indiscernibles in* $\mathfrak{A}$ *(with dimension* 1*).*

In the case of an infinite signature the counterexample to Conjecture 6.9 was obtained using the construction from [32] together with Theorem 6.12.

THEOREM 6.13 ([89]). *There is a sc-simple theory* T *of an infinite computable signature such that, for any uncountable* $\mathfrak{A} \models T$ *and any* $\mathfrak{L} \models T_{\mathrm{DLO}}$, *we have* $\mathfrak{A} \not\leqslant_\Sigma \mathfrak{L}$.

Using Theorem 6.13 and the construction of Hrushovski which allow us to interpret countably categorical structures with an infinite signature in countably categorical structures with a finite signature, we get the following result.

THEOREM 6.14 ([98]). *There is a c-simple theory* T *with a finite signature, such that, for any uncountable* $\mathfrak{A} \models T$ *and any* $\mathfrak{L} \models T_{\mathrm{DLO}}$, *we have* $\mathfrak{A} \not\leqslant_\Sigma \mathfrak{L}$.

It is known that Conjecture 6.9 is true for rather a "rich" class of c-simple theories (see Theorem 6.17 below).

DEFINITION 6.15. Let $n \in \omega$. A (first-order) theory T is called *n-discrete* if any finite type of T is uniquely determined by its n-subtypes.

A theory T is called *discrete* if it is n-discrete for some $n \in \omega$. If T is n-discrete and has a finite number of n-types then T is ω-categorical and submodel complete in some expansion by a finite number of definable predicates. Any regular n-discrete theory with a finite number of n-types is c-simple. Also, any submodel complete theory of a finite relational signature is n-discrete with a finite number of n-types, for some $n \in \omega$, and any ω-categorical submodel complete theory of a finite signature is n-discrete with a finite number of n-types, for some $n \in \omega$.

A theory T is called *sc-simple* [96] if it is ω-categorical, submodel complete, decidable, and has a decidable set of the complete formulas. Henceforth, a theory (of a finite signature) is *sc*-simple if it is *c*-simple and submodel complete.

As corollary of the Ehrenfeucht–Mostowski Theorem we get the next result.

PROPOSITION 6.16 ([97]). *If T is a sc-simple theory of a finite relational signature then, in any computable model of T, there exists an infinite computable set of order indiscernibles.*

Using this fact, we get a partial positive answer to Conjecture 6.9.

THEOREM 6.17 ([97]). *Let T be sc-simple theory of a finite relational signature. There exists an uncountable model $\mathfrak{A}$ of T such that $\mathfrak{A} \leqslant_\Sigma \mathfrak{L}$, $\mathfrak{L} \models T_{\mathrm{DLO}}$.*

We now present some examples of *sc*-simple theories constructed via Fraïssé limits.

Let *FinGraph* be the class of all finite symmetric graphs. It is easy to check that this class satisfies the properties HP, JEP, AP, and has a ULF-computable presentation.

DEFINITION 6.18. A symmetric graph $\mathfrak{A}$ is called *random* if, for any finite $X, Y \subseteq A$ such that $X \cap Y = \emptyset$, there is a vertex $v \in A \setminus (X \cup Y)$ such that v is adjacent with all vertices from X and not adjacent with vertices from Y.

PROPOSITION 6.19 ([19]). *If $\mathfrak{A}$ is the Fraïssé limit of the class FinGraph then $\mathfrak{A}$ is a random graph. As a corollary,* $\mathrm{Th}(\mathfrak{A})$ *is sc-simple.*

Let σ be a finite predicate signature. The class *Fin(σ)* of all finite structures of signature σ satisfies the properties HP, JEP, AP, and has a ULF-computable presentation.

DEFINITION 6.20. Let σ be a finite predicate signature. A *random structure* *Ran*(σ) of signature σ is the Fraïssé limit of the class *Fin(σ)*.

A structure $\mathfrak{A}$ is called *locally constructivizable* [10] if $\mathrm{Th}_\exists(\mathfrak{A}, \overline{a})$ is c.e. for any $\overline{a} \in A^{<\omega}$. It is easy to verify that a structure $\mathfrak{A}$ is locally constructivizable if and only if, for any $\overline{a} \in A^{<\omega}$, there exist a constructivizable structure $\mathfrak{B}$ and a tuple $\overline{b} \in B^{<\omega}$ such that $(\mathfrak{A}, \overline{a}) \equiv_1 (\mathfrak{B}, \overline{b})$ (or, equivalently, $\mathbb{HF}(\mathfrak{A}, \overline{a}) \equiv_1 \mathbb{HF}(\mathfrak{B}, \overline{b})$). The symbol $\equiv_\alpha$, here and later, denotes elementary equivalence with respect to the class of formulas with less than α groups of alternating groups of quantifiers in the prenex normal form ($0 \leqslant \alpha \leqslant \omega$). Hence, the following definition is a generalization of the notion of local constructivizability.

DEFINITION 6.21 ([94]). A structure $\mathfrak{A}$ is called *locally constructivizable of level* α ($0 < \alpha \leqslant \omega$) if for any $\overline{a} \in A^{<\omega}$ there exists a constructivizable structure $\mathfrak{B}$ and a tuple $\overline{b} \in B^{<\omega}$ such that

$$\mathbb{HF}(\mathfrak{A}, \overline{a}) \equiv_\alpha \mathbb{HF}(\mathfrak{B}, \overline{b}).$$

Local constructivizability of any level is preserved by Σ-definability.

PROPOSITION 6.22 ([94]). *Let $\mathfrak{A}$ and $\mathfrak{B}$ be such that $\mathfrak{A} \leqslant_\Sigma \mathfrak{B}$ and $\mathfrak{B}$ is locally constructivizable of level α, $0 < \alpha \leqslant \omega$. Then $\mathfrak{A}$ is also locally constructivizable of level α.*

A structure $\mathfrak{A}_0$ is called *saturated enough* [10] if there exists an ω-saturated structure $\mathfrak{A}_1$ such that $\mathfrak{A}_0 \preccurlyeq \mathfrak{A}_1$ and $\mathbb{HF}(\mathfrak{A}_0) \preccurlyeq \mathbb{HF}(\mathfrak{A}_1)$. Any structure with a c-simple theory is saturated enough and locally constructivizable of level ω. Moreover, its countable "computable simulation", in the terminology from [52], is unique up to the computable isomorphism. The situation is different in the case of regular theories: there are structures with a regular theory, which are not locally constructivizable even of level 1. For example, consider the fields $\mathbb{R}$ and $\mathbb{Q}_p$ of real and p-adic numbers.

COROLLARY 6.23 ([10]). *For any linear order $\mathfrak{L}$, fields $\mathbb{R}$ and $\mathbb{Q}_p$ are not Σ-definable in $\mathbb{HF}(\mathfrak{L})$.*

We conclude with the list of references which are relevant to the topics discussed in the paper.

REFERENCES.

[1] Ash, C., Knight, J.F., Manasse, M. and Slaman, T. (1989) Generic copies of countable structures, *Ann. Pure Appl. Logic* **42**, 195–205.

[2] Ashaev, I.V., Belyaev, V.Ya. and Myasnikov, A.G. (1993). Toward a generalized computability theory, *Algebra Logic* **32**, 4, 183–205.

[3] Baleva, V. (2006) The jump operation for structure degrees, *Arch. Math. Logic* **45**, 249–265.

[4] Barwise, J. (1975). *Admissible Sets and Structures*, Springer, Berlin–Heidelberg–New York.

[5] Blum, L., Shub, M. and Smale, S. (1989). On a theory of computation and complexity over the real numbers: NP-completeness, recursive functions and universal machines, *Bull. Am. Math. Soc.* **21**, 1, 1–46.

[6] Chisholm, J. (1990). Effective model theory vs. recursive model theory, *J. Symbolic Logic* **55**, 1168–1191.

[7] Ershov, Yu.L. (1980). *Decidability Problems and Constructivizable Models*, Nauka, Moscow.

[8] Ershov, Yu.L. (1985). Σ-definability in admissible sets, *Sov. Math. Dokl.* **32**, 767–770.

[9] Ershov, Yu.L. (1995). Definability in hereditarily finite manifolds, *Dokl. Math.* **51**, 1, 8–10.

[10] Ershov, Yu.L. (1996). *Definability and Computability*, Consultants Bureau, New York–London–Moscow.

[11] Ershov, Yu.L. (1998). Σ-definability of algebraic structures. In *Handbook of Recursive Mathematics, Vol. 1*, Yu.L. Ershov, S.S. Goncharov, A. Nerode, and J.B. Remmel (eds), Elsevier, 235–260.

[12] Ershov, Yu.L. (2000). *Definability and Computability*, Ekonomika, Nauch. Kniga, Moscow–Novosibirsk.

[13] Ershov,Yu.L. and Goncharov, S.S. (2000). *Constructive Models*, Consultants Bureau, New York–London–Moscow.

[14] Ershov, Yu.L., Goncharov, S.S. and Sviridenko, D.I. (1986). Semantic programming, *Inform. Processing* **86**, 1093–1100.

[15] Ershov, Yu.L., Puzarenko, V.G. and Stukachev, A.I. (2011). HF-computability, In *Computability in Context: Computation and Logic in the Real World*, S.B. Cooper and A. Sorbi (eds.), Imperial College Press/World Scientific, 173–248.

[16] Goncharov, S.S. and Sviridenko, D.I. (1985). Σ-programming (Russian), *Vychisl. Sist.* **107**, 3–29.

[17] Goncharov, S.S. and Khoussainov, B. (2004). Complexity of categorical theories with computable models, *Algebra Logic* **43**, 6, 365–373.

[18] Gordon, C. (1970). Comparisons between some generalizations of recursion theory, *Compos. Math.* **22**, 333–346.

[19] Hodges, W. (1993). *Model Theory*, Cambridge University Press, Cambridge.

[20] Kalimullin, I.S. (2006). The Dyment reducibility on the algebraic structures and on the families of subsets of ω, *Logical Approaches to Computational Bariers, CiE2006, Report Series*, Swansea, 150–159.

[21] Kalimullin, I.S. (2009). Uniform reducibility of representability problems for algebraic structures, *Sib. Math. J.* **50**, 2, 265–271.

[22] Khisamiev, A.N. (1996). Σ-enumeration and Σ-definability in $\mathrm{HF}_{\mathcal{M}}$ (Russian), *Vychisl. Sist.* **156**, 44–58.

[23] Khisamiev, A.N. (1997). Numberings and definability in the hereditarily finite superstructure of a model, *Sib. Adv. Math.* **7**, 3, 63–74.

[24] Khisamiev, A.N. (1998). On definability of a model in a hereditarily finite admissible set (Russian), *Vychisl. Sist.* **161**, 15–20.

[25] Khisamiev, A.N. (1998) Strong Δ_1-definability of a model in an admissible set, *Sib. Math. J.* **39**, 1, 168–175.

[26] Khisamiev, A.N. (1999). On resolvable and internally enumerable models (Russian), *Vychisl. Sist.* **165**, 31–35.

[27] Khisamiev, A.N. (2000). The intrinsic enumerability of linear orders, *Algebra Logic* **39**, 6, 423–428.

[28] Khisamiev, A.N. (2001). Quasiresolvable models and *B*-models, *Algebra Logic* **40**, 4, 272–280.

[29] Khisamiev, A.N. (2004). Quasiresolvable models, *Algebra Logic* **43**, 5, 346–354.

[30] Khisamiev, A.N. (2004). On the Ershov upper semilattice $\mathfrak{L}_E$, *Sib. Math. J.* **45**, 1, 173–187.

[31] Kierstead, H.A. and Remmel, J.B. (1983). Indiscernibles and decidable models, *J. Symb. Logic* **48**, 1, 21–32.

[32] Kierstead, H.A. and Remmel, J.B. (1985). Degrees of indiscernibles in decidable models, *Trans. Am. Math. Soc.* **289**, 1, 41–57.

[33] Korovina, M.V. (1992). Generalised computability of real functions, *Sib. Adv. Math.* **2**, 4, 1–18.

[34] Korovina, M.V. (1996). On the universal recursive function and on abstract machines on real numbers with the list superstructure (Russian), *Vychisl. Sist.* **156**, 24–43.

[35] Korovina, M.V. (2002). Fixed points on the real numbers without the equality test, *Electron. Notes Theor. Comput. Sci.* **66**, 1.

[36] Korovina, M.V. (2003). Computational aspects of Sigma-definability over the real numbers without the equality test. In CSL; *Lect. Notes Comput. Sci.* **2803**, 330–344.

[37] Korovina, M.V. (2003). Gandy's theorem for abstract structures without the equality test. In LPAR; *Lect. Notes Comput. Sci.* **2850**, 290–301.

[38] Korovina, M.V. (2003). Recent advances in S-definability over continuous data types. In *Ershov Memorial Conference*; *Lect. Notes Comput. Sci.* **2890**, 238–247.

[39] Korovina, M.V. and Kudinov, O.V. (1996). A new approach to computability of real-valued function (Russian), *Vychisl. Sist.* **156**, 3–23.

[40] Korovina, M.V. and Kudinov, O.V. (1998). Characteristic properties of majorant-computability over the reals. In CSL; *Lect. Notes Comput. Sci.* **1584**, 188–203.

[41] Korovina, M.V. and Kudinov, O.V. (1998). New approach to computability, *Sib. Adv. Math.* **8**, 3, 59–73.

[42] Korovina, M.V. and Kudinov, O.V. (1999). A logical approach to specification of hybrid systems. In *Ershov Memorial Conference*; *Lect. Notes Comput. Sci.* **1755**, 10–16.

[43] Korovina, M.V. and Kudinov, O.V. (2000). Formalisation of computability of operators and real-valued functionals via domain theory. In CCA; *Lect. Notes Comput. Sci.* **2064**, 146–168.

[44] Korovina, M.V. and Kudinov, O.V. (2001). Semantic characterisations of second-order computability over the real numbers. In CSL; *Lect. Notes Comput. Sci.* **2142**, 160–172.

[45] Korovina, M.V. and Kudinov, O.V. (2001). Generalised computability and applications to hybrid systems. In *Ershov Memorial Conference*; *Lect. Notes Comput. Sci.* **2244**, 494–499.

[46] Korovina, M.V. and Kudinov, O.V. (2005). Towards computability of higher type continuous data. In CiE2005; *Lect. Notes Comput. Sci.* **3526**, 235–241.

[47] Korovina, M.V. and Kudinov, O.V. (2008). Effectively enumerable topological spaces (Russian), *Vestn. Novosib. Gos. Univ., Ser. Mat. Mech. Inform.* **8**, 2, 74–83.

[48] Korovina, M.V. and Kudinov, O.V. (2008). *Basic principles of Σ-definability and abstract computability*, Bericht Nr. 08-01, Fachbereich Mathematik, D-57068, Siegen.

[49] Korovina, M.V. and Vorobjov, N. (2004). Pfaffian hybrid systems. In CSL; *Lect. Notes Comput. Sci.* **3210**, 430–441.

[50] Lacombe,D. (1964). Deux généralizations de la notion de récursivité relative, *C. R. Acad. Sci., Paris* **258**, 3410–3413.

[51] Miller, R. (2007). Locally computable structures. In CiE2007, *Lect. Notes Comput. Sci.* **4497**, 575–584.

[52] Miller, R. and Mulcahey, D. (2008). Perfect local computability and computable simulations. In CiE2008, *Lect. Notes Comput. Sci.* **5028**, 388–397.

[53] Montague, R. (1967). Recursion theory as a branch of model theory. In *Proceedings of the Third International Congress for Logic, Methodology and Philosophy of Science*, North-Holland, Amsterdam-London, 63–86.

[54] Montalban, A. (2009). Notes on the jump of a structure. In CiE2009, *Lect. Notes Comput. Sci.* **5635**, 372–378.

[55] Morozov, A.S. (2000). A Σ subset of natural numbers which is not enumerable by natural numbers, *Sib. Math. J.* **41**, 6, 1162–1165.

[56] Morozov, A.S. (2002). Presentability of groups of Σ-presentable permutaions over admissible sets, *Algebra Logic* **41**, 4, 254–266.

[57] Morozov, A.S. (2004). On the relation of Σ-reducibility between admissible sets, *Sib. Math. J.* **45**, 3, 522–535.

[58] Morozov, A.S. (2006). Elementary submodels of parametrizable models, *Sib. Math. J.* **47**, 3, 491–504.

[59] Morozov, A.S. (2008). On the index sets of Σ-subsets of the real numbers, *Sib. Math. J.* **49**, 6, 1078–1084.

[60] Morozov, A.S. and Korovina, M.V. (2008) Σ-definability of countable structures over real numbers, complex numbers and quaternions, *Algebra Logic* **47**, 3, 193–209.

[61] Morozov, A.S. and Puzarenko, V.G. (2004). Σ-subsets of natural numbers, *Algebra Logic* **43**, 3, 162–178.

[62] Moschovakis, Y.N. (1969). Axioms for computation theories – first draft. In *Logic Colloquium '69*, R.O. Gandy and C.E.M. Yates (eds), North-Holland, Amsterdam, 199–255.

[63] Moschovakis, Y.N. (1969). Abstract computability and invariant definability, *J. Symb. Log.* **34**, 605–633.

[64] Moschovakis, Y.N. (1969). Abstract first order computability I, *Trans. Am. Math. Soc.* **138**, 427–464.

[65] Moschovakis, Y.N. (1969). Abstract first order computability II. *Trans. Am. Math. Soc.* **138**, 465–504.

[66] Moschovakis, Y.N. (1974). *Elementary Induction on Abstract Structures*, North-Holland, Amsterdam–London.

[67] Puzarenko, V.G. (2000). On computability over models of decidable theories, *Algebra Logic* **39**, 2, 98–113.

[68] Puzarenko, V.G. (2002). On model theory in hereditarily finite superstructures, *Algebra Logic* **41**, 2, 111–122.

[69] Puzarenko, V.G. (2004). The Löwenheim–Skolem–Mal'tsev Theorem for $\mathbb{HF}$-structures, *Algebra Logic* **43**, 6, 418–423.

[70] Puzarenko, V.G. (2006). Definability of the field of reals in admissible sets, *Logical Approaches to Computational Barriers, CiE2006, Report Series*, Swansea, 236–240.

[71] Puzarenko, V.G. (2009). About a certain reducibility on admissible sets, *Sib. Math. J.* **50**, 2, 330–340.

[72] Richter, L. (1981). Degrees of structures, *J. Symb. Log.* **46**, 723–731.

[73] Romina, A.V. (1998). Hyperarithmetical stability of Boolean algebras (Russian), *Vychisl. Sist.* **161**, 21–27.

[74] Romina, A.V. (2000). Autostability of hyperarithmetical models, *Algebra Logic* **39**, 2, 114–118.

[75] Romina, A.V. (2000a). Definability of Boolean algebras in $\mathbb{HF}$-superstructures, *Algebra Logic* **39**, 6, 407–411.

[76] Rudnev, V.A. (1986). A universal recursive function on admissible sets, *Algebra Logic* **25**, 4, 267–273.

[77] Rudnev, V.A. (1988). Existence of an inseparable pair in the recursive theory of admissible sets, *Algebra Logic* **27**, 1, 33–39.

[78] Sacks, G.E. (1990). *Higher Recursion Theory*, Springer, Berlin–Heidelberg–New York.

[79] Schmerl, J.H. (1980). Decidability and $\aleph_0$-categoricity of theories of partially ordered sets, *J. Symb. Log.* **45**, 585–611.

[80] Slaman, T.A. (1998) Relative to any non-recursive set, *Proc. Am. Math. Soc.* **126**, 2117–2122.

[81] Sorbi, A. (1996). The Medvedev lattice of degrees of difficulty, In *Computability, Enumerability, Unsolvability: Directions in Recursion Theory*. LMS Lecture Notes, **24**, 289–312.

[82] Soskov, I.N. (2004). Degree spectra and co-spectra of structures, *Ann. Univ. Sofia* **96**, 45–68.

[83] Soskova, A.A. (2007). A jump inversion theorem for the degree spectra. In CiE2007, *Lect. Notes Comput. Sci.* **4497**, 716–726.

[84] Soskov, I.N. and Soskova, A.A. (2009). A jump inversion theorem for the degree spectra, *J. Log. Comput.* **19**, 199–215.

[85] Stukachev, A.I. (1996). Uniformization Theorem for HF(R) (Russian). In *Proceedings of the XXXIV International Scientific Student Conference 'Student i Nauchno-Tehnicheskij Progress: Matematica'*, Novosibirsk (1996), p. 83.

[86] Stukachev, A.I. (1997). Uniformization property in hereditary finite superstructures, *Sib. Adv. Math.* **7**, 1, 123–132.

[87] Stukachev, A.I. (1998). Uniformization property in hereditary finite superstructures (Russian), *Vychisl. Sist.* **161**, 3–14.

[88] Stukachev, A.I. (2002). Σ-admissible families over linear orders, *Algebra Logic* **41**, 2, 127–139.

[89] Stukachev, A.I. (2004). Σ-definability in hereditary finite superstructures and pairs of models, *Algebra Logic* **43**, 4, 258–270.

[90] Stukachev, A.I. (2005). On inner constructivizability of admissible sets (Russian), *Vestn. Novosib. Gos. Univ., Ser. Mat. Mech. Inform.* **5**, 1, 69–76.

[91] Stukachev, A.I. (2005). Presentations of structures in admissible sets. In CiE2005, *Lect. Notes Comput. Sci.* **3526**, 470–478.

[92] Stukachev, A.I. (2006). On mass problems of presentability. In TAMC2006, *Lect. Notes Comput. Sci.* **3959**, 774–784.

[93] Stukachev, A.I. (2007). Degrees of presentability of structures, I, *Algebra Logic* **46**, 6, 419–432.

[94] Stukachev, A.I. (2008). Degrees of presentability of structures, II, *Algebra Logic* **47**, 1, 65–74.

[95] Stukachev, A.I. (2009). A jump inversion theorem for the semilattices of Σ-degrees (Russian), *Sib. Elektron. Mat. Izv.* **6**, 182–190.

[96] Stukachev, A.I. (2010). A jump inversion theorem for the semilattices of Σ-degrees, *Sib. Adv. Math.* **20**, 1, 68–74.

[97] Stukachev, A.I. (2010a). Σ-definability of uncountable structures of c-simple theories, *Siberian Mathematical Journal* **51**, 515–524.

[98] Stukachev, A.I. (2013). On semilattices of Σ-degrees of structures. To appear in *Algebra Logic*.

[99] Vajtsenavichyus, R.Yu. (1989). On admissible sets with inner resolutions (Russian, English abstract), *Mat. Logika Primen.* **6**, 9–20.

[100] Vajtsenavichyus, R.Yu. (1989). On necessary conditions for the existence of a universal function on an admissible set (Russian, English abstract), *Mat. Logika Primen.* **6**, 21–37.

[101] *Van den Dries, L.* (1984). Algebraic theories with definable Skolem functions, *J. Symbolic Logic* **49**, 625–630.

[102] Wehner, S. (1998). Enumerations, countable structures and Turing degrees, *Proc. Am. Math. Soc.* **126**, 2131–2139.

Sobolev Institute of Mathematics,
Novosibirsk State University,
Novosibirsk, Russia.
E-mail: aistu@math.nsc.ru